Applied Probability and Statistics

BAILEY · The Elements of Stochastic Processes with Applications to the Natural Sciences

BARTHOLOMEW · Stochastic Models for Social Processes, *Second Edition*

BENNETT and FRANKLIN · Statistical Analysis in Chemistry and the Chemical Industry

BHAT · Elements of Applied Stochastic Processes

BOX and DRAPER · Evolutionary Operation: A Statistical Method for Process Improvement

BROWNLEE · Statistical Theory and Methodology in Science and Engineering, *Second Edition*

CHAKRAVARTI, LAHA, and ROY · Handbook of Methods of Applied Statistics, Vol. II

CHERNOFF and MOSES · Elementary Decision Theory

CHIANG · Introduction to Stochastic Processes in Biostatistics

CLELLAND, deCANI, BROWN, BURSK, and MURRAY · Basic Statistics with Business Applications, *Second Edition*

COCHRAN · Sampling Techniques, *Second Edition*

COCHRAN and COX · Experimental Designs, *Second Edition*

COX · Planning of Experiments

COX and MILLER · The Theory of Stochastic Processes

DANIEL and WOOD · Fitting Equations to Data

DAVID · Order Statistics

DEMING · Sample Design in Business Research

DODGE and ROMIG · Sampling Inspection Tables, *Second Edition*

DRAPER and SMITH · Applied Regression Analysis

DUNN and CLARK · Applied Statistics: Analysis of Variance and Regression

ELANDT-JOHNSON · Probability Models and Statistical Methods in Genetics

FLEISS · Statistical Methods for Rates and Proportions

GOLDBERGER · Econometric Theory

GUTTMAN, WILKS, and HUNTER · Introductory Engineering Statistics, *Second Edition*

HAHN and SHAPIRO · Statistical Models in Engineering

HALD · Statistical Tables and Formulas

HALD · Statistical Theory with Engineering Applications

HOEL · Elementary Statistics, *Third Edition*

HOLLANDER and WOLFE · Nonparametric Statistical Methods

HUANG · Regression and Econometric Methods

JOHNSON and KOTZ · Distributions in Statistics
- Discrete Distributions
- Continuous Univariate Distributions-1
- Continuous Univariate Distributions-2
- Continuous Multivariate Distributions

continued overleaf

Applied Probability and Statistics (Continued)

JOHNSON and LEONE · Statistics and Experimental Design: In Engineering and the Physical Sciences, Volumes I and II
LANCASTER · The Chi Squared Distribution
LANCASTER · An Introduction to Medical Statistics
LEWIS · Stochastic Point Processes
MILTON · Rank Order Probabilities: Two-Sample Normal Shift Alternatives
OTNES and ENOCHSON · Digital Time Series Analysis
RAO and MITRA · Generalized Inverse of Matrices and Its Applications
SARD and WEINTRAUB · A Book of Splines
SARHAN and GREENBERG · Contributions to Order Statistics
SEAL · Stochastic Theory of a Risk Business
SEARLE · Linear Models
THOMAS · An Introduction to Applied Probability and Random Processes
WHITTLE · Optimization under Constraints
WILLIAMS · Regression Analysis
WONNACOTT and WONNACOTT · Econometrics
YOUDEN · Statistical Methods for Chemists
ZELLNER · An Introduction to Bayesian Inference in Econometrics

Tracts on Probability and Statistics

BILLINGSLEY · Ergodic Theory and Information
BILLINGSLEY · Convergence of Probability Measures
CRAMER and LEADBETTER · Stationary and Related Stochastic Processes
JARDINE and SIBSON · Mathematical Taxonomy
KINGMAN · Regenerative Phenomena
RIORDAN · Combinatorial Identities
TAKACS · Combinatorial Methods in the Theory of Stochastic Processes

Stochastic Models for Social Processes

2nd Edition

Stochastic Models for Social Processes

2nd Edition

D. J. BARTHOLOMEW

Professor of Statistics
London School of Economics and
Political Science

JOHN WILEY & SONS
London · New York · Sydney · Toronto

Library of Congress Catalog Card Number: 73-2776

ISBN 0 471 05451 8

First edition 1967

Second edition 1973

Printed in Great Britain by
Unwin Brothers Limited, Old Woking, Surrey

Preface

This book is a contribution to the study of social phenomena by means of the theory of stochastic processes. Like the first edition, it is written with the needs of two groups of people in mind: first, for social scientists, in the broadest sense, who wish to have an account of what the theory can offer in their own fields; and secondly, for research workers, teachers and students of mathematics and statistics to whom it offers a field of application which is both mathematically stimulating and practically important. The viewpoint is that of a statistician who has developed a growing interest in the quantitative aspects of social processes. The range of topics covered is narrower than the title might suggest, having been deliberately restricted to those areas with which I am familiar. However, the book is intended to exemplify an approach to social phenomena which, I believe, has much wider validity.

Since the first edition appeared in 1967 there has been a rapid growth in the literature of stochastic modelling in the social sciences. This is reflected in a fourfold growth in the length of the present bibliography as compared with its predecessor. I have aimed to take account of new work in the area covered by the original book, but have resisted the temptation to introduce completely new topics. The opportunity has been taken to correct, clarify and up date the material at many points. The new material is, in every case, a direct outgrowth of topics treated in the first edition. A list of the principal changes, chapter by chapter, is as follows.

Chapter 1 is a revised and extended version of the original Chapter 1.

Chapter 2 covers the same ground as the original but has been much extended in length to take account of the considerable amount of recent work on social and occupational mobility.

Chapter 3 is similarly revised and contains new material on career patterns.

Chapter 4 is a new chapter containing material developed from the final section of the original Chapter 3.

Chapter 5 is an amalgam of the old Chapters 4 and 5 with the deletion of the material on labour turnover, which now appears in Chapter 6.

Chapter 6 on durations is new. It extends and generalizes material which appeared in the original Chapters 4 and 6.

Chapter 7 is a revision of the original Chapter 6 with the deletion of models for the leaving process and the addition of some new work on the distribution of the number of renewals.

Chapter 8 is substantially the same as the old Chapter 7.

Chapters 9 and 10 are based on the old Chapter 8. A division has been made into those models derived from the simple birth process and those which involve cessation of spreading. Most of the new material, principally on spatial diffusion, is in Chapter 9.

In concluding the preface to the first edition I referred to the hazards besetting those who enter an interdisciplinary no-man's land. I also expressed the hope that the next decade would see greater collaboration and understanding between mathematicians and social scientists. The sympathetic reception which the first edition received from social scientists, especially, encourages me to think that the hope was well founded and that a second excursion may help to bring its realization nearer.

Canterbury, Kent D. J. BARTHOLOMEW

Acknowledgments

It is a pleasure to acknowledge the help which I have received in preparing this edition. My debt remains, of course, to those acknowledged in the first edition; their contribution now has a new lease of life.

The research on control theory given in Chapter 4 was begun during a visit to the University of California at Berkeley during the summer of 1969. The stimulating environment created by Robert Oliver, Kneale Marshall and their colleagues was much appreciated. A visit to the Technion, Israel, early in 1972 at the invitation of Jonathan Halpern provided the much needed time to commit the new material of Chapters 4 and 6 to writing. Denis Mollison and Andrew Cliff, both of the University of Cambridge, have helped me considerably with the new material in Chapters 9 and 10. It is impossible at this stage to recall all who have contributed by comment or suggestion over the years, and to single out some would seem invidious. To all I am most grateful.

Most of the work was done at the University of Kent, where I have had the help of a number of research students and assistants. In the later stages Pauline Sales of the University of Kent and Ian Gardiner of the London School of Economics have been of particular assistance. My interest in applications to manpower planning has largely arisen out of my association with the Statistics Divisions of the Civil Service Department. The opportunity to see models carried through to implementation is a powerful stimulus and I am grateful to Tony Smith, Director of Statistics in the Department, for making this possible.

Permission to quote from their publications was given by D. Van Nostrand Company Inc. in respect of Kemeny and Snell's *Finite Markov Chains* and by the New York State School of Industrial and Labour Relations, Cornell University, in respect of Blumen, Kogan and McCarthy's *The Industrial Mobility of Labour as a Probability Process.*

Producing a book is a tedious business and no part more so than the preparation and checking of the final copy. In this labour I have had the cheerful and efficient help of Penny Brissenden, Statistics secretary at the University of Kent, and Marian, my wife. They have earnt the thanks of all who find this book useful.

Contents

CHAPTER 1

Introduction

1.1 STOCHASTIC THEORY AND SOCIAL PHENOMENA

A stochastic process is one which develops in time according to probabilistic laws. This means that we cannot predict its future behaviour with certainty; the most that we can do is to attach probabilities to the various possible future states. Such processes occur widely in nature and their study has provided the impetus for the rapid development of the theory of stochastic processes in the past few decades. Until relatively recently, most applications have been to problems in the physical and biological sciences, where the ideas of quantitative analysis were already familiar. Progress has been much slower in the social sciences for a number of reasons. One of these is undoubtedly the lack of sufficient people with training and interest in both the theory of stochastic processes and the relevant parts of social science. Much is being done to remedy this situation on a wider front and some interesting developments are indicated by Kruskal (1970) and Rosenbaum (1971). A more fundamental reason is that so many of the basic problems of measurement in the social sciences remain unsolved, and without measurement there can be little hope of using mathematical techniques. Indeed, there is still considerable argument among many social scientists about the relevance of quantitative thinking to substantive social problems. These debates are reminiscent of those which took place among biologists around the turn of the century.

In spite of the difficulties, progress has been rapid in recent years. In the first edition of this book we noted the major contribution of Coleman (1964b) in mathematical sociology and that of Steindhl (1965) who demonstrated the possibilities for stochastic analysis in the field of economics. Much earlier, Bush and Mosteller (1955) had developed stochastic learning models in psychology, where the statistical approach was more familiar. Kemeny and Snell's (1962) book, at a more elementary level, brought the stochastic study of social phenomena within the range of undergraduates. In the five years since the appearance of the first edition of this book there has been a rapid growth in the literature, as may be seen by the much greater length of the present bibliography. Among the more substantial contributions, Thonstad's (1969) book on educational planning makes extensive use of stochastic models. In the neighbouring area of demography, Keyfitz (1968) has provided a basic source of methodology for the mathematical analysis of population. White's (1970a) work on chains of opportunity is a detailed study of job mobility based on the idea of modelling the flow of vacancies through the system. Buying behaviour has been studied as a stochastic process and many references are included in the bibliography. A convenient point of entry into the literature is via the book by Massey and coworkers (1970). Quantitative geographers have shown a growing interest in stochastic methods, much of it stimulated by the pioneering work of Hägerstrand (1967) on the diffusion of innovations.

Most of the existing research has been done by subject specialists—psychologists, biologists, sociologists, management scientists, demographers, geographers, etc.—and published in the journals appropriate to these subjects. This has inhibited the flow of ideas across subject boundaries and statisticians, who are well placed to play a unifying role, have been slow to recognize the need. One of the objects of this book is to facilitate communication by bringing together models from different sources and describing them from a common viewpoint.

The developments referred to above which have taken place in recent years have been almost overshadowed by the rise of interest in manpower planning. This is an interdisciplinary activity involving much besides statistical analysis. Nevertheless, it has proved a fruitful field for the application of stochastic models where the gap between theory and application has been successfully bridged. Much of the new work in this area has been published in volumes of conference proceedings, including Wilson (1969), Smith (1971) and Bartholomew and Smith (1971). A considerable part of the first edition related to models of relevance to manpower planning; this emphasis remains in the present edition. Most of the processes discussed are sociological in the broadest sense, but in so far as they are relevant to control and decision problems they might be classed as management science or operational research. From the methodological point of view the material covered in the book is part of the subject matter of statistics, taken here to include probability.

In common with most branches of applied mathematics, we lay the foundations for the theoretical study by constructing mathematical models. The non-mathematical reader will be aware of the use made of actual physical models by engineers. For example, facts about the behaviour of an aircraft are deduced from the behaviour of the model under simulated flight conditions. The accuracy of these deductions will depend on the success with which the model embodies the features of the actual aircraft.

The aircraft and the model differ in many respects. They may be of different size and made of different materials; many of the detailed fittings not directly concerned with flight behaviour will be omitted from the model. The important thing about the model is not that it should be exactly like the real thing but that it should behave in a wind tunnel like the aircraft in flight. The basic requirement is thus that the aircraft and the model should be *isomorphous* in all *relevant* respects.

A mathematical model is used in an entirely analogous way. In the social systems which we shall discuss the constituent parts are interrelated. When one characteristic is changed there will be consequential changes in other parts of the system. Provided that the changes in question can be quantified, these interrelationships can be described, in principle at least, by mathematical equations. A set of equations which purports to describe the behaviour of the system is a mathematical model. Put in another way, such a model is a set of assumptions about the relationships between the parts of the system. Its adequacy is judged by the success with which it can predict the effects of changes in the social system which it describes and by whether or not it can account for changes which have occurred in the past. The model is an abstraction of the real world in which the relevant relations between the real elements are replaced by similar relations between mathematical entities.

Mathematical models may be deterministic or stochastic. If the effect of any change in the system can be predicted with certainty the system is said to be deterministic. In practice, especially in the social sciences, this is not the case. Either because the system is not fully specified or because of the unpredictable character of much human behaviour there is usually an element of uncertainty in any prediction. This uncertainty can be accommodated if we introduce probability distributions into the model in place of mathematical variables. More precisely, this means that the equations of the model will have to include random variables. Such a model is described as stochastic.

The necessity of using stochastic models in this book can be seen by considering some of the phenomena which we shall discuss. In the case of social mobility, for instance, no one can predict with certainty whether or not a son will follow in his father's footsteps. Similarly there is, in general, no certain means of telling when a man will decide to change his job or whether a student will pass his examination. In the same way the diffusion of news in a social group depends, to a large extent, on chance encounters between members of the group. It is the inherent uncertainty brought about by the freedom of choice available to the individual which compels us to formulate our models in stochastic terms. As we shall see later, there are often advantages in using deterministic methods to obtain approximations to stochastic models, but this does not affect the basic nature of the process. There has often been debate among sociologists about the rival merits of deterministic and stochastic methods. It should be clear from the above discussion that any model describing human behaviour should be formulated in stochastic terms. When it comes to the solution of the models it may be advisable to use a deterministic approximation. The greater simplicity of the deterministic version of a model may also make it easier to grasp the nature of the phenomena in question. Nevertheless, these are tactical questions which do not affect the basic principle which we have laid down.

The range of stochastic models covered in this book is somewhat narrower than the definition given above suggests. We shall be concerned here with what may be called 'explanatory' models as opposed to 'black-box' models. This division is somewhat arbitrary and it would be difficult to define either kind precisely. It is nevertheless worth making the distinction for practical purposes. A black-box model is concerned with the relationship between the output of a system and the input. It is stochastic if the output cannot be exactly predicted from the input. A model in this case usually takes the form of a regression equation (or equations) linking output variables to input variables. The form of the regression equation is purely empirical, being chosen from among a simple class of functions (linear, say) because it fits the output most closely. It makes no pretence of explaining how the input is turned into the output. An explanatory model, on the other hand, aims to describe the mechanism by which the input to the system becomes the output. A black-box model may be sufficient if our object is to predict the output or to control it by manipulating the input. If we wish to understand how the system works we shall need a model which attempts to explain what goes on inside the 'black box'.

To illustrate the point, consider recruitment and wastage in a firm. Over a long period we would probably observe a relationship between the series of recruitement and wastage figures. An estimate of this relationship would be a black-box model, and it could be used to predict future recruitment from past records of wastage and recruitment. An explanatory model would attempt to describe the movement of individuals within the firm and, in particular, the dependence of loss and recruitment on such things as grade and length of service. It will be clear from the next section that our purposes in this book are better served by explanatory models. (We are aware that what is an explanation at one level may be a mystery when viewed at a deeper level. The only claim made here is that our models aim to take us below the surface.) A comprehensive account of what we have called black-box modelling is contained in Box and Jenkins (1971).

1.2 FUNCTIONS OF A MODEL

Stochastic models of social phenomena have been constructed in the past with different objects in view. For our purposes it will be useful to distinguish four main functions of models. The first is to give *insight* into and *understanding* of the phenomenon in question. This is the activity characteristic of the pure scientist. The investigation begins with the collection of data on the process and the formulation of a model which embodies the observed features of the system. This we shall describe as the *model-building* stage. The next step is to use the model to make predictions about the system which can be tested by observation. This activity will require the use of mathematical reasoning to make deductions from the model and will be referred to as *model-solving*. The final step is to compare the deductions with the real world and to modify the original model if it proves to be inadequate; this is the procedure of *model-testing*. An investigation of this kind is not complete until the operation of the system has been accurately and comprehensively described in mathematical terms and when the solution of the equations which arise has been obtained.

The second and third objectives in model-building fall within the province of the applied social scientist. Of these the use of models for *prediction* is widely recognized. The social planner wants to know what is likely to happen if specific policies are implemented. The manager wishes to know, in advance, the consequences of various recruitment or promotion policies for the staffing of his firm. A model which adequately describes the behaviour of the system will be capable of providing answers to such questions. It is important to emphasize that no prediction is unconditional; we are always forecasting what will happen *if* certain conditions are met.

Closely related to the problem of prediction is the question of the *design* of social systems and their mode of operation. Some of the models for hierarchical organizations described in Chapters 3, 5 and 8 were first constructed to establish principles for the design of recruitment and promotion policies. The management structure of a firm has to be such that it provides the correct number of people with the requisite skills and experience at each level to carry out the functions of the organization efficiently. Whether or not this object is achieved depends on the manpower policies of the firm. Hence it is necessary to evaluate competing policies in terms of their success in attaining the stated objectives. In the natural sciences it is usually possible to conduct experiments to answer questions analogous to those above. In the social sciences this is rarely possible or desirable. For example, it is not feasible to alter the management structure of a firm or to change a person's occupation in the interests of pure research. Even if such an experiment were permissible it would not achieve the ends for which it was designed. The very fact that people know themselves to be the subject of experiment is sufficient to influence their behaviour and so vitiate the experiment. To some extent the model provides the social scientist with a substitute for the natural scientist's laboratory. It is possible to experiment in the 'world' generated by the model; if this faithfully mirrors reality the results of these experiments will be applicable to the real world.

The suggestion that mathematical methods may lead to the manipulation of social systems is often viewed with misgivings by sociologists and others who have a concern for individual freedom. Chapter 4 is, in fact, concerned with discovering what degree of control can be exercised over a manpower system and with how to exercise that control most effectively. This suggestion raises an important point, but the debate has to be conducted in terms of what is desirable. Here we are concerned with what is possible; it does not follow that what is possible must be done.

The fourth contribution which stochastic modelling can make to social research is in the field of measurement. This statement may appear paradoxical in view of our earlier remark that lack of suitable measures has itself hindered the development of stochastic models. Both statements are true, but at different levels of sophistication, as the following examples will show. In several applied fields research workers have become aware that the obvious but crude measures they were using were inadequate. For example, the success of a treatment for cancer is often assessed by calculating the proportion of patients who survive for a given period, say five years. However, this measure is influenced by irrelevant factors such as 'natural' mortality which should, ideally, be eliminated from the measure. Fix and Neyman (1951) showed how this could be done by constructing a stochastic model of the post-treatment period and estimating one of its parameters (see Chapter 5). Another example arose in the study of labour turnover. A widely used method of measuring turnover is to express the number of people leaving a firm per unit time as a percentage of the average labour force during the same period. Large values of the measure have often been taken as indicative of low morale. The measure is inadequate because it ignores the strong dependence of propensity to leave on length of service. This dependence is such that a firm with a large number of new recruits will have a misleadingly high figure of turnover. By constructing the renewal theory model of Chapter 7, it was possible to demonstrate the limitations of the usual index and to determine the conditions under which it could be meaningfully used. The results of this study emphasized the advantages of measures of turnover based on such things as the mean or median length of service.

1.3 OBJECTIONS TO THE USE OF STOCHASTIC THEORY IN THE SOCIAL SCIENCES

The use of mathematical methods in the social sciences is in its infancy. It is not surprising therefore that the advocates of the quantitative approach should find their methods subject to criticism. This is especially true in the case of the application of stochastic processes, where few social scientists have the competence to judge the claims which are made. Some of this criticism doubtless arises from reluctance to change established methods and habits of thought, but some deserves serious consideration. Our object in dealing with some substantive objections is not primarily to contribute to the debate but to clarify the nature and especially the limitations of our methods.

It is often argued that there is a fundamental discontinuity between the natural and social sciences. According to this view the attempt to extend the application of stochastic theory into the social sciences is doomed to failure because of the essential difference in the subject matter of the two branches of science. While it may be reasonable to use stochastic models for molecular or even animal behaviour it is absurd to suppose, the argument runs, that the same laws are applicable to human behaviour; to treat human beings as subject to 'laws' seems to be depriving them of freedom of choice. This objection rests on a misunderstanding of the role of probability theory in model-building. It is precisely because man is a free agent that his behaviour is unpredictable and hence must be described in probabilistic terms. The only alternative is to adopt a deterministic point of view; it is this and not the stochastic approach which is open to the charge of making man an automaton. An example may help to reinforce the point. In Chapter 2 we shall discuss a model of labour mobility. As part of this model we shall postulate a chance mechanism to describe the movement of workers from one job to another. The objection in this case would be that a man does not make this decision by resort to dice or fortune tellers. Instead, he considers the advantages and disadvantages of remaining in his present employment and so arrives at a responsible and rational decision. Our contention is not that the employee actually uses a chance device to make the decision but that the group behaves *as if* its individual members did use such a method. The function of probability theory is thus simply to describe observed variability; it carries no implications about the freedom, or otherwise, of human choice. It is a fact of experience that 'choice may mimic chance'.

A variation of this objection is that the quantitative study of society is in some way dehumanizing. Thus a writer in a recent issue of a personnel magazine headed his article 'Are manpower planners human?'. Another article in a university magazine spoke of statisticians as those who 'pretend to know the tricks of shortcutting reality by trivializing and dehumanizing, thus falsifying'. If the attempt to understand the aggregate behaviour of people in quantitative terms diverts attention from them as individuals then there is substance to this objection. If it is seen instead as adding a new perspective, or dimension, to the study of humanity then it enriches rather than impoverishes.

A second objection arises from the complexity of social phenomena. This is often expressed by saying that social situations are far too complicated to allow mathematical study and that to ignore this fact is to be led into dangerous oversimplification. The premiss of this objection must be accepted. Social phenomena are often exceedingly complex and our models are bound to be simplifications. Even if it is allowed that our model need only reproduce the relevant features of the real process the problem remains formidable. However, we would argue that there is no alternative to simplification. The basic limiting factor is not the mathematical apparatus available but the ability of the human mind to grasp a complex situation. There is no point in building models whose ramifications are beyond our comprehension. Perhaps the only safeguard against oversimplification is to use a battery of models instead of a single one. Any particular model will be a special case of the more complex model which would be needed to achieve complete realism. Greater confidence can be placed in conclusions which are common to several special cases than in those applicable to one arbitrarily selected model. In many places, especially Chapter 8, the strategy dictated by these considerations will be evident. Instead of making 'realistic' assumptions and accepting the complexity which goes with them we have made two, or more, 'unrealistic' but simple assumptions. Where possible these have been selected as extremes between which the true situation must lie. If the conclusions about a particular question are similar we may apply them to the real life situation without abandoning the framework of thought provided by the simple model.

The art of model-building is to know where and when to simplify. The object, as Coleman (1964b) has expressed it, is to condense as much as possible of reality into a simple model. Not only does this give us understanding of the system but it provides a convenient base-line with which to compare the behaviour of actual systems. These remarks made above should not be interpreted as justifying unlimited simplification. They apply, of course, to the model-building stage of the process, but this cannot be considered in isolation from the model-solving aspects. In the past the temptation to trim the model to a form having a tractable mathematical solution has been strong. The situation has been radically altered by the wide availability of high-speed computers, but the habits of thought of the old era linger on. It is no longer necessary for a solution to be obtainable in closed form in terms of simple functions in order to yield useful information about the process.

The presence of computers leads to a third and more subtle objection to the model-solving aspects of the subject. A parallel situation is found in the theory of queues, where a great volume of effort has been devoted to the solution of special models. It is argued that, from a practical point of view, the results can be obtained more easily and rapidly by simulation on a computer. The same argument applies in the social sciences. If our objective is simply to obtain a quick answer in a specific situation, simulation will usually be the best method of attack. In applied social science this will be the norm and the practitioner in the field will find that his interest is in the model-building aspects of the book. On the other hand, if our primary interest in stochastic models is to gain insight into the workings of social phenomena, simulation is less satisfactory. In essence, we are then searching for general solutions whereas simulation provides us with solutions in special cases only. The economy and clarity of a conclusion conveyed by a simple formula is such that the effort of problem-solving is well worthwhile even if we have to be content with an approximation.

It is necessary to distinguish clearly between two senses in which the term 'simulation'† is currently used. In the remarks made above we were speaking of simulating individual behaviour. In a stochastic model this involves determining each individual's changes of state according to the realization of random events as prescribed by the probabilistic assumptions of the model. By this means we generate artificial data on an individual basis of the kind that would be observed in practice. The term 'simulation' is also used by some writers to refer to the calculation of the expected values of random variables associated with the model. Such calculations tell us, in an average sense, what would happen if the model were allowed to operate and in this sense may be said to simulate the process. They provide no information about the stochastic behaviour of the system.

Extensive use has been made of simulation techniques, especially of the latter kind, in the study of social phenomena by Forrester and his colleagues at the Massachusetts Institute of Technology. Their work on industrial, urban and world dynamics is well known and one of its main aims is to gain insight into the behaviour of very complex systems. Our approach is complementary to theirs, though less ambitious, in that it seeks to achieve the same end through mathematical analysis.

† The term 'Monte Carlo method' is also used in the first sense given here.

1.4 PLAN OF THE BOOK

In a new and rapidly developing subject it would be premature to claim that a complete and systematic treatment of our subject can be given. It is not possible to cast the presentation of each model into a standard mould. Many of the models were first published in research journals covering many subjects and spanning a period extending over nearly forty years. The different backgrounds and interests of the authors gives the original material a fragmentary appearance. We have attempted to unify the notation, terminology and methodology and to extend the analysis where possible. In some cases this treatment has been very drastic, especially for that work which originated in an actuarial context. In spite of the extensive reworking which has been undertaken much unevenness remains. Some of the models are well established; they have passed through the stages of model-building, -solving, -testing and -rebuilding. Their adequacy can be tested and demonstrated by reference to published data. Other models are much more recent in origin and the data required to test their adequacy is not yet available. It is to be hoped that the incompleteness revealed by our presentation will serve to stimulate further research.

In spite of the variety of models and methods there is a common basic pattern underlying the discussion of each model. The members of the system being studied will, at any given time, be in one of a set of possible 'states'. These states may be discrete—for example the social or occupational classes—or they may form a continuum. An example of the latter is the number of people in a given occupational group with x years of service in that group. The principal object of the analysis is to find the expectations of the numbers in the classes at any time. After an introduction we shall usually begin by describing the basic model. This involves a description of the system and of the probabilistic laws which govern transitions between states. Next we proceed as far as possible with the model-solving aspect of the problem and conclude with any generalizations which are possible and a discussion of the light which the analysis throws on the general problem.

The line of development in the book is from the simple to the complex. This applies, broadly speaking, both to the social phenomena studied and to the mathematical theory. In order to help the reader trace this line of development we shall describe three methods of classifying the models.

One method of classification is according to whether the model is closed or open. These terms appear to have various connotations in sociology. They are used here in a direct and obvious sense which carries no judgement value. The membership of a closed system does not change over time; no persons leave and none are admitted. Models of this kind are described in Chapters 2, 5, 9 and 10. Our interest there centres upon the changing internal structure of the system. An open system has both gains and losses. We are still interested in internal changes but, in addition, there are changes in the input, output and total size to be considered. Some of the systems we shall study lose members but have no input of new members. Such systems are treated as closed by introducing a state to include all the losses and regarding it as part of the system. It is, of course, possible to convert open systems into closed ones by this kind of device, but we have not found this to be convenient and so prefer to retain the distinction.

A second method of classification depends on whether time is treated as discrete or continuous. The choice is partly a matter of realism and partly one of convenience. In some applications changes of state can only take place at fixed intervals of time. If the model of Chapter 3 is used to predict university enrolments then the natural unit of time is the term, quarter or year and a discrete time model is thus more realistic. In the case of the labour mobility models changes of job can take place at any time and a continuous time model is therefore appropriate. However, we can sometimes profitably take advantage of the fact that a discrete time model can be used to approximate a system in continuous time and vice versa. In applied research where it is necessary to compute solutions the discrete version of a model is usually easier to handle. But if we are interested mainly in the mathematical analysis of the model we shall often find that the techniques for handling the continuous time version are more readily available.

Finally, the models presented in this book can be dichotomized according to whether or not they are based on the Markov property. In essence this requires that it be possible to deduce the future development of the process from knowledge of its present state. Information about the history of the process has no predictive value. This is a severe and sometimes unrealistic restriction. The advantage of working with Markovian models is that the mathematical analysis is more tractable. In consequence the theory of Markov processes is much better developed. However, even when the Markov assumption is invalid it may not prove to be critical. In other cases the assumption can be made more realistic by redefining the states of the system. Where necessary we have used non-Markovian models, for example in Chapters 7, 8, 9 and 10, but it will usually be found in such cases that the results obtained are more limited in their scope.

Throughout the book we have concentrated on finding the mean values of the random variables of interest. In physical applications where, for example, the number of molecules in a system is very large the mean values are sufficient. In a social system the number of members may be relatively small and the mean values will then give an incomplete and perhaps misleading picture of the process. Wherever possible we have therefore also given results about variances and covariances and, where obtainable, about the complete distribution. Elsewhere we have sometimes quoted results of simulation studies but much more remains to be done.

At the model-testing stage problems of statistical inference arise. Methods of estimating the parameters of a model and of testing the goodness of fit are often required. The statistical theory which is available for this purpose in the theory of stochastic processes is still meagre, but some progress has been made since the first edition appeared. Where it has proved necessary to estimate a parameter we have used the most obvious method available without considering its efficiency. While such a practice is permissible in the early stages of the development of a subject in order to establish a bridgehead, it should not be taken as a pattern for future work.

1.5 NOTES FOR THE NON-MATHEMATICAL READER

It will be evident from a quick perusal of the following pages that a wide range of mathematical techniques has been used. The treatment is informal and non-rigorous but, even so, many social scientists who are interested in the subject matter may find themselves in difficulty. The only completely satisfactory solution to this problem is for the person concerned to learn more mathematics. In the long term we hope that mathematics will become a regular part of the social scientist's education in rather the same way as it is for the scientist or engineer today. Nevertheless, it should be possible to obtain a good idea of the problems, methods and results by reading between the formulae. We have tried to express the conclusions in words, but they are more adequately conveyed by the formulae and tables which follow from the analysis. Conclusions which seem to have considerable practical importance and generality have been printed in italics. The non-mathematical reader must appreciate that mathematics is a very condensed language and hence be prepared to spend the time necessary to extract the meaning of simple formulae. He would be well advised to compute numerical examples and draw graphs until the meaning conveyed by a formula is clear to him. In some cases this has already been done in the text but, to conserve space, we have usually been content to point out only the most important results of the analysis. The ability to write computer programs in a language such as BASIC, FORTRAN or ALGOL would be a great advantage for those who wish to fully explore the models.

The newcomer to the theory of stochastic processes will find that much of the treatment is self-contained. The remainder draws frequently upon standard results. There are now many textbooks and monographs on the theory of stochastic processes. Those most commonly referred to in this book are those by Bailey (1964), Parzen (1962), Cox and Miller (1965), Karlin (1966), Moran (1968) and Bartlett (1955), but this list is far from being exhaustive.

Notation is often a source of difficulty for those who lack mathematical experience. The problem is magnified when an attempt is made to weld material from diverse sources into a whole. It is clearly desirable to have a uniform notation throughout, but it is equally desirable to conform, as far as possible, with what may have become standard notation in the various fields of application. These two desiderata are sometimes in conflict and a compromise has to be made. We have adopted the following strategy. Fundamental quantities which occur throughout the book are denoted by the same symbol everywhere. Where applicable these symbols are those most widely used in the theory of stochastic processes. The most important of these are as follows:

T denotes time measured from the origin of the process and it may be discrete or continuous.

t and τ also refer to time, but measured from some other origin.

n refers to a number of people. A single subscript, as in n_i, means the number in a given state, i; where these numbers change with time the number at time T is denoted by $n_i(T)$; a double subscript, as in n_{ij}, indicates a flow from state i to state j in an interval of time or at a point of time, this interval or point being identified by adding the time in brackets.

N denotes the total number of people in a system.

p and q are used in a similar way to n except that they refer to probabilities (or proportions); thus p_{ij} is the probability of an individual moving from state i to state j.

k is the total number of states in a system.

g, h, i, j, m, r and s are often used as subscripts indexing the states of a system, occasionally with primes attached; i and j are used only in this connection.

Some other symbols have different meanings in different contexts, but these have been arranged to minimize the risk of confusion. A few symbols such as X, x, u and v are general purpose symbols with only a local meaning.

$f(\cdot)$ denotes a probability density function; $F(\cdot)$ is the corresponding distribution function and $G(\cdot) = 1 - F(\cdot)$.

Matrices and vectors are printed in bold face type—capitals for matrices and lower case for vectors. The same letter of the alphabet is used as for the elements; thus, for example, $\mathbf{P}$ is the matrix with elements $\{p_{ij}\}$. Transposition of rows and columns is denoted by a prime. Vectors are written as rows. This is a change from the first edition in which they were written as columns. The advantages of the new convention are most apparent in the new Chapter 4, where it leads to considerable typographical simplification.

CHAPTER 2

Models for Social and Occupational Mobility

2.1 INTRODUCTION

Human societies are often stratified into classes on the basis of such things as income, occupation, social status or place of residence. Members of such societies move from one class to another in what often seems to be a haphazard manner. For example, sons do not always follow in their fathers' footsteps and, even if they did, the varying numbers of offspring in different classes would lead to fluctuations in the class sizes. Similarly, in a free society a person has some degree of choice about changing his job or moving house. The inherent uncertainty of individual behaviour in these situations means that the future development of the mobility process cannot be predicted with certainty but only in terms of probability. The essential ingredient of any stochastic model of mobility is thus a probabilistic description of how movement takes place from one class to another. In the simplest type of model the assumption is made that the chance of moving depends only on the present class and not on the past. If movement can be regarded as taking place at discrete points in time the appropriate model becomes a simple Markov chain.

There are very many processes in the natural and social sciences which can be represented as a set of flows of objects or people between categories of some kind. The Markov chain model has been used in the study of many of them. In political studies, for example, Anderson (1954) and, more recently, Hawkes (1969) and Miller (1972) have attempted to describe changes in voting behaviour as a Markov process. Geographical mobility within a country has been investigated by Tarver and Gurley (1965), Morrison (1967), Land (1969) and Long (1970), while Berry (1971), among others, has used the model for the spread of negro ghettos in urban areas. In the latter case areas are classified according to the proportion of negro households and the changing structure reflects the changing composition of a region. Lui and Chow (1971) used a Markov chain model to study the spread of the use of interuterine devices in Taiwan. In much the same way Fuguitt (1965) used the model to study the growth and decline in the size of towns. Similar work on the size of firms may be found in the work of Hart and Prais (1956), Adelman (1958) and Preston and Bell (1961). There have been a number of applications of the model in marketing; see, for example, Styan and Smith (1964). Outside the social sciences an interesting application of the model has been made by Foley (1967) to the performance of computer executive systems.

The two applications which have received particular attention are the closely related fields of social and occupational mobility. Our main attention in this chapter will be focused on these two areas. A society is divided into classes on the basis of social status or occupation and the changing structure of society is studied with the aid of the model. Social class is a subject of considerable interest to sociologists. Attempts to classify and describe social groupings in quantitative terms have a long history, but it is only recently that the dynamics of social change have been the subject of stochastic analysis.

The earliest paper in which social mobility was viewed as a stochastic process appears to be that of Prais (1955a). More recent work is covered by Hodge (1966), Lieberson and Fuguitt (1967), McGinnis (1968), Henry and coworkers (1971), McFarland (1970), Henry (1971), Sabolo (1971) and Spilerman (1972a, 1972b). Closely related work on demographic models will be found in Matras (1960a, 1960b, 1967) and Sykes (1969). Social class is often defined in terms of occupation. The pioneering study in this field is that by Blumen and coworkers (1955), whose 'mover–stayer' model occupies a central place in this chapter. A recent application of this model to the British labour force is given in Wynn and Sales (1973b).

The basic form of the Markov chain model for social and occupational mobility is rarely adequate in practice. We shall therefore discuss a number of generalizations designed to introduce greater realism. All of the variants considered are closed, which means that they do not allow for movements in or out of the system. In practice this means that any losses are assumed to be made good by equivalent gains. An open system can always be turned into a closed one by introducing extra strata from which all new members come and to which old ones go. Adelman (1958) used this device, but we shall find it more convenient to dispense with it by treating closed models in this chapter and open ones in Chapter 3.

It is important to distinguish between intergenerational and intragenerational mobility. The first refers to changes of status from father to son through succeeding generations. Here the basic time unit is discrete and is measured in generations. Such models and their ramifications are the subject of Section 2.2. Intragenerational models are concerned with the movement of individuals between classes during their life span. One may still speak of social mobility if the classes chosen reflect social status and this usage is common in the literature. However, the main object of such analyses is usually centred on the movement between occupations, and we shall therefore emphasize the distinction by speaking of occupational mobility. If the categories are based on the industrial classification of jobs so that inter-industry movement is the subject of interest, it is more usual to speak of labour mobility. As far as thc modclling aspects are concerned labour mobility can be conveniently treated with intragenerational mobility. A characteristic feature of all intragenerational models is that changes of state can take place at almost any time rather than on a discrete time scale. For practical reasons it may only be possible to observed changes at discrete points, for example, annually. This fact may make it desirable to use a discrete time model. Models for occupational mobility will be discussed in Section 2.3.

In formulating and analysing our models we shall be using the theory of discrete time Markov chains. Accounts of the theory, at many levels, can be found in almost all modern texts on probability theory. Convenient references are Feller (1968), Kemeny and Snell (1960), Bailey (1964), Cox and Miller (1965), Karlin (1966) and Moran (1968). In Section 2.2 we shall encounter a branching process and the main model of Section 2.3 is a semi-Markov process. In all cases our treatment will be elementary and self-contained. The Markov chain models of this chapter also form the basis of the models for the open systems introduced in Chapter 3.

2.2 MODELS FOR SOCIAL MOBILITY

The basic model

First we consider a very simple model for the development of a single family line and then, in later sections, investigate the consequences of removing its more unrealistic features. The fundamental requirement in a model is that it must specify the way in which changes in social class occur. We shall assume that these are governed by transition probabilities which are independent of time. Let p_{ij} denote the probability that the son of a father in class i is in class j; since the system is closed

$$\sum_{j=1}^{k} p_{ij} = 1$$

where k is the number of classes. We denote by $\mathbf{P}$ the matrix of transition probabilities.† If we consider only family lines in which each father has

† When writing out tables of transition probabilities we shall follow the usual practice of putting the value of p_{ij} in the ith row and jth column.

exactly one son the class history of the family will be a Markov chain. By regarding society as composed of such family lines we could make deductions about the changing structure of society. In practice the requirement that each father shall have exactly one son is not met. As a result some lines become extinct and others branch. However, in a population whose size remains constant over a period of time, each father will have *on average* one son. We may expect our results for the simple model to apply in an average sense in such an actual society. Later we shall place this reasoning on a firmer footing and show that this expectation is fulfilled.

Suppose that the probability that the initial progenitor of a family line is in class j at time zero is $p_j(0)$. Let the probability that the line is in class j at time T $(T = 1, 2, 3, \ldots)$ be $p_j(T)$. The probabilities $\{p_j(T)\}$ can then be computed recursively from the fact that

$$p_j(T+1) = \sum_{i=1}^{k} p_i(T)p_{ij} \qquad (i = 1, 2, \ldots, k). \tag{2.1}$$

In matrix notation these equations may be written as

$$\mathbf{p}(T+1) = \mathbf{p}(T)\mathbf{P} \tag{2.2}$$

where $\mathbf{p}(T) = (p_1(T), p_2(T), \ldots, p_k(T))$. Repeated application of (2.2) gives

$$\mathbf{p}(T) = \mathbf{p}(0)\mathbf{P}^T. \tag{2.3}$$

The elements of $\mathbf{p}(T)$ may also be interpreted as the expected proportions of the population in the various classes at time T. If the original classes of the family lines are known the vector $\mathbf{p}(0)$ would then represent the initial class structure.

The matrix $\mathbf{P}^T$ plays a fundamental role in the theory of Markov chains. It can be used to obtain the 'state' probabilities from (2.3), but its elements also have a direct probabilistic interpretation. Let $p_{ij}^{(T)}$ denote the (i, j)th element in $\mathbf{P}^T$, then (2.3) may be written

$$p_j(T) = \sum_{i=1}^{k} p_i(0)p_{ij}^{(T)} \qquad (j = 1, 2, \ldots, k). \tag{2.4}$$

It is clear from this representation that $p_{ij}^{(T)}$ is the probability that a family line goes from class i to class j in T generations. The case $i = j$ is of special interest as the probabilities $p_{ii}^{(T)}$ can be made the basis of measures of mobility.

In many applications the population has been in existence for many generations so that the 'present' state corresponds to a large value of T. It is therefore of considerable practical interest to investigate the behaviour of the probabilities $\{p_i(T)\}$ and $\{p_{ij}^{(T)}\}$ as T tends to infinity. It is shown in the general theory of Markov chains that this limiting behaviour depends on the structure of the matrix **P**. Provided that the matrix **P** is *regular* it may be shown that these probabilities all approach limits as T tends to infinity. A regular (finite) Markov chain is one in which it is possible to be in any state (class) after some number, T, of generations, no matter what the initial state. More precisely, a necessary and sufficient condition for the chain to be regular is that all of the elements of $\mathbf{P}^T$ are non-zero for some T. All transition matrices which are likely to occur in the present context are regular but, later in the book, we shall meet examples which do not possess this property.

With the existence of the limits assured it is a straightforward matter to calculate them. Thus if we write $\lim_{T\to\infty} p_j(T) = p_j$ it follows from (2.2) that the limiting structure must satisfy

$$\mathbf{p} = \mathbf{pP} \tag{2.5}$$

with

$$\sum_{j=1}^{k} p_j = 1.$$

The limiting structure, or distribution, can thus be obtained by solving a set of simultaneous equations. An important property of the solution is that it does not depend on the initial state of the system. Since, by our assumptions, each family line extant will have reached the equilibrium given by (2.5), the vector **p** gives the expected structure of the population at the present time. If this structure is all that can be observed we have no means of reconstructing the transition matrix. Neither, in fact, can we deduce **P** from two consecutive observed structures $\mathbf{p}(T)$ and $\mathbf{p}(T+1)$, although White (1963) and Matras (1967) have considered what incomplete information can be obtained in these circumstances. The limiting value of $\mathbf{P}^T$, denoted by $\mathbf{P}^\infty$, can be deduced from (2.3). It must satisfy

$$\mathbf{p} = \mathbf{p}(0)\mathbf{P}^\infty$$

which can only be so if

$$\mathbf{P}^\infty = \begin{Bmatrix} p_1 & p_2 & \cdots & p_k \\ p_1 & p_2 & \cdots & p_k \\ \vdots & \vdots & & \vdots \\ p_1 & p_2 & \cdots & p_k \end{Bmatrix} \tag{2.6}$$

which implies that

$$\lim_{T \to \infty} p_{ij}^{(T)} = p_j. \tag{2.7}$$

The foregoing analysis shows that if our model provides an adequate description of actual societies then their future development depends only on their initial structure and the transition matrix. Of these two features the initial distribution has a diminishing influence as time passes. In the long run, therefore, the structure of the society is determined by its transition matrix. This conclusion implies that the study of mobility must be centred upon the transition probabilities. In particular, *measures* of mobility should be functions of the elements of **P**. Before pursuing this proposal we may test the adequacy of the model using actual data on mobility.

Test of the adequacy of the model

Three empirical studies of mobility have been published which give sufficient data to provide a partial test of the theory. One of these due to Glass and Hall (Glass, 1954) is based on a random sample of 3,500 pairs of fathers and sons in Britian. A second study made by Rogoff (1953) is based on data from marriage licence applications in Marion County, Indiana. Rogoff obtained data for two periods; one from 1905 to 1912, with a sample size of 10,253, and a second from 1938 to 1941, when the sample size was 9,892. Further data relating to Denmark are given in Svalagosta (1959).

The data obtained by Glass and Hall were used by Prais (1955a) and much of the material in this section is taken from his paper. Glass and Hall classified the members of their sample according to the seven occupational groups listed in Table 2.1, which also gives the estimated transition probabilities.†

The first two columns of Table 2.2 give the class structure of the population tion in two succeeding generations. If the Markov model is adequate and if the society has reached equilibrium we would expect these distributions to be the same, apart from sampling fluctuations. We would also expect them both to agree with the equilibrium distribution obtained from (2.5). Prais (1955a) made the calculations necessary for this comparison and his results are given in Table 2.2.

The differences between the three distributions are not large although there does appear to be a shift towards the lower classes as we move across the table. If this trend is genuine and not merely the result of sampling fluctuations it might be taken to indicate that the process had not reached

†The question of sampling populations with a view to valid estimation of transition probabilities requires further investigation.

Table 2.1 Estimated transition probabilities for England and Wales in 1949 (Glass and Hall's data)

	Son's class						
Father's class	1	2	3	4	5	6	7
1. Professional and higher administrative	0·388	0·146	0·202	0·062	0·140	0·047	0·015
2. Managerial and executive	0·107	0·267	0·227	0·120	0·206	0·053	0·020
3. Higher grade supervisory and non-manual	0·035	0·101	0·188	0·191	0·357	0·067	0·061
4. Lower grade supervisory and non-manual	0·021	0·039	0·112	0·212	0·430	0·124	0·062
5. Skilled manual and routine non-manual	0·009	0·024	0·075	0·123	0·473	0·171	0·125
6. Semi-skilled manual	0·000	0·013	0·041	0·088	0·391	0·312	0·155
7. Unskilled manual	0·000	0·008	0·036	0·083	0·364	0·235	0·274

Table 2.2 Actual and equilibrium distributions of the social classes in England and Wales (1949), estimated from Glass and Hall's data

Class	*Fathers*	*Sons*	*Predicted equilibrium*
1	0·037	0·029	0·023
2	0·043	0·046	0·042
3	0·098	0·094	0·088
4	0·148	0·131	0·127
5	0·432	0·409	0·409
6	0·131	0·170	0·182
7	0·111	0·121	0·129

equilibrium. Another possible explanation is discussed below. A complete answer to the question of sampling error is not available, but a first step towards the solution of the problem is provided by the sampling theory derived later. Although we cannot obtain a complete test of the model using these data it does appear that there is a broad compatibility between the data and the predictions of the theory.

Rogoff's data lead to similar conclusions about the applicability of Markov chain theory to social mobility. She obtained information at two dates separated by thirty years and so we can see whether there had been any significant change in the transition probabilities during that period. The two transition matrices are given in Table 2.3.

Table 2.3 Transition probabilities estimated from Rogoff's data for Marion County, Indiana[a]

	1905–1912			1938–1941		
	1	2	3	1	2	3
1. Non-manual	0·594	0·396	0·009	0·622	0·375	0·003
2. Manual	0·211	0·782	0·007	0·274	0·721	0·005
3. Farm	0·252	0·641	0·108	0·265	0·694	0·042

[a] Taken from Kemeny and Snell's *Finite Markov Chains*, Copyright 1960, D. Van Nostrand Company, Inc., Princeton, N.J.

We have followed Kemeny and Snell (1960) in adopting a coarse grouping for the classes in place of Rogoff's very fine breakdown. Allowing for sampling error, it appears that changes did take place in the transition matrix over the period in question. However, the changes are not large and suggest that we shall not be involved in gross error if we treat them as constant over moderately short periods.

Neither set of data allows us to make a direct test of the Markov property. This property requires that a son's class should depend only on that of his father and not on that of his grandfather. To test this assumption we need records of family history over at least three generations. Indirect support for the assumption is provided by the close agreement between the equilibrium class structure predicted by Markov theory and the observed class structure.

There are theoretical grounds for believing that the Markov property will not hold exactly for a social mobility process. These grounds arise from the fact that the class boundaries are drawn arbitrarily. Thus, for example, we could subdivide the seven categories used by Glass and Hall. Alternatively, some classes could be amalgamated to give a smaller number of categories. It is known (see Kemeny and Snell, 1960, Chapter 6) that if the states of a Markov chain are pooled then the new chain does not, in general, have the Markov property. In the present context this means that we cannot arbitrarily rearrange the classes and retain the Markov property. Even if there is one system of classification for which the property holds this may not be the one which we happen to have chosen. However, as we have pointed out in Chapter 1, it is sufficient if our model embodies the main features of the process without being correct in every detail. To summarize: although the assumptions on which the theory depends are not completely realistic the model is sufficiently near to reality to justify its further use and development.

The measurement of mobility

It is convenient, especially when comparing two societies, to have a measure of mobility. According to the Markov model the society is characterized by the transition matrix and hence we are led to construct measures of mobility from the elements of the matrix. Numerous descriptive measures of mobility which depend on the transition proportions have been devised for empirical work. Some examples are listed in Matras (1960b) and the subject is discussed in great detail in Boudon (1973). It is not our intention to give a full account of this important topic but rather to show the bearing which stochastic theory has upon the problem of measurement. We may begin by considering the form of the transition matrix under extreme degrees of mobility. In a completely immobile society sons will have the same class as their fathers and **P** will have 1's along its principal diagonal and 0's elsewhere. It is less easy to decide on the other end-point of our scale of mobility. Prais (1955a) defined a perfectly mobile society as one in which the son's class is independent of his father's class. For such a society the rows of **P** will be identical. This may be a desirable social norm with which to compare existing societies but it does not coincide with the maximum amount of movement which can take place between classes. The point may be illustrated by considering the following transition matrices for $k = 2$:

$$\text{(a)}\quad \begin{pmatrix} 1 & 0 \\ 0 & 1 \end{pmatrix} \qquad \text{(b)}\quad \begin{pmatrix} \frac{1}{2} & \frac{1}{2} \\ \frac{1}{2} & \frac{1}{2} \end{pmatrix} \qquad \text{(c)}\quad \begin{pmatrix} 0 & 1 \\ 1 & 0 \end{pmatrix}.$$

Case (a) represents the immobile society in which sons follow fathers; case (b) is one example of the perfectly mobile society as defined by Prais; and case (c) represents the extreme of movement in which every son has a different class from his father. We must therefore decide whether we are primarily interested in movement as such, or in deviation from some ideal such as that represented by case (b). In practice it is likely that transition matrices will be intermediate between (a) and (b), but it could be important not to confuse (a) and (c).

One of the simplest measures of mobility when $k = 2$ is obtained by taking the determinant of **P**. This takes the value 1 for an immobile society, 0 for Prais's perfectly mobile society and -1 when the maximum amount of movement takes place. This measure is less satisfactory when $k > 2$. It is zero whenever two rows of **P** are equal, which implies perfect mobility for only two classes. Also, there is some ambiguity in deciding on the appropriate form of matrix to take the place of case (c). If we select the matrix

$$\begin{pmatrix} 0 & 0 & 1 \\ \frac{1}{2} & 0 & \frac{1}{2} \\ 1 & 0 & 0 \end{pmatrix}$$

for $k = 3$, the determinant is zero and the measure has the same value as for the very different situation when the rows are identical.

A more direct and meaningful measure is obtained by counting the class boundaries crossed in passing from one generation to the next. If $f_1, f_2, \ldots, f_k$ are the numbers in the classes of a given generation then the expected number of class boundaries crossed in moving to the next generation is

$$\sum_{i=1}^{k} \sum_{j=1}^{k} f_i p_{ij} |i - j|.$$

This measure depends on which generation we choose as our base-line. A more satisfactory version is obtained by replacing the f's by the equilibrium class structure. Our measure would then be defined as

$$D = \sum_{i=1}^{k} \sum_{j=1}^{k} p_i p_{ij} |i - j|. \tag{2.8}$$

This is near to zero if the p_{ii}'s are close to unity for all i†, and increases in value with the amount of movement. The behaviour of D in the case $k = 2$ is easily investigated. Here we have

$$p_1 = \frac{p_{21}}{p_{12} + p_{21}} \qquad p_2 = \frac{p_{12}}{p_{12} + p_{21}}$$

$$0 < p_{12}, p_{21} < 1$$

and hence

$$D = \frac{2p_{12}p_{21}}{p_{12} + p_{21}}.$$

D varies between zero for the immobile society and unity for the extreme degree of movement represented by case (c) above. For Prais's perfectly mobile society $p_{12} = 1 - p_{21}$ and hence $D = 2p_{12}(1 - p_{12}) \leq \frac{1}{2}$. Our proposed measure is thus a genuine measure of movement rather than of closeness to the ideal represented by case (c). A disadvantage of D for $k > 2$

† If $p_{ii} = 1$ for all i, **P** is not regular and the equilibrium distribution **p** does not exist. In this case we may define $D = 0$.

is that it cannot be used for comparing two societies with different groupings for the classes.

The following example illustrates the use of the index when $k = 3$. Suppose that population I has transition matrix

$$\begin{pmatrix} 0{\cdot}4 & 0{\cdot}3 & 0{\cdot}3 \\ 0{\cdot}2 & 0{\cdot}5 & 0{\cdot}3 \\ 0{\cdot}2 & 0{\cdot}2 & 0{\cdot}6 \end{pmatrix}.$$

It may easily be verified that the limiting structure is (7/28, 9/28, 12/28). Population II has transition matrix

$$\begin{pmatrix} 0{\cdot}5 & 0{\cdot}3 & 0{\cdot}2 \\ 0{\cdot}2 & 0{\cdot}5 & 0{\cdot}3 \\ 0{\cdot}2 & 0{\cdot}3 & 0{\cdot}5 \end{pmatrix}$$

with limiting structure (16/56, 21/56, 19/56). For population I

$$D = \tfrac{1}{28}[7\{0{\cdot}3 + 2(0{\cdot}3)\} + 9\{0{\cdot}2 + 0{\cdot}3\} + 12\{2(0{\cdot}2) + 0{\cdot}2\}] = \tfrac{18}{28}$$

and for population II

$$D = \tfrac{1}{56}[16\{0{\cdot}3 + 2(0{\cdot}3)\} + 21\{0{\cdot}2 + 0{\cdot}3\} + 19\{2(0{\cdot}2) + 0{\cdot}3\}] = \tfrac{35}{56}.$$

Mobility in population I is thus marginally higher than in II, but overall the two populations are almost equally mobile as judged by the D-index.

Any measure which attempts to summarize the contents of the transition matrix into a single number is bound to result in oversimplification. A more detailed picture of the process can be obtained if we replace the single number by a set of numbers. Two proposals were made by Prais (1955a), and these will now be described.

The first set of measures is obtained by considering only the main diagonal of $\mathbf{P}$. Thus we consider only whether or not a transition takes place and we ignore the kind of transition. It is easy to see that the duration of stay in the jth class has a geometric distribution with mean $\mu_j = 1/(1 - p_{jj})$. In a mobile society these means will be small and in an immobile society they will be large. Prais suggested that the means might be standardized by expressing them in terms of the means for a standard population. He chose the perfectly mobile society with the same equilibrium structure as the given population. The measures are thus

$$\mu_j^* = (1 - p_j)/(1 - p_{jj}) \qquad (j = 1, 2, \ldots, k). \tag{2.9}$$

The values of the μ_j^* for the data given in Table 2.1 are given in Table 2.4.

Table 2.4 The expected stay in each social class for Glass and Hall's data

Class (j)	$(1 - p_{jj})^{-1}$	$(1 - p_j)^{-1}$	$\mu_j^* = (1 - p_j)/(1 - p_{jj})$
1	1·63	1·02	1·59
2	1·36	1·04	1·30
3	1·23	1·10	1·12
4	1·27	1·15	1·11
5	1·90	1·69	1·12
6	1·45	1·22	1·19
7	1·38	1·15	1·20

Most of the figures in the last column of the table are close to 1, indicating that a high degree of mobility existed in England and Wales in 1949. The relatively high values for classes 1 and 2 are suggestive but may not be significant. Prais calculated standard errors for the μ_i and found that they lay between 0·54 and 1·30. It is thus clear that differences between entries in the last column must be treated with caution.

The measures $\{\mu_j^*\}$ depend only the diagonal elements of $\mathbf{P}$. The information contained in the off-diagonal elements can be utilized by enlarging the set of measures still further. A convenient and meaningful way to do this is to consider the diagonal elements of the matrices $\mathbf{P}^T$ for $T = 2, 3, \ldots$. The quantity $p_{jj}^{(T)}$ is the probability that a family line which is initially in class j is also in class j, T generations later. This probability will be near to 1 in an immobile society and relatively small in a mobile society. If we standardize this probability by dividing by $p_j = \lim_{T\to\infty} p_{jj}^{(T)}$ we obtain Prias's 'immobility ratios'. However, a point is rapidly reached where our set of measures contains more elements than the transition matrix itself. The immobility ratios are best regarded as an alternative way of presenting the information contained in $\mathbf{P}$ rather than as summary measures of mobility.

We have already emphasized that the foregoing discussion depends for its validity on the adequacy of the Markov model. If this model is not correct then the measures proposed will be open to criticism. It seems very doubtful whether the simple Markov model is satisfactory, even when it appears to describe the overall pattern on mobility reasonably well. We shall shortly turn to several extensions and generalizations in an attempt to overcome this difficulty, but these are also subject to an important limitation. They all treat flows between classes as being generated in the class from which movement starts. Thus we can speak of them as 'push' flows because the member of the family line can be thought of as being pushed from one class to another. In practice it is not always reasonable to think of movements in this way. Some are more accurately described as 'pulls'. This is particularly true if the classes are based on occupation. For example, a decrease in the proportion of manual jobs available will inevitably lead to movements to the non-manual classes. The individual is then pulled into a new class by the changing occupation structure. The real issue is concerned with what causes movement. Do the flows change the structure or does the changing structure generate the flows? It is not always possible to make the distinction clearly since there may be elements of both push and pull in a system. Nevertheless these remarks serve to emphasize the value of a stochastic model in clarifying the issues involved in choosing a measure. One which is appropriate for pushes will not necessarily be adequate for pulls. The distinction between push and pull flows is a fundamental one in modelling social systems; it underlies the difference between the Markov models of Chapters 3, 4 and 5 and the renewal models of Chapters 7 and 8.

The variability of the class sizes

The theory developed so far enables us to calculate expected numbers, or proportions, in each class at any time in the future. So far we have no means of determining the variances and covariances of our predictions. A method for obtaining the moments and product moments of the class numbers has been given by Pollard (1966), and we shall now describe its application to this problem.

Let N denote the number of family lines in the population; this remains constant through time. Let the size of the jth class at time T be $n_j(T)$ and let the number of transitions between class i and class j between T and $T + 1$ be $n_{ij}(T)$. It follows from the definitions that

$$n_j(T + 1) = \sum_{i=1}^{k} n_{ij}(T) \qquad (j = 1, 2, \ldots, k). \tag{2.10}$$

If we take expectations on each side of this equation we arrive at (2.1) because $En_j(T + 1) = Np_j(T + 1)$ and $En_{ij}(T) = Np_i(T)p_{ij}$. Consider the covariance of $n_j(T + 1)$ with $n_l(T + 1)$. It will simplify the presentation of the theory if we adopt the convention that $\text{cov}(x_i, x_j) = \text{var}(x_i)$ if $i = j$. We then have

$$\begin{aligned}\text{cov}\,\{n_j(T + 1), n_l(T + 1)\} \\ &= E\{n_j(T + 1)n_l(T + 1)\} - En_j(T + 1)En_l(T + 1) \\ &= \sum_{i=1}^{k}\sum_{i'=1}^{k} [E\{n_{ij}(T)n_{i'l}(T) - En_{ij}(T)En_{i'l}(T)\}]\end{aligned} \tag{2.11}$$

by (2.10). In order to evaluate the expectations in (2.11), we make use of a well-known result about conditional expectations to the effect that $E(x) =$

$E_X(x|X)$. In the present case we obtain the expectations conditional upon $n_i(T)$. These follow from the fact that, given $n_i(T)$, $n_{ij}(T)$ $(j = 1, 2, \ldots, k)$ are multinomially distributed with probabilities p_{ij} $(j = 1, 2, \ldots, k)$. Hence

$$E\{n_{ij}(T)|n_i(T)\} = n_i(T)p_{ij} \tag{2.12}$$

and

$$E\{n_{ij}(T)n_{i'l}(T)|n_i(T), n_{i'}(T)\} = n_i(T)n_{i'}(T)p_{ij}p_{i'l} \qquad (i \neq i'),$$
$$E\{n_{ij}(T)n_{il}(T)|n_i(T)\} = n_i(T)\{n_i(T) - 1\}p_{ij}p_{il} + \delta_{jl}n_i(T)p_{ij} \tag{2.13}$$

where $\delta_{jl} = 1$ if $j = l$ and is zero otherwise. The unconditional expectations are now obtained from (2.12) and (2.13) by taking the expectations of the right-hand sides with respect to $n_i(T)$. Substituting these expressions in (2.11) then gives

$$\operatorname{cov}\{n_j(T+1), n_l(T+1)\} = \sum_{i=1}^{k} \sum_{i'=1}^{k} p_{ij}p_{i'l} \operatorname{cov}\{n_i(T), n_{i'}(T)\}$$
$$+ \sum_{i=1}^{k} (\delta_{jl}p_{ij} - p_{ij}p_{il})En_i(T). \tag{2.14}$$

We have thus obtained a recurrence relation between the expectations and covariances at time T and the covariances at time $T + 1$. Since the covariances at $T = 0$ are zero the complete set can be computed from (2.14) and (2.1). If we require them, the same method can be extended to yield the higher moments and product moments.

Since the expectations and covariances at $T + 1$ are linear functions of the corresponding quantities at T we can express the relationship in matrix notation. To do this we introduce the vector of means and covariances and denote it by $\boldsymbol{\mu}(T)$. In this vector we first list the k means followed by the k^2 covariances in dictionary order, that is

$$\boldsymbol{\mu} = [E(n_1), \ldots, E(n_k), \operatorname{cov}(n_1, n_1), \operatorname{cov}(n_1, n_2), \ldots, \operatorname{cov}(n_1, n_k),$$
$$\operatorname{cov}(n_2, n_1), \operatorname{cov}(n_2, n_2), \ldots, \operatorname{cov}(n_2, n_k), \ldots, \operatorname{cov}(n_k, n_k)],$$

where we have omitted the arguments of the n's and $\boldsymbol{\mu}$ for brevity. There is some redundancy in this listing because, for example, the covariance between n_1 and n_2 appears as $\operatorname{cov}(n_1, n_2)$ and as $\operatorname{cov}(n_2, n_1)$, but by retaining this the symmetry of the expressions is preserved. Equations (2.1) and (2.14) may now be combined and written in the form

$$\boldsymbol{\mu}(T+1) = \boldsymbol{\mu}(T)\boldsymbol{\Pi} \tag{2.15}$$

where the elements of the $k(k + 1) \times k(k + 1)$ matrix $\boldsymbol{\Pi}$ are functions of the p_{ij}'s. The matrix $\mathbf{\Pi}$ may be partitioned as follows:

$$\begin{array}{c|c} \mathbf{P} & \mathbf{X} \\ \hline \mathbf{O} & \mathbf{Y} \end{array}$$

where $\mathbf{P}$ is the $k \times k$ transition matrix, $\mathbf{O}$ is a $k^2 \times k$ zero matrix, $\mathbf{X}$ is a $k \times k^2$ matrix with elements of the form $\delta_{jl}p_{ij} - p_{ij}p_{il}$ and $\mathbf{Y}$ is a $k^2 \times k^2$ matrix with elements of the form $p_{ij}p_{i'l}$.† In the case of $\mathbf{X}$ and $\mathbf{Y}$, (j, l) indexes the column and i or (i, i') the row.

Equation (2.15) is a generalization of (2.2) and it can be used in the same way. It is no longer true that $\mathbf{\Pi}$ is a stochastic matrix because it has negative elements but the rows still sum to one. This fact can be used to deduce that the elements of $\boldsymbol{\mu}(T)$ tend to limits as T tends to infinity. Consequently, these limits must satisfy the equations

$$\left.\begin{aligned} \boldsymbol{\mu}(\infty) &= \boldsymbol{\mu}(\infty)\mathbf{\Pi} \\ \sum_{i=1}^{k} \bar{n}_i(\infty) &= N \end{aligned}\right\} \tag{2.16}$$

where here, and subsequently, the bar denotes the expectation of the random variable. A computer program is available to generate the vector $\boldsymbol{\mu}(T)$ for all T. We shall give some results of calculations in Chapter 3 for a more general model. We can get an idea of the kind of solution obtained in the present application by considering a special case.

Suppose that we have a perfectly mobile society in Prais's sense with $p_{ij} = p_j$ $(j = 1, 2, \ldots, k)$. Equation (2.14) then simplifies to give

$$\operatorname{cov}\{n_j(T+1), n_l(T+1)\} = N\delta_{jl}p_j - Np_jp_l + p_jp_l \sum_{i=1}^{k} \sum_{i'=1}^{k} \operatorname{cov}\{n_i(T), n_{i'}(T)\}. \tag{2.17}$$

Now

$$\sum_{i=1}^{k} \sum_{i'=1}^{k} \operatorname{cov}\{n_i(T), n_{i'}(T)\} = E\left\{\sum_{i=1}^{k} (n_i - \bar{n}_i)\right\}^2 = 0$$

so that the last term in (2.17) vanishes. The part which remains will be recognized as giving the variances and covariances of the multinomial distribution. In fact, it is easy to see directly that $\{n_j(T)\}$ has a multinomial distribution with parameters N and p_j $(j = 1, 2, \ldots, k)$ for all T. The class

†$\mathbf{Y}$ is the *direct matrix product* $\mathbf{P} \times \mathbf{P}$. The direct matrix product can be used to facilitate the solution of the more general problem of finding the higher moments and product moments of the $\{n_j\}$. See Pollard (1966).

distribution at time $T + 1$ is independent of that at time T. Hence we may regard the N family lines as being allocated independently to classes with probabilities p_j ($j = 1, 2, \ldots, k$), which is the condition for the distribution to be multinomial. In this case, therefore, the standard error of the predicted proportion p_j will be $\{p_j(1 - p_j)/N\}^{\frac{1}{2}}$.

The underlying assumption in the derivation of these results is that the individuals moving out of a given class are distributed according to a multinomial distribution over all classes. The probabilities governing these flows are those in the row of the transition matrix corresponding to the class in question. As far as the second moments are concerned, this assumption is embodied in the second part of (2.13) when $i = i'$, and this leads on to the result in (2.14). Wynn and Sales (1973a) have considered a more general error structure in which the covariances between any pair of flows out of a given class are allowed to be arbitrary subject to the overall requirement that the total flow is fixed. Thus Wynn and Sales (1973a) suppose that

$$\text{cov}\,\{n_{ij}(T), n_{i'l}(T)\} = 0 \quad \text{if } i \neq i' \text{ (as before)}$$

and

$$\text{cov}\,\{n_{ij}(T), n_{il}(T)\} = \alpha_{ijl} n_i(T) \qquad (i, j = 1, 2, \ldots, k) \tag{2.18}$$

conditional on $n_i(T)$. The α's are constrained by the fact that, for fixed $n_i(T)$,

$$\sum_{j=1}^{k} n_{ij}(T) = n_i(T)$$

and hence by

$$\begin{aligned} \text{var} \sum_{j=1}^{k} n_{ij}(T) &= \sum_{j=1}^{k} \sum_{l=1}^{k} \text{cov}\,\{n_{ij}(T), n_{il}(T)\} \\ &= \sum_{j=1}^{k} \sum_{l=1}^{k} \alpha_{ijl} = \text{var}\,\{n_i(T)\} = 0. \end{aligned} \tag{2.19}$$

Subject to this restraint the values of $\{\alpha_{ijl}\}$ can have any values, except that when $j = l$ the covariance becomes a variance and so must be positive. In the multinomial case

$$\begin{aligned} \alpha_{ijl} &= -p_{ij}p_{il} \qquad &\text{if } j \neq l \\ &= p_{ij}(1 - p_{ij}) \qquad &\text{if } j = l. \end{aligned} \tag{2.20}$$

The argument leading to (2.14) still holds, but the last term on the right-hand side has to be replaced by

$$\sum_{i=1}^{k} \alpha_{ijl} En_i(T).$$

In the matrix version of the equations the submatrix $\mathbf{X}$ becomes a $k \times k^2$ matrix with elements $\{\alpha_{ijl}\}$. Wynn and Sales (1973a) established the existence of the limiting form (2.16) and discussed the estimation of the errors; they applied the method to data relating to occupational mobility.

Time-dependent models

In this and the following sections we shall examine the effect of relaxing some of the assumptions on which the foregoing model depends. The first of these is that the transition probabilities are time-homogeneous. Rogoff's data suggest that this assumption may be reasonable in some societies, but it is of interest to point out that the generalization is straightforward.

Suppose that the transition matrix for the Tth generation is $\mathbf{P}(T)$. Equation (2.2) still holds with $\mathbf{P}$ replaced by $\mathbf{P}(T)$. Solving this equation recursively then gives, in place of (2.3),

$$\mathbf{p}(T) = \mathbf{p}(0) \prod_{i=1}^{T} \mathbf{P}(i). \tag{2.21}$$

It is thus possible to investigate the effect of any specified changes in the transition matrix. In general, there will not be an equilibrium class structure and the complexity of the model makes it difficult to devise useful measures of mobility. Matras (1967) has suggested that the transition probabilities relating to movements between times T and $T + 1$ might be made functions of the class structure at time T. This would seem especially plausible if the classes were based on occupational categories. The movement in or out of a class might then depend on the demand for the skills possessed by its members. The future development of the system would then be completely determined by the initial structure $\mathbf{p}(0)$ and the function relating the transition matrix to the structure on which it operates. The development of this model should prove to be very worthwhile.

Another kind of time-dependence was envisaged by Prais (1955a). He postulated that the discrepancy between the class structure for fathers and sons revealed by Table 2.2 might be due to a change in the definitions of the classes. Thus, suppose that, in the past, the process was time-homogeneous and had attained an equilibrium structure like that observed for fathers. The changes which occur in the next generation are then supposed to take place in two parts. The first is a change of 'true' class governed by the time-homogeneous matrix **P**; the second is a change in 'apparent' class resulting from a change in the system of classification. If the second transition has an associated matrix $\mathbf{R}(T)$ then the observed transition matrix would be $\mathbf{PR}(T)$, If it were possible to estimate $\mathbf{R}(T)$ from census or other information, it would be possible, in turn, to estimate **P** and so predict the development of the process. Prais (1955a) showed that a matrix $\mathbf{R}(T)$ could be constructed that would account for the observed difference in the class distributions of fathers and sons in Glass and Hall's data. The reader will recognize this as an attempt to incorporate a pull aspect into the model.

An apparent dependence of the transition matrix on time can arise due to heterogeneity in the population or to the non-Markovian character of the transitions as will shortly appear. Hodge (1966), who discussed the time-dependent model given above, obtained results illustrating this point by investigating the n-step transition matrix.

A model allowing for a differential birth-rate

The most restrictive and unrealistic of our assumptions is that each father has exactly one son. We have used a rough argument to suggest that our model will be a reasonable approximation if each father has, on average, one son. In this section we shall develop the theory necessary to place these remarks on a firmer footing.

Suppose that the distribution of the number of sons born to a member of the jth class is $P_j(i)$, $(i = 0, 1, 2, \ldots)$ and assume that these probabilities are time-homogeneous. The dependence of the distribution on j enables us to introduce differential birth rates between classes. We assume also that family sizes are independent. Let the mean number of male offspring in the jth class be denoted by v_j. Under this model the population size will not remain constant from one generation to the next so we must work in terms of class numbers instead of proportions. Using a simple conditional probability argument based on conditional expectations we have

$$\bar{n}_j(T+1) = \sum_{i=1}^{k} \bar{n}_i(T) v_i p_{ij} \qquad (j = 1, 2, \ldots, k;\, T \geq 0) \tag{2.22}$$

where $\bar{n}_i(0) = n_i(0)$ is the initial number in class i. This equation generates the expected class sizes in a manner analogous to (2.1) In order to make comparisons with the simple model we shall find it necessary to revert to proportions, writing

$$p_j(T) = \bar{n}_j(T) \Big/ \sum_{i=1}^{k} \bar{n}_i(T) \qquad (j = 1, 2, \ldots, k).$$

It is not strictly accurate to use this notation because the ratio of expected values is not, in general, equal to the expected value of the proportion. However, the distinction is not important for the heuristic reasoning which follows and we shall refer to $p_j(T)$ as defined above as an expected proportion.

If we divide both sides of (2.22) by

$$\sum_{i=1}^{k} \bar{n}_i(T)$$

and use the fact that

$$\sum_{i=1}^{k} \bar{n}_i(T+1) = \sum_{i=1}^{k} \bar{n}_i(T)v_i$$

we obtain the following expression:

$$p_j(T+1) = \sum_{i=1}^{k} p_i(T)v_i p_{ij}/\bar{v}(T) \qquad (j = 1, 2, \ldots, k) \tag{2.23}$$

where

$$\bar{v}(T) = \sum_{i=1}^{k} p_i(T)v_i \bigg/ \sum_{i=1}^{k} p_i(T).$$

This equation was given by Matras (1960b). When the classes have the same birth rate, v_i is independent of i and (2.23) reduces to (2.1). We therefore conclude that, as far as expectations are concerned, the theory developed at the beginning of this section is still applicable when the single family line is replaced by a more realistic branching process.

The effect of a differential birth rate can most easily be seen by writing $p_j(T)$ in the following form:

$$p_j(T) = \sum_{i=1}^{k} p_i(T-1)v_i p_{ij} \bigg/ \sum_{i=1}^{k} p_i(T-1)v_i. \tag{2.24}$$

Consider first the perfectly mobile society in which $p_{ij} = p_j$. In this case (2.24) gives

$$p_j(T) = p_j.$$

Expressed in words, this means that the social structure of a perfectly mobile society is unaffected by a differential birth rate between classes. This conclusion is intuitively obvious. By contrast the birth rates are of crucial importance in an immobile society. In that case $p_{ij} = 1$ if $i = j$ and zero otherwise. From (2.22) we find

$$\begin{aligned} En_j(T+1) &= v_j En_j(T) \\ &= v_j^{T+1} n_j(0) \qquad (j = 1, 2, \ldots, k;\ T \geq 0). \end{aligned}$$

The corresponding expression for $p_j(T)$ is

$$p_j(T) = v_j^T n_j(0) \Big/ \sum_{i=1}^{k} v_i^T n_i(0) \qquad (j = 1, 2, \ldots, k). \tag{2.25}$$

It is clear from the form of this equation that the class with the largest birth rate will eventually dominate the population. In the limit as T tends to infinity we shall have $p_j(\infty) = 1$ if j refers to the class with the highest birth rate and zero otherwise. For degrees of mobility intermediate between the two extremes which we have considered we would expect to find some tendency for classes with high birth rates to increase in size relative to the others. Some idea of the extent to which this is possible can be deduced from (2.24). The right-hand side of this equation is a weighted average of the probabilities p_{ij} $(j = 1, 2, \ldots, k)$ with positive weights. Hence it follows that

$$\min_i p_{ij} \leq p_j(T) \leq \max_i p_{ij} \qquad (j = 1, 2, \ldots, k) \tag{2.26}$$

for all T. Thus, however much the birth rates of the classes may differ, the class structure is bounded by the inequalities (2.26). For example, using the data of Table 2.1, the expected proportion in class 5 (skilled manual and routine non-manual) can never fall below 0·140 or exceed 0·473 as long as the model remains valid.

The theory presented above is clearly capable of further development. Matrix methods have been used in demographic work for many years. Matras (1967) has pointed out that these can easily be adapated to include changes in social class as well as changes in the age structure of a population. Now that Pollard's method for obtaining the moments of class sizes is available, considerable progress should be possible.

Heterogeneity in the transition probabilities

The Markov model of mobility provides a probabilistic description of the progress of a single family line. If all family lines have the same transition probabilities the theory will also describe the aggregate behaviour of a society, as we noted when analysing empirical data on mobility. However, in spite of the modest success we had in fitting the model, it seems rather unlikely that all family lines would have the same transition matrix. It is therefore of interest to see what the effects of differing matrices would be on the aggregate behaviour of the system. The same problem arises in intra-generational mobility, and it is in that connection that most of the work has been done. A basic reference is McFarland (1970); his work was followed by a numerical investigation of the effects of heterogeneity on the limiting structure by Morrison and coworkers (1971).

One way in which the effects of heterogeneity show themselves is in the structure of the T-step transition matrix. In a homogeneous Markov population this will be the Tth power of the one-step matrix. The nature of the departure from this expection is the subject of the following analysis.

Suppose that a proportion $x_i(h)$ of the members of class i at time zero move according to the transition matrix $\mathbf{P}(h)$ $(h = 1, 2, 3, \ldots)$. Then the expected proportion of the whole population in class i who will move to class j between time 0 and time 1 is

$$p_{ij} = \sum_h x_i(h)p_{ij}(h) \qquad (i, j = 1, 2, \ldots, k). \tag{2.27a}$$

In matrix notation this may be written

$$\mathbf{P} = \sum_h \mathbf{X}(h)\mathbf{P}(h) \tag{2.27b}$$

where $\mathbf{X}(h)$ is a square diagonal matrix with (i, i)th element equal to $x_i(h)$. After T steps those who have the transition matrix $\mathbf{P}(h)$ will have moved according to $\mathbf{P}^T(h)$, and therefore the overall transition matrix for the T-step transition will be

$$\mathbf{P}^{(T)} = \sum_h \mathbf{X}(h)\mathbf{P}^{(T)}(h). \tag{2.28}$$

This matrix will approach a limit as $T \to \infty$ because each of the matrices $\mathbf{P}^T(h)$ does so. But unlike the single Markov chain the rows of $\mathbf{P}^{(\infty)}$ will not, in general, be equal because of the unequal weighting introduced by the matrices $\{\mathbf{X}(h)\}$.

If we were only able to observe the one-step matrix given by (2.27b) then the Markov model would lead us to expect that $\mathbf{P}^{(T)}$ would be

$$\mathbf{P}^T = \left\{\sum_h \mathbf{X}(h)\mathbf{P}(h)\right\}^T. \tag{2.29}$$

The question which now arises has to do with how (2.28) differs from (2.29). A partial answer can be found numerically. To take a simple example, suppose that $k = 3$ and that h takes just two values with

$$\mathbf{P}(1) = \begin{pmatrix} 0{\cdot}7 & 0{\cdot}2 & 0{\cdot}1 \\ 0{\cdot}2 & 0{\cdot}6 & 0{\cdot}2 \\ 0{\cdot}2 & 0{\cdot}3 & 0{\cdot}5 \end{pmatrix}, \qquad \mathbf{P}(2) = \begin{pmatrix} 0{\cdot}5 & 0{\cdot}3 & 0{\cdot}2 \\ 0{\cdot}3 & 0{\cdot}4 & 0{\cdot}3 \\ 0{\cdot}3 & 0{\cdot}4 & 0{\cdot}3 \end{pmatrix}.$$

Further, suppose that $x_i(h) = \frac{1}{2}$ $(h = 1, 2; i = 1, 2, \ldots, k)$. Then

$$\mathbf{P} = \tfrac{1}{2}[\mathbf{P}(1) + \mathbf{P}(2)] = \begin{pmatrix} 0{\cdot}60 & 0{\cdot}25 & 0{\cdot}15 \\ 0{\cdot}25 & 0{\cdot}50 & 0{\cdot}25 \\ 0{\cdot}25 & 0{\cdot}35 & 0{\cdot}40 \end{pmatrix}$$

and

$$\mathbf{P}^2 = \begin{pmatrix} 0{\cdot}460 & 0{\cdot}328 & 0{\cdot}213 \\ 0{\cdot}338 & 0{\cdot}400 & 0{\cdot}263 \\ 0{\cdot}338 & 0{\cdot}378 & 0{\cdot}285 \end{pmatrix}, \qquad \mathbf{P}^4 = \begin{pmatrix} 0{\cdot}394 & 0{\cdot}362 & 0{\cdot}244 \\ 0{\cdot}379 & 0{\cdot}370 & 0{\cdot}252 \\ 0{\cdot}379 & 0{\cdot}369 & 0{\cdot}252 \end{pmatrix}.$$

The two- and four-step matrices calculated from (2.28) for this example are

$$\mathbf{P}^{(2)} = \begin{pmatrix} 0{\cdot}475 & 0{\cdot}320 & 0{\cdot}205 \\ 0{\cdot}330 & 0{\cdot}415 & 0{\cdot}255 \\ 0{\cdot}330 & 0{\cdot}370 & 0{\cdot}300 \end{pmatrix}, \qquad \mathbf{P}^{(4)} = \begin{pmatrix} 0{\cdot}407 & 0{\cdot}357 & 0{\cdot}236 \\ 0{\cdot}375 & 0{\cdot}375 & 0{\cdot}250 \\ 0{\cdot}375 & 0{\cdot}371 & 0{\cdot}254 \end{pmatrix}.$$

For this particular population the Markov model underestimates the proportions who will remain in the same grade over both two and four time periods. Is this true in general or is it peculiar to this example?

We shall show that this characteristic is typical but not universal and try to give some insight into the features of the matrices $\{\mathbf{P}(h)\}$ which produce this phenomenon.

To simplify the analysis let $x_i(h) = x(h)$ for all i; it will be sufficient to consider the case $T = 2$. We shall compare the (j, j)th element of (2.28) with that of (2.29). From (2.28)

$$p_{jj}^{(2)} = \sum_h x(h) \sum_{l=1}^{k} p_{jl}(h) p_{lj}(h) \tag{2.30}$$

and from (2.29)

$$p_{jj}^{2*} = \sum_{l=1}^{k} \left\{\sum_h x(h) p_{jl}(h)\right\} \left\{\sum_h x(h) p_{lj}(h)\right\} \tag{2.31}$$

where p_{jj}^{2*} denotes the (j, j)th element of $\mathbf{P}^2$. The difference $p_{jj}^{(2)} - p_{jj}^{2*}$ may thus be written

$$\sum_{l=1}^{k} \sum_h x(h) \left\{p_{jl}(h) - \sum_m x(m) p_{jl}(m)\right\} \left\{p_{lj}(h) - \sum_m x(m) p_{lj}(m)\right\}$$

$$= \sum_{h=1}^{k} \operatorname{cov}(p_{jl}, p_{lj}) \tag{2.32}$$

where the covariances are calculated with respect to the distribution $\{\mathbf{x}(m)\}$. The conditions under which the diagonal elements of $\mathbf{P}^{(2)}$ will be larger than those of $\mathbf{P}^2$ are thus related to the correlation between the flows in opposite directions between each pair of classes. Thus if a large flow (proportionately speaking) from class j to l tends to be accompanied by a large flow in the reverse direction for most pairs of classes the phenomenon observed will occur. One set of conditions under which (2.32) is certainly positive is when the matrices are symmetric, that is with $p_{jl}(h) = p_{lj}(h)$ $(j, l = 1, 2, \ldots, k)$. Matrices having this property are called doubly stochastic because their columns as well as their rows add up to one. The limiting structure for such matrices has equal proportions in each category. Another example, which arises in the next section, occurs when $p_{jl}(h) = p_{lj}(h)p_l(h)/p_j(h)$ $(l, j = 1, 2, \ldots, k)$ where $p_l(h)$ and $p_j(h)$ are the limiting probabilities associated with $\mathbf{P}(h)$; the Glass and Hall matrix very nearly satisfies this condition.

An example in which the diagonal elements do not differ in this systematic way is provided by

$$\mathbf{P}(1) = \begin{pmatrix} 0{\cdot}5 & 0{\cdot}4 & 0{\cdot}1 \\ 0{\cdot}2 & 0{\cdot}5 & 0{\cdot}3 \\ 0{\cdot}2 & 0{\cdot}3 & 0{\cdot}5 \end{pmatrix}, \qquad \mathbf{P}(2) = \begin{pmatrix} 0{\cdot}5 & 0{\cdot}1 & 0{\cdot}4 \\ 0{\cdot}3 & 0{\cdot}5 & 0{\cdot}2 \\ 0{\cdot}3 & 0{\cdot}2 & 0{\cdot}5 \end{pmatrix}$$

with $x_i(h) = \frac{1}{2}$ $(h = 1, 2; i = 1, 2, 3)$. In this case

$$\mathbf{P} = \begin{pmatrix} 0{\cdot}50 & 0{\cdot}25 & 0{\cdot}25 \\ 0{\cdot}25 & 0{\cdot}50 & 0{\cdot}25 \\ 0{\cdot}25 & 0{\cdot}25 & 0{\cdot}50 \end{pmatrix} \quad \text{and} \quad \mathbf{P}^2 = \begin{pmatrix} 0{\cdot}375 & 0{\cdot}312 & 0{\cdot}312 \\ 0{\cdot}312 & 0{\cdot}375 & 0{\cdot}312 \\ 0{\cdot}312 & 0{\cdot}312 & 0{\cdot}375 \end{pmatrix}.$$

But

$$\mathbf{P}^{(2)} = \frac{1}{2}\begin{pmatrix} 0{\cdot}35 & 0{\cdot}43 & 0{\cdot}22 \\ 0{\cdot}26 & 0{\cdot}42 & 0{\cdot}32 \\ 0{\cdot}26 & 0{\cdot}38 & 0{\cdot}36 \end{pmatrix} + \frac{1}{2}\begin{pmatrix} 0{\cdot}40 & 0{\cdot}18 & 0{\cdot}42 \\ 0{\cdot}36 & 0{\cdot}32 & 0{\cdot}32 \\ 0{\cdot}36 & 0{\cdot}23 & 0{\cdot}41 \end{pmatrix} = \begin{pmatrix} 0{\cdot}375 & 0{\cdot}305 & 0{\cdot}320 \\ 0{\cdot}310 & 0{\cdot}370 & 0{\cdot}320 \\ 0{\cdot}310 & 0{\cdot}305 & 0{\cdot}385 \end{pmatrix}.$$

Even with matrices having such different structures as the two in this example, the two-step matrices do not differ very much. The reader will find it quite difficult to construct examples in which $\mathbf{P}^{(2)}$ has smaller diagonal elements than $\mathbf{P}^2$.

One particular special case of the model incorporating differing transition matrices has received a great deal of attention. This is the so-called 'mover–stayer' model which arises if $h = 1, 2$ and $\mathbf{P}(1) = \mathbf{I}$, the unit matrix, and $\mathbf{P}(2) = \mathbf{P}$, an arbitrary transition matrix. According to this model some members of the population never move whereas the remainder move as in the simple Markov model. The 'mover–stayer' model is of more relevance to occupational mobility and we shall return to it in Section 2.3. It was first proposed by Blumen and coworkers (1955) and has since been fitted to British mobility data by Wynn and Sales (1973b). McCall (1971) has used it as a model in the study of income dynamics.

Time reversal in the Markov chain model

The basic difference equation of the Markov model for intergenerational mobility arose by considering the expected class in the next generation conditional upon present class. By this means it has proved possible to predict future class structures and to deduce their limiting form. It is sometimes interesting to proceed in the reverse direction and consider what happens if we try to go backwards in time. The relevant transition probabilities are then $Pr\{\text{family line now in class } i \textit{ came from} \text{ class } j\} = \tilde{p}_{ij}$, say. These probabilities are simply related to the forward transition probabilities as follows. Let $X(T)$ temporarily denote the number of the class in which a particular family line is at time T. Then

$$Pr\{X(T+1) = j, X(T) = i\} = Pr\{X(T) = i\}Pr\{X(T+1) = j|X(T) = i\}$$
$$= Pr\{X(T+1) = j\}Pr\{X(T) = i|X(T+1) = j\}.$$

Equating the two alternative forms of the right-hand side,

$$Pr\{X(T) = i|X(T+1) = j\} = p_{ij}p_i(T)/p_j(T+1) = \tilde{p}_{ji}. \qquad (2.33)$$

Although the reversed process is Markovian it is not in general time-homogeneous. However, if the system has reached the steady state $p_i(T)$ and $p_j(T+1)$ will have their limiting values. In that case, exchanging i and j,

$$\tilde{p}_{ij} = p_{ji}p_j/p_i \qquad (i, j = 1, 2, \ldots, k). \qquad (2.34)$$

It may happen that the backward and forward transition matrices turn out to be the same, that is $\tilde{\mathbf{P}} = \mathbf{P}$. This means that the process appears the same in whichever direction it is viewed in time: we shall call such a process, reversible. One consequence of reversibility is that if we consider someone in class i then the chance that his son is in class j is the same as the chance that his father was in class j. Yet another way of expressing this property is in terms of the expected numbers of changes of class between one generation and the next. Suppose that we have a population of N family lines; then the expected number of changes from class i to class j from one generation to the next will be

$$Np_ip_{ij}.$$

The expected flow in the opposite direction is

$$Np_j p_{ji}.$$

Under the reversibility condition, obtained by putting $\tilde{p}_{ij} = p_{ij}$ in (2.34), these two flows are equal. Reversibility therefore implies an equal exchange between classes. Since the system is in equilibrium there is no *net* change in the class sizes from one generation to the next. But reversibility requires something much stronger; it is as though a family line can only move to a new class if there is a compensatory move in the reverse direction. This does not seem very plausible in the case of social mobility but, as Kemeny and Snell (1960) noted, it does seem to be the case for the Glass and Hall data. The point can be examined by computing the matrix with (i, j)th element equal to $p_i p_{ij}$; this matrix will be symmetric for a reversible process. If we form the matrix for Glass and Hall's data given in Tables 2.1 and 2.2 we obtain

$$\begin{pmatrix} 0{\cdot}0089 & 0{\cdot}0034 & 0{\cdot}0046 & 0{\cdot}0014 & 0{\cdot}0032 & 0{\cdot}0011 & 0{\cdot}0003 \\ 0{\cdot}0045 & 0{\cdot}0112 & 0{\cdot}0095 & 0{\cdot}0050 & 0{\cdot}0087 & 0{\cdot}0022 & 0{\cdot}0008 \\ 0{\cdot}0031 & 0{\cdot}0089 & 0{\cdot}0165 & 0{\cdot}0168 & 0{\cdot}0314 & 0{\cdot}0059 & 0{\cdot}0054 \\ 0{\cdot}0027 & 0{\cdot}0050 & 0{\cdot}0142 & 0{\cdot}0269 & 0{\cdot}0546 & 0{\cdot}0157 & 0{\cdot}0079 \\ 0{\cdot}0037 & 0{\cdot}0098 & 0{\cdot}0307 & 0{\cdot}0503 & 0{\cdot}1935 & 0{\cdot}0699 & 0{\cdot}0511 \\ 0{\cdot}0000 & 0{\cdot}0024 & 0{\cdot}0075 & 0{\cdot}0160 & 0{\cdot}0712 & 0{\cdot}0568 & 0{\cdot}0282 \\ 0{\cdot}0000 & 0{\cdot}0010 & 0{\cdot}0046 & 0{\cdot}0107 & 0{\cdot}0470 & 0{\cdot}0303 & 0{\cdot}0353 \end{pmatrix}. \quad (2.35)$$

The near symmetry of this matrix implies that there is almost equal exchange between all pairs of class. It is difficult to think of any sociological reason why this should be so, and the matter obviously requires further investigation.

Cumulative inertia models

We have already seen how heterogeneity in the transition matrix could explain the occurrence of larger values on the main diagonal of a T-step transition matrix than predicted by Markov theory. This is not the only type of generalization which is capable of producing this phenomenon. An alternative explanation is provided by a model due originally to McGinnis (1968). He introduced the idea of 'cumulative inertia', which describes the situation when an individual's chance of changing class declines with increasing length of stay in that class. Empirical evidence in support of such a hypothesis has been provided by Myers and coworkers (1967), Morrison (1967) and Land (1969), for the case of intragenerational mobility. It has some plausibility in the intergenerational case also. At first sight this assumption destroys the Markov property since the chance of moving now has to depend on past history as well as the present state. This difficulty can be circumvented by redefining the states in a manner which is widely used in the theory of stochastic processes.

Instead of defining the states of the Markov chain in terms of classes (or occupational groups) they are now defined in terms of both class and length of stay in the class. Length of stay is measured in units of the time interval between changes of state and the members of a given class are subclassified according to their length of stay. Each state is thus described by a pair of numbers—the class and the duration of service. The process defined on this set of states is now a Markov chain because we have made length of stay part of the description of the current state. There is nothing in the specification so far to prevent an indefinitely long stay in a given grade, which means that we must allow for an infinite number of states within each class. This takes us out of the realm of finite Markov chains but, at the theoretical level, this presents no insuperable problems. In practice, it is convenient to work with finite Markov chains having a fairly small number of states. Henry and coworkers (1971) suggested a modification of the McGinnis model which achieves this objective. Each class is divided into groups according to length of service, as before, but now all length of service groups above some chosen level are combined into a single group. Thus, for example, the length of service categories might be one, two, three, four and 'more than four' years.

A transition matrix for such a modified model would have the following form, where X stands for a non-zero entry:

$$
\left[\begin{array}{ccccc|ccccc|ccccc}
O & X & \cdots & \cdot & O & X & O & \cdots & \cdot & O & X & O & \cdots & \cdot & O \\
\cdot & O & X & & \cdot & X & O & \cdots & \cdot & O & X & O & \cdots & \cdot & O \\
\vdots & & \ddots & \ddots & \vdots & \vdots & \vdots & & & \vdots & \vdots & \vdots & & & \vdots \\
\cdot & & & O & X & \cdot & \cdot & & & \cdot & \cdot & \cdot & & & \cdot \\
O & \cdots & \cdots & O & X & X & O & \cdots & \cdot & O & X & O & \cdots & \cdot & O \\
\hline
X & O & \cdots & \cdot & O & O & X & \cdots & \cdot & O & X & O & \cdots & \cdot & O \\
X & O & \cdots & \cdot & O & \cdot & O & X & & \cdot & X & O & \cdots & \cdot & O \\
\vdots & \vdots & & & \vdots & \vdots & & \ddots & \ddots & \vdots & \vdots & \vdots & & & \vdots \\
\cdot & \cdot & & & \cdot & \cdot & & & O & X & \cdot & \cdot & & & \cdot \\
X & O & \cdots & \cdot & O & O & \cdots & \cdots & O & X & X & O & \cdots & \cdot & O \\
\hline
X & O & \cdots & \cdot & O & X & O & \cdots & \cdot & O & O & X & \cdots & \cdot & O \\
X & O & \cdots & \cdot & O & X & O & \cdots & \cdot & O & \cdot & O & X & & \cdot \\
\vdots & \vdots & & & \vdots & \vdots & \vdots & & & \vdots & \vdots & & \ddots & \ddots & \vdots \\
\cdot & \cdot & & & \cdot & \cdot & \cdot & & & \cdot & \cdot & & & O & X \\
X & O & \cdots & \cdot & O & X & O & \cdots & \cdot & O & O & \cdots & \cdots & O & X
\end{array}\right]
$$

In a typical row we have the probabilities that an individual who has been in a given class for a certain length of time will move to another state. He either stays in the same class and moves up to the next higher length of service group or he moves to the lowest length of service category of another class. The exception to this pattern is in the last length of service category of each class. In this case someone remaining in the class stays in the same category and so there is a non-zero entry on the diagonal. Various assumptions are possible about the value of these entries. They can be unity, meaning that a person reaching that length of stay remains in the grade for ever; they can be zero if a person must move on reaching that level; or they can have some intermediate value. This last possibility is equivalent to a constant staying probability for all lengths of service beyond the lower limit of the final category. McGinnis' postulate of cumulative inertia requires that the super-diagonal elements in each diagonal submatrix form a non-decreasing sequence. The idea of incorporating the effect of increasing seniority into the Markov model in this way is widely used in manpower planning, using an extended version of the model to be discussed in Chapter 3. For further discussion of this point see, also, Bartholomew (1971).

The long-run behaviour of the cumulative inertia model depends on the structure of the extended transition matrix. If absorption in any class is possible (that is if there is a non-zero chance that someone in a class will never move out) the ultimate structure will depend on the initial structure. The theory of absorbing Markov chains can be used to deal with such cases. In the contrary case the matrix is regular and a limiting structure exists which is independent of the initial structure.

In order to demonstrate that cumulative inertia can produce larger diagonal elements in the T-step transition matrix than would be predicted on Markov theory we consider a simple example. In a two-class system the members of each class are divided up according to whether they have less or more than one year of service. The four states of the system are then (Class 1, <1 year's service), (Class 1, >1 year's service), (Class 2, <1 year's service) and (Class 2, >1 year's service). Suppose that the transition matrix for annual movements is

$$\mathbf{P} = \left(\begin{array}{cc|cc} 0 & 0{\cdot}6 & 0{\cdot}4 & 0 \\ 0 & 0{\cdot}8 & 0{\cdot}2 & 0 \\ \hline 0{\cdot}3 & 0 & 0 & 0{\cdot}7 \\ 0{\cdot}1 & 0 & 0 & 0{\cdot}9 \end{array}\right). \tag{2.36}$$

The dotted lines indicate the class boundaries. This matrix has the same form as the general case shown above, and the chance of staying in a class also increases as length of stay increases. Suppose further that, initially, there are equal numbers in each state. Then someone observing the transitions over one time period, but ignoring length of stay, will record the one-step matrix for the two classes as

$$\mathbf{P}^* = \begin{pmatrix} \frac{1}{2}(0{\cdot}6 + 0{\cdot}8) & \frac{1}{2}(0{\cdot}4 + 0{\cdot}2) \\ \frac{1}{2}(0{\cdot}3 + 0{\cdot}1) & \frac{1}{2}(0{\cdot}7 + 0{\cdot}9) \end{pmatrix} = \begin{pmatrix} 0{\cdot}7 & 0{\cdot}3 \\ 0{\cdot}2 & 0{\cdot}8 \end{pmatrix}.$$

His estimate of the two-step matrix will then be

$$(\mathbf{P}^*)^2 = \begin{pmatrix} 0{\cdot}55 & 0{\cdot}45 \\ 0{\cdot}30 & 0{\cdot}70 \end{pmatrix}. \tag{2.37}$$

However, the actual transitions will be governed by (2.36) for which

$$\mathbf{P}^2 = \left(\begin{array}{cc|cc} 0{\cdot}12 & 0{\cdot}48 & 0{\cdot}12 & 0{\cdot}28 \\ 0{\cdot}06 & 0{\cdot}64 & 0{\cdot}16 & 0{\cdot}14 \\ \hline 0{\cdot}07 & 0{\cdot}18 & 0{\cdot}12 & 0{\cdot}63 \\ 0{\cdot}07 & 0{\cdot}06 & 0{\cdot}04 & 0{\cdot}81 \end{array}\right). \tag{2.38}$$

For someone who fails to distinguish the length of service categories the observed two-step matrix between classes will be

$$(\mathbf{P}^2)^* = \begin{pmatrix} \frac{1}{2}(0{\cdot}12 + 0{\cdot}48 + 0{\cdot}06 + 0{\cdot}64) & \frac{1}{2}(0{\cdot}12 + 0{\cdot}28 + 0{\cdot}16 + 0{\cdot}14) \\ \frac{1}{2}(0{\cdot}07 + 0{\cdot}18 + 0{\cdot}09 + 0{\cdot}06) & \frac{1}{2}(0{\cdot}12 + 0{\cdot}63 + 0{\cdot}04 + 0{\cdot}81) \end{pmatrix}$$

$$= \begin{pmatrix} 0{\cdot}65 & 0{\cdot}35 \\ 0{\cdot}20 & 0{\cdot}80 \end{pmatrix}. \tag{2.39}$$

When compared with (2.37) we see that the incorporation of cumulative inertia has the effect of producing the higher diagonal elements which the theory was designed to explain.

2.3 MODELS FOR OCCUPATIONAL MOBILITY

A basic model

There are obvious affinities between the movement of family lines between social classes and the movement of employees between jobs or of households from one area to another. In fact, the simple Markov model with which we began this chapter has often been proposed as a model of occupational or geographical mobility. However, these types of mobility are intragenerational and so there is no fixed time interval between successive moves. People can change their jobs or move their homes at any time and so different individuals will have made differing numbers of moves in any given time. It is not surprising, therefore, that the simple Markov model has been less successful in describing intragenerational mobility. An adequate model must not only specify which changes of state take place but, also, when they take place. A full analysis thus requires continuous time models of the kind which we shall discuss in Chapter 5. However, in practice, it is often impossible to observe a mobility process continuously and we have to be content with observations of its state at fixed intervals of time. As far as this aspect is concerned the process can be modelled in terms of a matrix of transition probabilities and a probability distribution of the number of transitions which take place in the intervals between the observations.

The stochastic models to be described are based on the classic study of Blumen, Kogan and McCarthy (1955) (referred to hereafter as BKM). They investigated the flows of employees between occupational categories in the United States, using data for the whole population. Their models are also relevant to the study of flows between individual employers in a given locality and to geographical mobility. In fact, occupational and geographical mobility are closely linked because a change of job is often accompanied by a change of residence. For the sake of clarity we shall refer to occupational mobility throughout. The original BKM work has been developed in a number of directions and applied to other sets of mobility data. Some references are given in the course of the discussion.

We shall follow the same pattern of development as in Section 2.2, beginning with a very simple model and then generalizing it in various ways suggested by the data on mobility. It will be assumed throughout that we have a closed system of employers (or industries) and employees. An equivalent assumption is that any losses from the system are made good by new members who are similar in all relevant respects to those that they replace. In practice it will usually be necessary to introduce 'unemployed' as an occupational category if the system is to be closed. Our model must specify the means by which changes of state are determined and also the timing of these changes. We do this as follows.

In the employment history of a given individual we postulate a number of 'decision' points at which he considers changing his employment. At each of these points he may move to another job or remain where he is. We assume that the changes of employment are governed by time-homogeneous transition probabilities p_{ij} $(i, j = 1, 2, \ldots, k)$, where k is the number of employment categories. Thus the diagonal elements $\{p_{ii}\}$ of the transition matrix give the probabilities that no change is made. The model can be given a slightly different interpretation as follows. Suppose that the k categories represent fairly broad industrial classifications. The decision points could be replaced by 'change' points at which the employee leaves his present employer. Under this interpretation the probabilities $\{p_{ii}\}$ would relate to a move to a different employer within the same industrial classification. We shall use the terminology appropriate to the first variant of the model, but our results can easily be translated into the language of the second. The specification of the model is completed by supposing that the decision points for each individual are realizations of the same time-homogeneous point process. The number of decision points in the interval $(0, T)$ will be a random variable which we shall denote by $m(T)$. Let $Pr\{m(T) = m\}$ be denoted by $P_m(T)$. In some applications in which employees can only leave their jobs at the end of a week or month it would be more realistic to treat the intervals between decision points as discrete variables. This refinement does not seem justified at the present stage and we shall confine the discussion to the continuous case. A process in which changes of state occur according to a Markov chain and in which the time intervals between changes are random variables is sometimes known as a semi-Markov process or a Markov renewal process. The theory of such processes has been discussed, for example, by Pyke (1961a, 1961b) and Taga (1963).

If a complete job history were available for each employee in the system, then its two main features could be studied separately. The data on job changes, or decisions on whether to change, could be used to estimate the transition probabilities and hence predict the equilibrium occupational structure in the same way as the model for social mobility. Similarly, the record of decision points could be made the basis for studying the stochastic process $m(T)$. In the BKM study, complete information was lacking because the records only gave the current occupations at quarterly intervals. There was no record of changes that had taken place during these intervals. This is a common situation and the particular interest of the BKM model is that it shows what can be learned from such limited data.

We suppose that a job census is taken at regular intervals, recording the numbers of employees in each occupational category. For convenience we shall refer to the interval between censuses as a quarter, as in the BKM investigation. In the formulae we shall retain full generality, denoting it by τ. Let $p_{ij}(\tau)$ be the probability that a man who is in category i at the beginning of a quarter is in category j at the beginning of the next quarter. We shall assume that the process $\{m(T)\}$ is time-homogeneous and hence that the transition probabilities $p_{ij}(\tau)$ do not depend on T. If there are exactly m decision points during the quarter we have $p_{ij}(\tau) = p_{ij}^{(m)}$. However, m is itself a random variable with distribution $P_m(\tau)$. Hence

$$p_{ij}(\tau) = \sum_{m=0}^{\infty} P_m(\tau) p_{ij}^{(m)} \qquad (i, j = 1, 2, \ldots, k) \tag{2.40}$$

where we define $p_{ii}^{(0)} = 1$ and $p_{ij}^{(0)} = 0$ $(i \neq j)$. The matrix $\mathbf{P}(\tau) = \{p_{ij}(\tau)\}$ can be observed in practice, whereas $\mathbf{P}$ cannot. Our analysis will therefore be devoted to discovering how much about the future of the process can be deduced from $\mathbf{P}(\tau)$ alone.

The answer to this question would present no problems if the states of the system at successive quarters could be treated as a Markov chain. We would then be in a situation identical with that of the last section with the 'generations' replaced by 'quarters'. Unfortunately this is the case only in rather special circumstances. An obvious condition for the system to be a Markov chain, which is both necessary and sufficient, is that

$$p_{ij}(r\tau) = p_{ij}^{(r)}(\tau) \qquad (r = 1, 2, 3, \ldots) \tag{2.41}$$

where $p_{ij}^{(r)}(\tau)$ denotes the r-quarter transition probability for the Markov chain with transition matrix $\mathbf{P}(\tau)$. We now ask for what kinds of stochastic process $\{m(T)\}$, if any, does (2.41) hold. The probability on the left-hand side of (2.41) is

$$\sum_{m=0}^{\infty} P_m(r\tau) p_{ij}^{(m)}.$$

That on the right-hand side is the (i, j)th element in the matrix

$$\left\{ \sum_{m=0}^{\infty} P_m(\tau) \mathbf{P}^m \right\}^r$$

where we define $\mathbf{P}^0$ to be the unit matrix $\mathbf{I}$. The condition expressed by (2.41) is thus equivalent to

$$\sum_{m=0}^{\infty} P_m(r\tau) \mathbf{P}^m = \left\{ \sum_{m=0}^{\infty} P_m(\tau) \mathbf{P}^m \right\}^r. \tag{2.42}$$

There are only two stochastic processes $\{m(T)\}$ of practical interest for which this condition holds. The first is the degenerate case when decision points occur at regular intervals. Under these circumstances

$$P_m(\tau) = 1 \text{ if } m = m_0$$
$$= 0 \text{ otherwise}$$

and

$$P_m(r\tau) = 1 \text{ if } m = rm_0,$$

where m_0 is the fixed number of decisions made in an interval of length τ. The second case when (2.42) is satisfied is when $\{m(T)\}$ is a Poisson process. Then we have, when the parameter of the process is λ,

$$\left.\begin{aligned} P_m(\tau) &= \frac{(\lambda\tau)^m}{m!}\mathrm{e}^{-\lambda\tau} \\ P_m(r\tau) &= \frac{(\lambda\tau r)^m}{m!}\mathrm{e}^{-\lambda r\tau} \end{aligned}\right\} \quad (m = 0, 1, 2, \ldots)$$

and (2.42) is satisfied. The Poisson process implies that the intervals between decision points are independently and exponentially distributed.

The foregoing theory shows that the one-quarter transition matrix only defines a Markov chain if the decision points occur either regularly or randomly in time. If, therefore, we observe a discrepancy in practice between the predicted and observed 'r-quarter' transition matrices, one explanation could be that the stochastic process governing when movements take place is neither regular nor random. Such a discrepancy did, in fact, occur in the BKM study. Using the observed one-quarter matrix the authors estimated those for four and eight quarters using Markov chain theory. The result was that the actual matrices for the longer periods did not agree with the predictions. A typical example is given in Table 2.5, which shows that the Markov prediction underestimates the elements on the main diagonal.

The letters A–K denote industry codes and U is the unemployed category. It is thus clear that, within the framework of the model, the hypothesis of regular or completely random spacing of decision points is untenable. We shall therefore present a more general model and show that it is capable of explaining the discrepancies which we have observed.

A model allowing for individual differences

The direct approach to the problem of reconciling the differences between theory and observation would be to search for a stochastic process $\{m(T)\}$ which gives a better fit. We shall find such a process but it will be arrived at indirectly. According to our model the number of decision points in each quarter for any individual will vary. On the simple model the pattern of variation will be the same for all individuals in the population. Although we have no direct evidence on this point it would be remarkable if there were no individual differences in the rate of occurrence of decision points. We shall therefore generalize the original model in such a way that each individual may have a different 'decision rate'. Let us assume that the random variable

Table 2.5 Comparison of observed and predicted eight-quarter transition probabilities for males aged 20–24. The upper figure is the observed proportion and the lower that predicted by Markov theory [a]

Industry code	A	B	C	D	E	F	G	H	J	K	U
A	0·000	0·062	0·062	0·000	0·125	0·156	0·312	0·000	0·000	0·000	0·281
	0·002	0·086	0·105	0·042	0·116	0·053	0·181	0·016	0·058	0·004	0·337
B	0·003	0·449	0·039	0·020	0·048	0·035	0·079	0·014	0·023	0·006	0·284
	0·002	0·144	0·087	0·040	0·104	0·050	0·163	0·018	0·052	0·004	0·336
C	0·002	0·037	0·461	0·023	0·046	0·021	0·101	0·007	0·022	0·002	0·278
	0·002	0·077	0·176	0·039	0·103	0·046	0·163	0·106	0·050	0·004	0·324
D	0·000	0·064	0·044	0·459	0·083	0·024	0·091	0·011	0·030	0·002	0·192
	0·001	0·070	0·080	0·218	0·099	0·046	0·141	0·015	0·047	0·003	0·279
E	0·002	0·045	0·042	0·034	0·489	0·031	0·094	0·010	0·023	0·002	0·227
	0·001	0·072	0·075	0·040	0·276	0·046	0·147	0·013	0·044	0·004	0·279
F	0·003	0·056	0·033	0·022	0·054	0·440	0·090	0·020	0·026	0·010	0·245
	0·002	0·081	0·076	0·038	0·097	0·166	0·152	0·017	0·050	0·004	0·316
G	0·002	0·047	0·051	0·025	0·046	0·038	0·491	0·020	0·044	0·002	0·235
	0·002	0·080	0·084	0·039	0·098	0·049	0·261	0·017	0·053	0·004	0·314
H	0·000	0·044	0·007	0·015	0·026	0·085	0·096	0·439	0·074	0·000	0·214
	0·001	0·077	0·077	0·035	0·090	0·048	0·170	0·158	0·052	0·004	0·287
J	0·002	0·061	0·033	0·018	0·054	0·035	0·145	0·019	0·339	0·000	0·294
	0·002	0·084	0·085	0·038	0·105	0·049	0·178	0·018	0·105	0·004	0·333
K	0·000	0·113	0·097	0·032	0·121	0·048	0·137	0·032	0·024	0·048	0·347
	0·002	0·089	0·096	0·047	0·130	0·048	0·179	0·023	0·056	0·006	0·325
U	0·001	0·069	0·068	0·035	0·077	0·040	0·153	0·018	0·055	0·004	0·482
	0·002	0·090	0·095	0·042	0·112	0·052	0·179	0·019	0·058	0·004	0·346

[a] Data from Blumen and coworkers (1955).

$m(T)$ has a distribution $P_m(T; \lambda)$, where λ is a parameter which characterizes the individual. If the variation of λ in the population is described by a distribution function $F(\lambda)$ then the observed probability law of the process will be

$$P_m(T) = \int_0^\infty P_m(T; \lambda)\, \mathrm{d}F(\lambda). \tag{2.43}$$

Without loss of generality we may define λ in such a way that it is proportional to $\overline{m}(T)$ and hence its range of variation may be limited to $(0, \infty)$. This is a different kind of heterogeneity to that which we introduced in Section 2.2. There we supposed that all people changed state at the same time but according to different transition probabilities. Here we suppose they all have the same transition matrix but that some people change more frequently than others.

It will be shown below that the probabilities given by (2.43) do not satisfy the condition of (2.42), even if $P_m(T; \lambda)$ is a Poisson distribution. Hence there is no immediate need to take $m(T)$ as anything other than a Poisson process. We must therefore first consider whether it is possible to find an $F(\lambda)$ which, when combined with a Poisson process, leads to a tenable hypothesis. No general results have been obtained but the following example, due to Blumen and coworkers (1955), answers the question we have raised. Suppose that the transition matrix $\mathbf{P}$, which would be unknown in practice, is

$$\begin{pmatrix} 0{\cdot}70 & 0{\cdot}15 & 0{\cdot}15 \\ 0{\cdot}20 & 0{\cdot}60 & 0{\cdot}20 \\ 0{\cdot}25 & 0{\cdot}25 & 0{\cdot}50 \end{pmatrix}. \tag{2.44}$$

Let $P_m(T; \lambda)$ be a Poisson distribution with mean λT. We suppose that for half of the employees $\lambda = 1/10$ and that for the other half $\lambda = 7/10$. Then

$$P_m(\tau) = \frac{1}{2m!}\left\{\left(\frac{\tau}{10}\right)^m e^{-\tau/10} + \left(\frac{7\tau}{10}\right)^m e^{-7\tau/10}\right\} \qquad (m = 0, 1, 2, \ldots). \tag{2.45}$$

If the unit of time between censuses is unity we shall assume that

$$P_0(1) = 0{\cdot}701, \qquad P_1(1) = 0{\cdot}219, \qquad P_2(1) = 0{\cdot}063,$$
$$P_3(1) = 0{\cdot}014, \qquad P_4(1) = 0{\cdot}002.$$

The mean of this distribution is 0·395 and the variance is 0·473, indicating a higher dispersion than the Poisson distribution having the same mean. Substitution of these values and the matrix (2.44) in (2.40) gives

$$\mathbf{P}(\tau = 1) = \begin{pmatrix} 0{\cdot}90 & 0{\cdot}05 & 0{\cdot}05 \\ 0{\cdot}07 & 0{\cdot}87 & 0{\cdot}06 \\ 0{\cdot}08 & 0{\cdot}08 & 0{\cdot}84 \end{pmatrix}. \tag{2.46}$$

This is the matrix which would be calculated from the data collected for two successive quarters. Treating it as the transition matrix of a Markov chain we would predict the eight-quarter matrix by raising the matrix (2.46) to the eighth power. Thus

$$\{\mathbf{P}(\tau = 1)\}^8 = \begin{pmatrix} 0{\cdot}54 & 0{\cdot}25 & 0{\cdot}21 \\ 0{\cdot}33 & 0{\cdot}44 & 0{\cdot}24 \\ 0{\cdot}35 & 0{\cdot}30 & 0{\cdot}36 \end{pmatrix}.$$

However, the eight-quarter matrix which we would expect from our present model is

$$\mathbf{P}(\tau = 8) = \begin{pmatrix} 0{\cdot}63 & 0{\cdot}21 & 0{\cdot}17 \\ 0{\cdot}26 & 0{\cdot}55 & 0{\cdot}19 \\ 0{\cdot}28 & 0{\cdot}23 & 0{\cdot}48 \end{pmatrix}.$$

The heterogeneity which we have introduced thus leads us to underestimate the diagonal elements. This is exactly what Blumen and coworkers (1955) found in practice.

It is not necessary to assume that there are individual differences in the population. The same result would be obtained if the decision points occurred according to (2.45) for each person. This might be the case if decision points were of two kinds, one kind occurring at rate $\lambda = 1/10$ and the other at rate $\lambda = 7/10$. It is an unfortunate fact that we cannot, on the data available, distinguish between these two models for the decision point process. For predictive purposes they are identical; if we require a full explanation of the process, data on individual case histories is necessary.

Spilerman (1972b) has examined the case when λ varies continuously by assuming that

$$\mathrm{d}F(\lambda) = \frac{c^\nu}{\Gamma(\nu)}\lambda^{\nu-1}\,\mathrm{e}^{-c\lambda}\,\mathrm{d}\lambda \qquad (\lambda \geq 0;\, c, \nu > 0).$$

This is a flexible form capable of describing patterns of variability ranging from a J-shaped distribution at one extreme to a unimodal symmetric distribution at the other. We shall use it again in this connexion in Chapter 6. Substituting in (2.43) we have, for the time interval τ,

$$\begin{aligned} P_m(\tau) &= \int_0^\infty \frac{(\lambda\tau)^m}{m!}\,\mathrm{e}^{-\lambda\tau}\frac{c^\nu}{\Gamma(\nu)}\lambda^{\nu-1}\,\mathrm{e}^{-c\lambda}\,\mathrm{d}\lambda \\ &= \frac{c^\nu\tau^m}{m!\,\Gamma(\nu)}\int_0^\infty \lambda^{m+\nu-1}\,\mathrm{e}^{-\lambda(c+\tau)}\,\mathrm{d}\lambda \\ &= \binom{m+\nu-1}{m}\left(\frac{c}{c+\tau}\right)^\nu\left(\frac{\tau}{c+\tau}\right)^m \end{aligned} \tag{2.47}$$

where we define $x! = \Gamma(x + 1)$ for all x. Substitution of $P_m(\tau)$ in the matrix form of (2.40) now gives

$$\mathbf{P}(\tau) = \left(\frac{c}{c+\tau}\right)^{\nu} \sum_{m=0}^{\infty} \binom{m+\nu-1}{m} \left(\frac{\tau}{c+\tau}\right)^{m} \mathbf{P}^{m}. \tag{2.48}$$

If $\mathbf{P}$ were a scalar this would be a negative binomial series with sum

$$\left(\frac{c}{c+\tau}\right)^{\nu} \left\{\mathbf{I} - \frac{\tau}{c+\tau}\mathbf{P}\right\}^{-\nu}, \tag{2.49}$$

assuming convergence. We can, of course, define (2.49) as the infinite series (2.48), but this leaves us with the problem of how to compute (2.49) bearing in mind that ν need not be an integer. To avoid the problem we use the fact that $\mathbf{P}$ can be written in the form

$$\mathbf{P} = \mathbf{H}\mathbf{D}\mathbf{H}^{-1}$$

where $\mathbf{D}$ is a matrix having the eigenvalues of $\mathbf{P}$ on its main diagonal and zeros elsewhere, and $\mathbf{H}$ is a matrix whose ith column is the eigenvector associated with the ith eigenvalue of $\mathbf{P}$. It is easily verified that

$$\mathbf{P}^m = \mathbf{H}\mathbf{D}^m\mathbf{H}^{-1}.$$

Hence

$$\mathbf{P}(\tau) = \left(\frac{c}{c+\tau}\right)^{\nu} \mathbf{H}\left\{ \sum_{m=0}^{\infty} \binom{m+\nu-1}{m} \left(\frac{\tau}{c+\tau}\right)^{m} \mathbf{D}^{m}\right\}\mathbf{H}^{-1}. \tag{2.50}$$

The infinite matrix sum in brackets is a diagonal matrix whose (i, i)th element is the scalar sum

$$\sum_{m=0}^{\infty} \binom{m+\nu-1}{m} \left(\frac{\tau}{c+\tau}\right)^{m} \alpha_i^m = \left(1 - \frac{\tau\alpha_i}{c+\tau}\right)^{-\nu} \tag{2.51}$$

where α_i is the ith eigenvalue of $\mathbf{P}$. These series converge because, for a stochastic matrix, $\alpha_i \leq 1$ for all i and $\tau/(c + \tau) < 1$ for $\tau > 0$, implying that $\tau\alpha_i/(c + \tau) < 1$. An equivalent way of summing the series (2.48) is to use the spectral representation of $\mathbf{P}^m$ given in Chapter 3, equation (3.5).

Spilerman (1972b) made some numerical comparisons between the actual T-step matrix for this process and $\{\mathbf{P}(\tau)\}^T$. He found that the former had larger diagonal elements, just as with the two-point distribution for λ illustrated above.

The 'mover-stayer' model

Having shown that heterogeneity in the process $\{m(T)\}$ is a desirable ingredient of the model we must consider what form this should take. There is a certain plausibility in what Blumen and coworkers (1955) termed the 'mover-stayer' model. According to this model a certain proportion, $F(0)$, of employees do not change their jobs at all. These are the 'stayers'. The 'movers' are those who change employment in the way postulated by our model. If we retain the assumption of heterogeneity for the movers the one-quarter transition matrix will now be

$$\mathbf{P}(\tau) = F(0)\mathbf{I} + \int_0^\infty \sum_{m=0}^\infty P_m(\tau;\lambda)\mathbf{P}^m \, dF(\lambda) \tag{2.52}$$

where $\mathbf{I}$ is the unit matrix. We could allow the proportion of stayers and the distribution $F(\lambda)$ to vary from one category to another, but the simple assumption will suffice for the present purpose. Blumen and coworkers (1955) assumed homogeneity and supposed that the decision process was Poisson with the same λ for all movers. In this case (2.52) becomes

$$\mathbf{P}(\tau) = F(0)\mathbf{I} + \{1 - F(0)\} \sum_{m=0}^\infty \frac{(\lambda\tau)^m}{m!} e^{-\lambda\tau}\mathbf{P}^m. \tag{2.53}$$

Even with this crude model the predictions obtained were a marked improvement over the simple Markov version in the case of the eight-quarter matrix. The improvement was less marked for the four- and eleven-quarter matrices. Table 2.6 shows the observed and predicted values of the diagonal elements of the two matrices for the example used in Table 2.5.

Table 2.6 Comparison of observed and predicted values of the diagonal elements in the eight-quarter transition matrices for males aged 20–24[a]

Industry code	A	B	C	D	E	F	G	H	J	K	U
Observed	0·000	0·449	0·461	0·459	0·489	0·440	0·491	0·439	0·339	0·048	0·482
Predicted	0·003	0·442	0·464	0·474	0·512	0·444	0·489	0·446	0·338	0·049	0·536

[a] Data from Blumen and coworkers (1955).

Other generalizations of the basic model are possible. We could allow the matrix $\mathbf{P}$ to be time-dependent as we did in the social mobility model. In addition, we could introduce more elaborate stochastic processes to account for the decision points. Since the models developed above are capable of describing mobility patterns which have been observed in practice there is no immediate need for such generalizations. However, if data can be obtained on such things as individual job histories the scope for useful model-building will be enlarged.

Limiting behaviour of the models

When we turn to the limiting occupational structure the differences between all but one of our models vanish. First we consider the basic model and show that the limiting behaviour does not depend on the stochastic process $m(T)$. More precisely we shall show that, if $\mathbf{P}$ is regular, then

$$\lim_{r\to\infty} \mathbf{P}^r(\tau) = \lim_{r\to\infty} \mathbf{P}^r. \tag{2.54}$$

This is an extremely useful result. It means that we can compute the limiting occupational structure from $\mathbf{P}(\tau)$ by treating the quarterly structures *as if* they were realizations of a Markov chain. To prove (2.54) we find the limit of the left-hand side as follows.

$$\mathbf{P}(\tau) = \sum_{m=0}^{\infty} P_m(\tau)\mathbf{P}^m \tag{2.55}$$

by (2.40). We may thus think of $\mathbf{P}(\tau)$ as a matrix-valued generating function of the probability distribution $P_m(\tau)$. On raising the right-hand side of (2.55) to the rth power we thus obtain the generating function of the r-fold convolution of the distribution of $m(\tau)$. Hence we may write

$$\mathbf{P}^r(\tau) = \sum_{m=0}^{\infty} Pr\{S(r) = m\}\mathbf{P}^m$$

where $S(r)$ is the sum of r independent observations on $m(\tau)$. By applying Chebychev's inequality to the distribution of $S(r)$ we may show that

$$\lim_{r\to\infty} Pr\{S(r) \leq m\} = 0$$

for any finite m and therefore

$$\lim_{r\to\infty} \sum_{m=0}^{\infty} Pr\{S(r) = m\}\mathbf{P}^m = \lim_{m\to\infty} \mathbf{P}^m.$$

If we assume that the system we are studying has reached equilibrium we can test whether the changes of state follow a Markov chain. This test can easily be made with the BKM data. Taking the observed quarterly transition matrix we can predict the equilibrium distribution using (2.5).

A comparison of an observed and predicted structure taken from Blumen and coworkers is given in Table 2.7.

Table 2.7 Actual and predicted occupation structure using the Markov model and the quarterly transition matrix [a]

Occupational group	(C, D, E)	G	(F, H)	(A, B, J, K)	U	*Total*
Average percentage of workers observed	28·2	17·0	6·8	13·7	34·3	100·0
Predicted percentage using Markov model	27·0	18·0	8·0	15·0	32·0	100·0

[a] Data from Blumen and coworkers (1955).

The close agreement supports the hypothesis that the changes of occupation are accounted for by the Markov chain model. The discrepancies between the predicted and observed four- and eight-quarter transition matrices, noted earlier, point to the need for $\{m(T)\}$ to be a non-Poissonian point process. We must now consider whether the agreement observed in Table 2.7 is compatible with the models involving individual differences described earlier in this section. This is obviously the case because we have proved the result for any distribution $P_m(T)$ and hence, in particular, for those having the form of (2.43).

The mover-stayer model requires special treatment. Our theory as developed above applies only to the movers. They will reach an equilibrium structure $\mathbf{p}$ while the stayers remain in their original categories. The matrix giving the probabilities of transitions from state i to state j in r steps will thus have the limiting form

$$F(0)\mathbf{I} + \{1 - F(0)\}\mathbf{P}^{\infty}.$$

This is not the same form as we would get by finding $\lim_{r\to\infty} \{\mathbf{P}(\tau)\}^r$ from (2.53). The matrix $\mathbf{P}(\tau)$ for the mover-stayer model can be written in the form of (2.55) and hence it follows that

$$\lim_{r\to\infty} \{\mathbf{P}(\tau)\}^r = \mathbf{P}^{\infty}.$$

Therefore, if we treat $\mathbf{P}(\tau)$ as the transition matrix of a Markov chain the limiting structure that we shall predict is that of the movers only. The reason for this result is worth elaborating. On the basis of the one-quarter transition matrix alone we cannot distinguish between the mover-stayer model and one in which *all* members are movers with transition matrix

$F(0)\mathbf{I} + \{1 - F(0)\}\mathbf{P}$. The limiting result obtained is the one appropriate to the second model rather than the first. We cannot therefore treat the mover-stayer model as a Markov process, in discrete time with transition matrix $\mathbf{P}(\tau)$. This fact raises interesting estimation problems which have been discussed by Blumen and coworkers (1955) and Goodman (1961). A more recent discussion has been given by Wynn and Sales (1973b) who have fitted the model to British data on occupational mobility.

Concluding remarks

We have assumed that the system we have been studying is closed. This is a reasonable assumption in the short term but in view of our emphasis on limiting results it requires further comment. We remarked in the introduction that our model would cover open systems in which all losses were made good by 'equivalent' replacements. In a labour force whose size remains constant over long periods we may expect this requirement to be met in an average sense. A full justification of this remark must await the development of a theory of open systems to which we turn in the following chapter.

When discussing social mobility we included a section on the measurement of social mobility. It was argued that any such measures should be based on the matrix of transition probabilities. In the case of labour mobility we would similarly argue that the transition matrix and the stochastic process $\{m(T)\}$ need to be taken into account. This is a problem which would repay further research.

A fuller exploration of the application of semi-Markov processes to mobility processes has been made by Ginsberg (1971) and reviewed by him in Ginsberg (1972a). In his general formulation the lengths of time between transitions are allowed to depend on both the origin and the destination of the move. The long-run behaviour of such a system will depend only on the transition matrix, but the transient behaviour will also depend on k^2 probability distributions—one for each possible type of transition. Unless very large quantities of data are available the prospect of adequately estimating so many parameters and hence of judging the goodness of fit of the model is remote. The number of parameters can be reduced by, for example, making all the intermove distributions the same, which brings us back to the basic models of this section.

The chief value of the general semi-Markov model is that it can be used to examine the qualitative effects of the departures from the rather simple assumptions underlying the models of this chapter since it includes them all as special cases. In practice, however, the mathematical analysis is difficult and the potential of the model has yet to be realized.

CHAPTER 3

Markov Models for Educational and Manpower Systems

3.1 INTRODUCTION

The social systems described in Chapter 2 were closed. This meant either that no members moved in or out of the system or that any losses were replaced immediately by identical recruits. Our interest in that chapter was in the changing internal structure of the system. The assumption of a closed system was reasonable for the applications to social class and labour mobility for reasons which have already been explained. Nevertheless, there are many systems in which gains and losses are an important feature of the process. In this chapter we shall, therefore, give two generalizations of the closed Markov model in which recruitment and loss appear explicitly.

A feature common to each model is the stochastic mechanism governing losses from the system. This can easily be incorporated into the Markov chain model by associating a time-homogeneous loss probability with each grade or stratum. The difference between the two models lies in the factors which are supposed to control the input. For clarity we shall describe the situation by reference to two applications without intending thereby to place any limitation on the applicability of the models.

In the first model the number of recruits to the system at time T is either a known quantity or a realization of a known stochastic process. Under these conditions the sizes of the individual grades will be random variables with distributions determined by the stochastic nature of the loss and input. One example of a situation in which such a model is appropriate is provided by an educational system. The model described in Section 3.2, was, in fact, proposed by Gani (1963) for the university system of Australia. He wished to predict the total enrolment in universities and the numbers of degrees to be awarded in the future. The system thus consisted of the total university population of the country. The strata, or grades, were the four undergraduate and three postgraduate years of study. At the end of each year three alternatives were available to any student. He could move into the next higher year, by passing an examination; he could repeat the year if he failed the examination; or he could leave the university. Past data showed that it was reasonable to suppose that each kind of transition had a fixed probability. The input to the educational system consists of the total number of qualified students reaching the age for university entrance. This number could be accurately predicted up to eighteen years ahead from the known numbers in each age cohort. In practice there might be constraints affecting the size of the system but this does not detract from the usefulness of the model for predicting the demand for university places.

The same model has been widely used in other educational applications. Gani extended his treatment, in unpublished work, to accommodate the American university system, where progress is by 'credits' rather than years. Applications to the student population of the University of California have been made by Marshall (1973) and by Oliver and his coworkers in several unpublished reports. Clough and McReynolds (1966) have modelled the secondary school system in Ontario by extending the basic idea of Gani's model. Thonstad's (1969) work on the Norwegian educational system includes an extensive discussion of the Markov model, and the capacity models of Menges and Elstermann (1971) incorporate a Markovian component. Armitage, Smith and Alper (1969) have discussed the applicability of the model in educational planning and Armitage, Phillips and Davies (1970) dealt with it in relating to a model of the English secondary school system. Kamat (1968b, 1968c) proposed a special case of the model suitable for describing the progress of a cohort through the educational system.

The second field of application is in manpower planning. The states of the system then denote grades, age groups or other relevant classifications of the employees in a firm. To be applicable in the form described above it is necessary for the recruitment to the firm to be determined exogeneously. In practice this will not usually be the case because recruitement is often directly linked to the occurrence of vacancies. However, there are two sets of circumstances in which it is meaningful to consider a model in which recruitment numbers are given. One is where the demand for labour exceeds the supply; in this case it is the supply which controls the input and this is determined externally. The other is when, as a planning exercise, we wish to forecast the consequences of a fixed pattern of recruitment. An example of the latter use is given in Sales (1971) and Forbes (1971a).

The second of the two models to be considered in this chapter is particularly relevant in the manpower planning context. In it we assume that the total size of the system is fixed rather than the total number of recruits. The recruitment needs are then determined by the losses together with any change which is planned in the size of the system. If this version of the model is to be realistic there must always be a sufficient number of recruits available to fill any vacancies. In other words, the supply must exceed the demand.

Using the terminology of the last chapter, recruitment is a 'push' flow in Gani's version of the model and a 'pull' flow in the present case. The version of the model with a fixed total size is due to Young and Almond (1961), who developed the theory of the model and applied it to the staff structure of a British university. Since then the model has been widely used in manpower planning—see Young (1971) for further examples. In manpower applications, where the states are grades, the internal transitions will correspond to promotion, demotion or transfers. Young and Almond (1961) intended their model to be used for expanding systems but, with minor modifications, it can be adopted for use with any pattern of change.

In both versions of the model our main interest is in the 'stocks' of people in the various 'grades'. These stocks change over time as a result of the operation of fixed probabilities of flow between grades. The main emphasis in this chapter will be on the stochastic behaviour of the stock numbers and, in particular, on their means and variances. Of even greater interest in some applications is the question of the control of the stock numbers by manipulating the flows. This aspect of the model will be pursued in the next chapter in relation to manpower planning.

The model with given input is dealt with first in Section 3.2 and that with given total size follows in Section 3.3. Here, as elsewhere, we shall not discuss the statistical problems of estimation and hypothesis testing, but the interested reader may pursue the question in Sales (1971), Forbes (1971a) and Mahoney and Milkovich (1971).

3.2 A MODEL FOR A SYSTEM WITH GIVEN INPUT

The basic model

We consider a population whose members are divided into k strata. In many applications these will be ranked according to seniority, but this feature will not be considered explicitly until later. As far as possible the notation of Chapter 2 will be retained. Thus $n_j(T)$ denotes the number of people in grade j at time T $(T = 0, 1, 2, \ldots)$. The initial grade sizes, $n_j(0)$ $(j = 1, 2, \ldots, k)$, are assumed to be given and we define

$$N(T) = \sum_{j=1}^{k} n_j(T).$$

For $T > 0$ the grade sizes are random variables and we shall be concerned mainly with their expectations. These will be denoted as before by placing a bar over the symbol representing the random variable; thus $\bar{n}_j(T)$ is the expected number in grade j at time T. The number of new entrants to the system at time T will be written as $R(T)$. As remarked above, this may be a constant or a random variable. In the latter case $R(T)$ is to be understood as the expected number of entrants. The assumptions and notation for the transitions between grades used in the model of social mobility are retained. A member of grade i moves to grade j with probability p_{ij}, but it is no longer true that

$$\sum_{j=1}^{k} p_{ij} = 1.$$

In general

$$\sum_{j=1}^{k} p_{ij} < 1$$

because, in an open system, transitions out of the system are possible. The probability of loss from the ith grade at time T is denoted by $p_{i,k+1}$ and we note that

$$p_{i,k+1} = 1 - \sum_{j=1}^{k} p_{ij}.$$

To complete the specification of the model we must say how new entrants are allocated to the various grades. In many applications all recruits are placed in the lowest grade but we shall make the more general assumption that a proportion p_{0j} enter the jth grade. Obviously

$$\sum_{j=1}^{k} p_{0j} = 1$$

and the distribution $\{p_{0j}\}$ will be referred to as the 'recruitment distribution'. An alternative assumption about entry to the system is that a new recruit is allocated to grade j with *probability* p_{0j}. The actual numbers entering the various grades will then be multinomially distributed instead of fixed. Our theory covers both cases as long as it refers only to expectations because it then makes no difference whether p_{0j} is regarded as an actual or an expected proportion. The difference becomes important for the discussion of variances and covariances given later.

The probabilities which specify the process can be conveniently set out in standard form as follows:

$$\begin{array}{ccccc}
p_{01} & p_{02} & \cdots & p_{0k} & \\
\hline
p_{11} & p_{12} & \cdots & p_{1k} & p_{1,k+1} \\
p_{21} & p_{22} & \cdots & p_{2k} & p_{2,k+1} \\
\vdots & \vdots & & \vdots & \vdots \\
p_{k1} & p_{k2} & \cdots & p_{kk} & p_{k,k+1} \\
\hline
\end{array}$$

As before $\mathbf{P}$ will be the matrix with elements $\{p_{ij}\}$ $(i, j = 1, 2, \ldots, k)$; $\mathbf{p}_0$ denotes the recruitment distribution and $\bar{\mathbf{n}}(T)$ the vector of expected grade sizes at T.

Now that the operation of the system has been specified it is easy to write down equations relating the expected grade sizes at successive points in time. A straightforward argument using conditional expectations gives

$$\bar{n}_j(T+1) = \sum_{i=1}^{k} p_{ij}\bar{n}_i(T) + R(T+1)p_{0j} \qquad (T = 1, 2, 3, \ldots) \qquad (3.1a)$$
$$(j = 1, 2, \ldots, k)$$

or

$$\bar{\mathbf{n}}(T+1) = \bar{\mathbf{n}}(T)\mathbf{P} + R(T+1)\mathbf{p}_0. \qquad (3.1b)$$

Note that the recruits are treated as though they all arrive in the system at $T + 1$. Since $R(T)$ is known for all T this equation may be used to compute the expected grade sizes recursively. Repeated application of equation (3.1b) gives

$$\bar{\mathbf{n}}(\tau) = \bar{\mathbf{n}}(0)\mathbf{P}^T + \mathbf{p}_0\left\{\sum_{\tau=0}^{T-1} R(T-\tau)\mathbf{P}^\tau\right\} \qquad (3.2)$$

where $\mathbf{P}^0$ is again defined to be the unit matrix $\mathbf{I}$. The probabilities $\{p_{i,k+1}\}$ do not appear explicitly in these formulae but, because they are the complements of the row sums of $\mathbf{P}$, $\mathbf{P}^\tau$ and hence $\bar{\mathbf{n}}(T)$ depend upon their values. If $R(T)$ has a suitable mathematical form it may be possible to sum the matrix series appearing in (3.2) and so obtain the solution in closed form. This is the case if $R(T)$ is constant for all T or, more generally, if

$$R(T) = Rx^T \qquad (R > 0; x > 0; T \geq 1). \qquad (3.3)$$

We then have

$$\bar{\mathbf{n}}(T) = \bar{\mathbf{n}}(0)\mathbf{P}^T + Rx\mathbf{p}_0(x\mathbf{I} - \mathbf{P})^{-1}(x^T\mathbf{I} - \mathbf{P}^T), \qquad (3.4)$$

provided that the inverse of $x\mathbf{I} - \mathbf{P}$ exists. This will be so unless x is equal to any of the k eigenvalues of the matrix $\mathbf{P}$. For a matrix with non-negative elements and row sums strictly less than unity, the eigenvalues lie between 0 and 1. In particular $x = 1$ is not an eigenvalue and hence (3.4) applies for the case of constant input. However, as we shall see below, the chief value of this form is for the light which it throws on the limiting behaviour of the system.

The analysis of the following sections is designed to yield general results about the form of the solutions and their implications for the social process in question.

The spectral representation of $\bar{\mathbf{n}}(T)$

An alternative representation of $\bar{\mathbf{n}}(T)$ which has some advantages may be obtained by using a standard result in matrix theory known as Sylvester's theorem.† The theorem states that the Tth power of $\mathbf{P}$ can be expressed in the form

$$\mathbf{P}^T = \sum_{r=1}^{k} \lambda_r^T \mathbf{A}_r. \tag{3.5}$$

The constants $\lambda_1, \lambda_2, \ldots, \lambda_k$ are the eigenvalues, assumed distinct, of the matrix $\mathbf{P}$, and the matrices $\{\mathbf{A}_r\}$ have the following properties:

$$\begin{aligned} \mathbf{A}_r\mathbf{A}_s &= 0 \quad \text{if } r \neq s \\ &= \mathbf{A}_r \quad \text{if } r = s \end{aligned}$$

and

$$\sum_{r=1}^{k} \mathbf{A}_r = \mathbf{I}.$$

Equation (3.5) is described as the spectral representation of $\mathbf{P}^T$ and the matrices $\{\mathbf{A}_r\}$ are known as the 'spectral set'. If there are multiplicities among the eigenvalues, $\mathbf{P}^T$ can still be expressed in terms of powers of the eigenvalues but the expressions are more complicated.

If we substitute the expression for $\mathbf{P}^T$ given by (3.5) in (3.2) we shall have

$$\bar{\mathbf{n}}(T) = \sum_{r=1}^{k} \left[\lambda_t^T \bar{\mathbf{n}}(0)\mathbf{A}_r + \left\{ \sum_{\tau=0}^{T-1} R(T-\tau)\lambda_r^\tau \right\} \mathbf{p}_0 \mathbf{A}_r \right]. \tag{3.6}$$

This representation will be particularly useful if the series

$$\sum_{\tau=0}^{T-1} R(T-\tau)\lambda_r^\tau$$

can be expressed in closed form. If this is possible $\bar{\mathbf{n}}(T)$ will be a sum of k terms for all T; it can be found without having to calculate the structure for intermediate values of T. Equation (3.6) is also a good starting point for the investigation of the limiting behaviour of $\bar{\mathbf{n}}(T)$. In general, the determination of the eigenvalues and the spectral set involves extensive calculations if k is large, but these are routine operations for which computer

† A basic reference is Frazer, Duncan and Collar (1946); an application of the theorem to powers of stochastic matrices is given in Bailey (1964), pages 47–56. It has also been used by Young (1971).

programs are available. However, the analysis is simplified considerably in certain special cases of practical interest. One such example arises when the organization is hierarchical in form with downward movement not permitted. We shall discuss this special case and illustrate the foregoing theory in the following section.

Hierarchical structure with no demotions

Let us suppose that the k grades are arranged in increasing order of seniority and assume that transitions within the organization are to a higher grade only. This was certainly the case in the application to the Australian university system and is sufficiently common to warrant special attention. The matrix $\mathbf{P}$ is now upper triangular, thus

$$\mathbf{P} = \begin{pmatrix} p_{11} & p_{12} & \cdots & p_{1k} \\ 0 & p_{22} & \cdots & p_{2k} \\ \vdots & \vdots & & \vdots \\ 0 & 0 & \cdots & p_{kk} \end{pmatrix}.$$

The eigenvalues of such a matrix are obtained at once as $\lambda_r = p_{rr}$. Hence the eigenvalues will be distinct if the diagonal elements of $\mathbf{P}$ are distinct. In this case it is also possible to give an explicit expression for the elements of the matrices $\{\mathbf{A}_r\}$. If we denote the (i, j)th element of $\mathbf{A}_r$ by $A_{r,ij}$ then

$$\left.\begin{aligned} A_{r,ij} &= \sum_{m=1}^{j-i} \left(\prod_{s=i}^{j-m} p_{s,s+m} \bigg/ \prod_{\substack{s=i+m-1 \\ s \neq r}}^{j} (p_{rr} - p_{ss}) \right) \quad \left\{ \begin{matrix} j > r > i \\ j \geq r > i \\ j > r \geq i \end{matrix} \right\} \\ &= 1 \text{ if } i = r = j \\ &= 0 \text{ otherwise.} \end{aligned}\right\} \quad (3.7)$$

Substitution of the eigenvalues and the spectral set in (3.6) gives an expression from which $\bar{\mathbf{n}}(T)$ can be calculated.

Further simplification is possible if we place additional restraints on the kind of promotions which can occur. Suppose, for example, that promotion can only occur into the next highest grade so that $p_{s,s+m} = 0$ for $m > 1$. The first expression given in (3.7) can then be replaced by

$$A_{r,ij} = \prod_{s=i}^{j-1} p_{s,s+1} \bigg/ \prod_{\substack{s=i \\ s \neq r}}^{j} (p_{rr} - p_{ss}). \quad (3.8)$$

If $R(T)$ has the geometric form Rx^T

$$\bar{\mathbf{n}}(T) = \sum_{r=1}^{k} \left\{ p_{rr}^T \mathbf{n}(0)\mathbf{A}_r + Rx\left(\frac{x^T - p_{rr}^T}{x - p_{rr}}\right)\mathbf{p}_0\mathbf{A}_r \right\} \quad (x \neq p_{rr}). \tag{3.9}$$

It is thus clear that the diagonal elements of $\mathbf{P}$ and the value of x play a crucial role in determining the development of the process in time.

To illustrate the calculations required and to give some insight into the behaviour of hierarchical systems we shall discuss a numerical example with $k = 5$. The table of transition probabilities which we assume is set out in standard form below:

0·75	0·25	0	0	0	
0·65	0·20	0	0	0	0·15
0	0·70	0·15	0	0	0·15
0	0	0·75	0·15	0	0·10
0	0	0	0·85	0·10	0·05
0	0	0	0	0·95	0·05

The figures in this table have been chosen to reflect the kind of conditions which one might find in a typical management hierarchy. In an educational system the diagonal elements would tend to be much smaller. Three-quarters of new recruits enter the lowest grade and one-quarter the next lowest. The wastage probabilities, $\{p_{i,k+1}\}$, decrease as we move up the hierarchy since mobility between firms is usually more common at the lower levels. Loss from grade 5 would include retirements as well as transfers and so might well have a higher probability than the figure of 0·05 used in this example. The average time spent in grade i is $(1 - p_{ii})^{-1}$; in the present case this is between three and four years for the lowest three grades. The expected time to reach the highest grade is almost seventeen years.

The matrix $\mathbf{P}$ is triangular and hence the eigenvalues are equal to the diagonal elements. It therefore remains to calculate the spectral set $\{\mathbf{A}_r\}$. To do this we have to substitute numerical values given above in (3.8). For example,

$$A_{1\cdot 11} = 1, \quad A_{1\cdot 12} = p_{12}/(p_{11} - p_{22}) = (0{\cdot}20)/(-0{\cdot}05) = -4,$$

$$A_{1\cdot 13} = p_{12}p_{23}/(p_{11} - p_{22})(p_{11} - p_{33}) = 0{\cdot}20)(0{\cdot}15)/(-0{\cdot}05)(-0{\cdot}10) = 6,$$

etc. Written out in full the matrices are

$$\mathbf{A}_1 = \begin{pmatrix} 1 & -4 & 6 & -4\cdot5 & 1\cdot5 \\ 0 & 0 & 0 & 0 & 0 \\ 0 & 0 & 0 & 0 & 0 \\ 0 & 0 & 0 & 0 & 0 \\ 0 & 0 & 0 & 0 & 0 \end{pmatrix}$$

$$\mathbf{A}_2 = \begin{pmatrix} 0 & 4 & -12 & 12 & -4\cdot8 \\ 0 & 1 & -3 & 3 & -1\cdot2 \\ 0 & 0 & 0 & 0 & 0 \\ 0 & 0 & 0 & 0 & 0 \\ 0 & 0 & 0 & 0 & 0 \end{pmatrix}$$

$$\mathbf{A}_3 = \begin{pmatrix} 0 & 0 & 6 & -9 & 4\cdot5 \\ 0 & 0 & 3 & -4\cdot5 & 2\cdot25 \\ 0 & 0 & 1 & -1\cdot5 & 0\cdot75 \\ 0 & 0 & 0 & 0 & 0 \\ 0 & 0 & 0 & 0 & 0 \end{pmatrix}$$

$$\mathbf{A}_4 = \begin{pmatrix} 0 & 0 & 0 & 1\cdot5 & -1\cdot5 \\ 0 & 0 & 0 & 1\cdot5 & -1\cdot5 \\ 0 & 0 & 0 & 1\cdot5 & -1\cdot5 \\ 0 & 0 & 0 & 1 & -1 \\ 0 & 0 & 0 & 0 & 0 \end{pmatrix}$$

$$\mathbf{A}_5 = \begin{pmatrix} 0 & 0 & 0 & 0 & 0\cdot3 \\ 0 & 0 & 0 & 0 & 0\cdot45 \\ 0 & 0 & 0 & 0 & 0\cdot75 \\ 0 & 0 & 0 & 0 & 1 \\ 0 & 0 & 0 & 0 & 1 \end{pmatrix}.$$

Of the quantities appearing in (3.9) the initial grade structure $\bar{\mathbf{n}}(0)$ and the input sequence $\{R(T)\}$ remain to be specified. Let us assume that $\mathbf{n}(0) = N(0)$ $(0\cdot40, 0\cdot30, 0\cdot15, 0\cdot10, 0\cdot05)$ and $R(T) = R$. If we compute the total population size and the individual grade sizes at time T ($T > 0$) as multiples of $N(0)$ we need only express R as a fraction of $N(0)$. For the illustrative calculations we have chosen $N(0) = 9{\cdot}8333R$. The reason for this particular choice is that it makes $N(\infty) = N(0)$. This ensures that there is no long-term trend in the overall size and so enables us to concentrate on the changes in structure which take place. Under the assumptions we have made

$$\bar{\mathbf{n}}(T) = \sum_{r=1}^{5} \left\{ p_{rr}^{T}\mathbf{n}(0)\mathbf{A}_r + R\left(\frac{1 - p_{rr}^{T}}{1 - p_{rr}}\right)\mathbf{p}_0\mathbf{A}_r \right\} \tag{3.10}$$

and, in the limit,

$$\bar{\mathbf{n}}(\infty) = R \sum_{r=1}^{5} (1 - p_{rr})^{-1}\mathbf{p}_0\mathbf{A}_r. \tag{3.11}$$

In the example the vectors $\{\mathbf{p}_0\mathbf{A}_r\}$ are found to be

$$\mathbf{p}_0\mathbf{A}_1 = (0{\cdot}75 \quad -3{\cdot}00 \quad 4{\cdot}50 \quad -3{\cdot}375 \quad 1{\cdot}125),$$
$$\mathbf{p}_0\mathbf{A}_2 = (0 \quad 3{\cdot}25 \quad -9{\cdot}75 \quad 9{\cdot}75 \quad -3{\cdot}90),$$
$$\mathbf{p}_0\mathbf{A}_3 = (0 \quad 0 \quad 5{\cdot}25 \quad -7{\cdot}875 \quad 3{\cdot}9375),$$
$$\mathbf{p}_0\mathbf{A}_4 = (0 \quad 0 \quad 0 \quad 1{\cdot}50 \quad -1{\cdot}50),$$
$$\mathbf{p}_0\mathbf{A}_5 = (0 \quad 0 \quad 0. \quad 0 \quad 0{\cdot}3375).$$

Substitution in (3.11) thus gives the limiting structure as

$$\bar{\mathbf{n}}(\infty) = N(0)(0{\cdot}218 \quad 0{\cdot}230 \quad 0{\cdot}138 \quad 0{\cdot}138 \quad 0{\cdot}275). \tag{3.12}$$

The difference between this and the initial structure is quite striking. The highest grade has increased more than fivefold in size and the sizes of the lower grades have become more nearly equal. This feature is primarily due to the long average stay of twenty years in grade 5. In order to counteract this excessive growth at the top we might be able to increase p_{rr} for $r < 5$ by reducing the promotion probabilities. *This example illustrates the fact that a promotion policy which seems reasonable in itself may lead to undesirable consequences for the structure of the organization. In particular, it suggests that the pressure to grow at the top which is exhibited in many organizations may be the direct consequence of a fixed promotion policy.*

To calculate the intermediate structures we must first calculate the vectors $\{\mathbf{n}(0)\mathbf{A}_r\}$. We omit the details and proceed directly to Table 3.1 giving values of $\bar{\mathbf{n}}(T)$ for selected values of T.

***Table 3.1** Values of $\bar{n}_j(T)/N(0)$ for the example discussed in the text*

	T					
Grade	0	1	2	5	10	∞
1	0·400	0·336	0·295	0·239	0·220	0·218
2	0·300	0·315	0·313	0·280	0·243	0·230
3	0·150	0·158	0·165	0·174	0·159	0·138
4	0·100	0·107	0·115	0·136	0·154	0·138
5	0·050	0·058	0·065	0·091	0·137	0·275
$\sum_j \bar{n}_j(T)/N(0)$	1.000	0·974	0·953	0·920	0·913	1·000

The approach to equilibrium exhibits a pattern which we have observed in other instances. In the lower grades the limiting expectations are attained relatively quickly but the approach is very slow in grade 5. After ten years the expected size of grade 5 is still only half its equilibrium value while grades 1 and 2 have almost attained theirs. The total expected size of the population shows a steady decrease over the period up to ten years, but ultimately recovers its original value. The slowness of the approach to equilibrium in the upper grades and in the total size is accounted for by the term involving $(0{\cdot}95)^T$ which occurs in (3.10). With a coefficient of $-0{\cdot}6864$ its value does not become negligible until T is of the order of 100. In cases like this the limiting structure of the system is of little direct practical interest, although it does indicate the general direction of change. For some purposes it might be of greater interest to look at the expected proportions in each grade instead of the expected numbers. These can easily be obtained from Table 3.1, but they do not materially alter the general picture.

Application to a cohort

If we put $R(T) = 0$, for all T, in the model set out above, the various formulae describe the movement of the original members through the system. The stock numbers at time T are now given by

$$\bar{\mathbf{n}}(T) = \mathbf{n}(0)\mathbf{P}^T \tag{3.13}$$

and they will clearly tend to zero as T tends to infinity. The importance of this special case lies in its application to a cohort and the information which this gives about career patterns.

A cohort is a group of individuals joining the organization at the same time. Suppose a group is of size N and that they all enter grade i. (Here, as elsewhere in this chapter, we use the word grade instead of state because the main application is in manpower planning, but the results have a wider application.) In this case

$$\mathbf{n}(0) = N\,\mathbf{e}_i$$

where $\mathbf{e}_i$ is a vector having a one in the ith position and zeros elsewhere. The expected distribution of the cohort over the grades at various stages in its history can then be obtained from (3.13). One variable of practical interest is the expected proportion of the original cohort who survive in the organization for different lengths of time. This can easily be obtained by summing the elements of $\bar{\mathbf{n}}(T)$ and dividing by N. This line of approach was developed by Young (1971) and will be pursued further in Chapter 6, along with a number of other questions relating to durations. For the present we shall concentrate on the determination of the means and, where possible, the variances of various quantities of relevance in career planning.

Our analysis depends on the theory of absorbing Markov chains, a good account of which can be found in Kemeny and Snell (1960). The $(k + 1)$th grade, consisting of all those who have left, can be treated as a single absorbing state, the remaining states being transient. The (stochastic) transition matrix of such a chain is thus of the form

$$\left[\begin{array}{c|c} \mathbf{P} & \mathbf{p}'_{k+1} \\ \hline 0 & 1 \end{array}\right]$$

where $\mathbf{p}_{k+1} = (p_{1,k+1}, p_{2,k+1}, \ldots, p_{k,k+1})$. It turns out that a great deal of information about individual career histories can be deduced from the so-called fundamental matrix $(\mathbf{I} - \mathbf{P})^{-1}$. This matrix will occur again in Chapter 4 where it will be used to determine whether or not a given grade structure can be maintained.

Consider first a person who enters the system in grade i: what is the average length of time he will spend in grade j? To answer questions such as this we introduce random variables $\{X_{ij}^{(r)}\}$ defined as follows:

$$\begin{aligned} X_{ij}^{(r)} &= 1 \text{ if an entrant to grade } i \text{ is in grade } j \text{ after } r \text{ time units} \\ &= 0 \text{ otherwise } (i, j = 1, 2, \ldots, k;\ r = 0, 1, 2, \ldots). \end{aligned}$$

The total time spent by such an individual in grade j is thus

$$X_{ij} = \sum_{r=0}^{\infty} X_{ij}^{(r)} \qquad (i, j = 1, 2, \ldots, k).$$

(Note that $X_{ij}^{(0)} = 0$ if $j \neq i$ and $= 1$ otherwise.) We first find the expectation of X_{ij} which is given by

$$E(X_{ij}) = \sum_{r=0}^{\infty} E(X_{ij}^{(r)}). \tag{3.14}$$

It is well known from the general theory of Markov chains that

$$Pr\{X_{ij}^{(r)} = 1\} = p_{ij}^{(r)}$$

where $p_{ij}^{(r)}$ is the (i, j)th element of $\mathbf{P}^r$. Hence

$$E(X_{ij}^{(r)}) = p_{ij}^{(r)}$$

and therefore

$$E(X_{ij}) = \sum_{r=0}^{\infty} p_{ij}^{(r)}. \tag{3.15}$$

If we introduce the matrix $\mathbf{X} = \{X_{ij}\}$, (3.15) yields

$$E(\mathbf{X}) = \sum_{r=0}^{\infty} \mathbf{P}^{(r)} = \sum_{r=0}^{\infty} \mathbf{P}^r = (\mathbf{I} - \mathbf{P})^{-1}. \tag{3.16}$$

This result establishes the relationship between the fundamental matrix and the expected lengths of service in the grades. The expected stay of an entrant to grade i in the whole system is

$$E(X_i) = \sum_{j=1}^{k} E(X_{ij}) = d_i, \text{ say}, \tag{3.17}$$

where d_i is thus the sum of the ith row of $(\mathbf{I} - \mathbf{P})^{-1}$. Notice that all of these expectations apply equally to a new entrant to the organization and to someone newly promoted from inside. This fact is an immediate consequence of the Markov assumption which treats all entrants to a grade alike regardless of their past history. We might well doubt the validity of such an assumption in practice, but in organizations where all, or most, recruitment is at the bottom of a hierarchy the difficulty does not arise.

Stone (1972) has discussed the application of these results to a manpower system and has made the calculations for the five-grade matrix used above. For this matrix we find

$$(\mathbf{I} - \mathbf{P})^{-1} = \begin{pmatrix} 2{\cdot}86 & 1{\cdot}90 & 1{\cdot}14 & 1{\cdot}14 & 2{\cdot}29 \\ 0 & 3{\cdot}33 & 2{\cdot}00 & 2{\cdot}00 & 4{\cdot}00 \\ 0 & 0 & 4{\cdot}00 & 4{\cdot}00 & 8{\cdot}00 \\ 0 & 0 & 0 & 6{\cdot}67 & 13{\cdot}33 \\ 0 & 0 & 0 & 0 & 20{\cdot}00 \end{pmatrix}. \qquad \begin{matrix} \textit{Row totals} \\ 9{\cdot}33 \\ 11{\cdot}33 \\ 16{\cdot}00 \\ 20{\cdot}00 \\ 20{\cdot}00 \end{matrix}$$

Inspection of the row totals shows how the expectation of service increases as we move up the hierarchy. This result reflects the decrease of wastage with increasing seniority—once a recruit has survived the high wastage rates of the lowest grades his prospects of a long stay are improved.

The individual elements in the rows of $(\mathbf{I} - \mathbf{P})^{-1}$ show how the total expected service is divided among the grades. Thus on entering grade 1 an individual will expect to spend 2·86 years in the first grade, 1·90 years in the second, and so on. On moving up to the second grade the pattern changes, reflecting the fact that the individual has survived through grade 1. The diagonal elements are the expected stays in each grade at the time of entry to that grade.

Care must be taken in interpreting the off-diagonal elements. For example, 2·29 years is the time an entrant to grade 1 expects to spend in grade 5. Of all entrants to grade 1 most will never reach grade 5 and so will contribute nothing to the average. By the time a successful entrant reaches the threshold of grade 5 his expectation has increased to 20 years. The set of numbers in the ith row can thus be thought of as a typical career expectation for an entrant to that grade but, like all averages, they conceal considerable variation.

We have mentioned the chance that an entrant at the bottom will reach the top. Such chances can be calculated from the fundamental matrix. Let π_{ij} denote the probability that an entrant to grade i spends some time in grade j before leaving (In general, an individual may pass through j more than once; in a simple hierarchy with one-step promotion he can pass through it no more than once.) If μ_{ij} is the (i, j)th element of $(\mathbf{I} - \mathbf{P})^{-1}$ then clearly

$$\mu_{ij} = \pi_{ij}\mu_{jj} + (1 - \pi_{ij}) \times 0$$

or

$$\pi_{ij} = \mu_{ij}/\mu_{jj} \quad (i, j = 1, 2, \ldots, k). \tag{3.18}$$

Thus to obtain the set of probabilities $\{\pi_{ij}\}$ we must divide the elements in each column of $(\mathbf{I} - \mathbf{P})^{-1}$ by the diagonal element of that column. For the example,

$$\{\pi_{ij}\} = \begin{pmatrix} 1 & 0{\cdot}57 & 0{\cdot}29 & 0{\cdot}17 & 0{\cdot}11 \\ 0 & 1 & 0{\cdot}50 & 0{\cdot}30 & 0{\cdot}20 \\ 0 & 0 & 1 & 0{\cdot}60 & 0{\cdot}40 \\ 0 & 0 & 0 & 1 & 0{\cdot}67 \\ 0 & 0 & 0 & 0 & 1 \end{pmatrix}.$$

The diagonal elements must obviously be unity; the off-diagonal elements give the chance of reaching the grade corresponding to the column, given that we enter that corresponding to the row. Thus, for example, on entering grade 2 an individual has a 50 per cent chance of reaching the next grade, a 30 per cent chance of rising two grades and a 20 per cent chance of reaching the top.

In the case of a matrix like that just illustrated, in which $\mathbf{P}$ has non-zero elements only on the diagonal and super-diagonal, $(\mathbf{I} - \mathbf{P})^{-1}$ can be found algebraically (see equation 4.8). We then easily find

$$\pi_{ij} = \prod_{r=i}^{j-1} p_{r,r+1}/(1 - p_{r,r}) \qquad (j > i). \tag{3.19}$$

This result could have been deduced directly from the fact that $p_{r,r+1}/(1 - p_{r,r})$ is the probability of promotion from grade r, given that the individual does not leave. These formulas can, of course, be used in reverse to deduce the transition matrix required to achieve desired career prospects as specified by the π_{ij}'s or the μ_{ij}'s.

Another set of indices which may be of use in career planning are the probabilities of leaving from particular grades. Let a_{ij} be the probability that someone in grade i will leave when he is in grade j. The determination of the a_{ij}'s is discussed in another context in Chapter 6. The method is to introduce one absorbing state corresponding to the leavers from each grade so that the transition matrix of the chain has the form

$$\begin{array}{c|c} \mathbf{P} & \mathbf{Q} \\ \hline \mathbf{O} & \mathbf{I} \end{array}$$

where $\mathbf{Q}$ has diagonal elements equal to those of the vector $\mathbf{p}_{k+1}$ and zeros elsewhere. It is also shown in Chapter 6 how to make deductions about the behaviour of the system conditional on absorption in a particular state. This makes it a simple matter to carry out analyses like those discussed above for those who terminate their employment in a specified grade.

The method used to find the expected length of stay in each group (the μ_{ij}'s) can easily be extended to find the variances. Thus

$$E(X_{ij}^2) = E\left(\sum_{r=0}^{\infty} X_{ij}^{(r)}\right)^2 = \sum_{r=0}^{\infty}\sum_{s=0}^{\infty} E(X_{ij}^{(r)}X_{ij}^{(s)}). \tag{3.20}$$

If $r = s$,

$$E(X_{ij}^{(r)})^2 = E(X_{ij}^{(r)}) = p_{ij}^{(r)}.$$

Otherwise

$$\begin{aligned} E(X_{ij}^{(r)}X_{ij}^{(s)}) &= Pr\{X_{ij}^{(r)} = 1, X_{ij}^{(s)} = 1\} \\ &= Pr\{X_{ij}^{(r)} = 1\}Pr\{X_{ij}^{(s)} = 1 | X_{ij}^{(r)} = 1\} \\ &= p_{ij}^{(r)}p_{jj}^{(s-r)} \text{ if } s \geq r. \end{aligned} \tag{3.21}$$

If $s < r$ we simply interchange r and s. Substituting in (3.20)

$$\begin{aligned} E(X_{ij}^2) &= \sum_{r=0}^{\infty} p_{ij}^{(r)} + 2 \sum_{r=0}^{\infty} \sum_{s=r+1}^{\infty} p_{ij}^{(r)}p_{jj}^{(s-r)} \\ &= \mu_{ij} + 2 \sum_{r=0}^{\infty} p_{ij}^{(r)}(p_{jj}^{(1)} + p_{jj}^{(2)} + \dots) \\ &= \mu_{ij} + 2\mu_{ij}(\mu_{jj} - 1) = 2\mu_{ij}\mu_{jj} - \mu_{ij}. \end{aligned} \tag{3.22}$$

Hence

$$\begin{aligned} \text{var}(X_{ij}) &= 2\mu_{ij}\mu_{jj} - \mu_{ij} - \mu_{ij}^2 \\ &= \mu_{jj}^2(2\pi_{ij} - \pi_{ij}^2) - \mu_{jj}\pi_{ij} \qquad (i, j = 1, 2, \dots, k). \end{aligned} \tag{3.23}$$

In the important special case when $i = j$, $\pi_{ii} = 1$ and

$$\text{var}(X_{ii}) = \mu_{ii}^2 - \mu_{ii} = \mu_{ii}(\mu_{ii} - 1) \qquad (i = 1, 2, \dots, k). \tag{3.24}$$

This result, in particular, shows that the lengths of stay in each grade will be highly variable. If μ_{ii} is fairly large the standard deviation of length of stay will be almost as large as the expectation.

Limiting behaviour of $\bar{\mathbf{n}}(T)$

We have already considered the limiting behaviour of the model for a special case with constant input. Although the limits may be approached too slowly to be attained in practice they give valuable information about the inherent tendencies in the system and so we shall pursue the question in the general case. For this purpose we use the spectral representation given by (3.6) in conjunction with a theorem due to Perron and Frobenius (see, for example, Cox and Miller, 1965, page 120). When applied to the matrix **P** this states that

$$0 < \max_i \lambda_i \leq \max_i \sum_{j=1}^{k} p_{ij}$$

where the equality sign holds only if all of the row sums are equal. The largest eigenvalue will thus be strictly less than one unless all the row sums are equal to one. This cannot be so for an open system which requires that some, at least, of the inequalities

$$\sum_{j=1}^{k} p_{ij} < 1 \qquad (i = 1, 2, \ldots, k)$$

must hold. These results ensure that

$$\lim_{T\to\infty} \lambda_r^T = 0$$

for all T and hence that the leading term of (3.6) vanishes. The limiting solution thus depends on the behaviour of the series

$$S_{T,r} = \sum_{\tau=0}^{T-1} R(T-\tau)\lambda_r^\tau$$

as $T \to \infty$ for $0 < \lambda_r < 1$ and $r = 1, 2, \ldots, k$. Although we cannot obtain further detailed results without specifying the function $R(T)$ we can usefully distinguish three kinds of behaviour as follows.

(a) $\lim_{T\to\infty} R(T) = 0$. In this case $S_{T,r}$ tends to zero for all r and hence $\bar{\mathbf{n}}(\infty) = \mathbf{0}$. This result expresses the obvious fact that if the input vanishes the population will ultimately become extinct. We might ask what happens to the *relative* grade sizes as $T \to \infty$, but no general results are available.

(b) $\lim_{T\to\infty} R(T) = R$, a constant. Under this condition

$$\lim_{T\to\infty} S_{T,r} = R \sum_{j=0}^{\infty} \lambda_r^j = R(1-\lambda_r)^{-1}$$

which gives

$$\bar{\mathbf{n}}(\infty) = R \sum_{r=1}^{k} (1-\lambda_r)^{-1}\mathbf{p}_0\mathbf{A}_r. \tag{3.25}$$

A special case of this result was given earlier in (3.11). It should be noted that the limiting size and structure do not depend on $\mathbf{n}(0)$. The ultimate size is, in fact, proportional to R, the input per unit time. The same result holds if the eigenvalues are not all distinct.

(c) $\lim_{T\to\infty} R(T) = \infty$. If the input increases without limit it is obvious that the size of the organization will grow in like manner. Again it would be interesting to have information on the limiting values of the relative grade sizes. No general results are available, but a special case will be considered later.

The three cases listed above are not exhaustive because $R(T)$ need not tend to a limit; it might fluctuate indefinitely in a more or less erratic manner.

If this is the case there will be no limiting grade structure in the ordinary sense but it will still be true that the initial structure will exert a diminishing influence as T increases. The general results we have obtained about $\bar{\mathbf{n}}(\infty)$ could have been derived, though in a less complete form, by allowing T to approach infinity on both sides of the difference equation, (3.1b).

We shall illustrate the foregoing discussion by supposing that $R(T) = Rx^T (x, R > 0)$. By varying the value of x we can generate examples falling into each of the above three categories. If $x < 1$

$$\lim_{T \to \infty} R(T) = 0$$

and the total input is finite. Under these conditions $N(T)$ and the individual grade sizes tend to zero as T increases. If $x = 1$ case (b) above obtains, and an expression for the expected limiting grade structure has been given in (3.25). In certain special cases explicit expressions can be found for the limiting structure. For example, if recruitment is into the lowest grade only and if promotion is restricted to the next highest grade then $\lambda_r = p_{rr} (r = 1, 2, \cdots, k)$; the spectral set is given by (3.8) and hence

$$\bar{n}_j(\infty) = R \prod_{r=1}^{j-1} p_{r,r+1} \Big/ \prod_{r=1}^{j} (1 - p_{rr}) \qquad (j = 1, 2, \ldots, k) \tag{3.26}$$

where the first product is defined to be one when $j = 1$. This formula clearly demonstrates how grade sizes depend on the values of the p_{rr}'s and, in particular, on their closeness to one.

If $x > 1$ the term in x^T in (3.9) becomes dominant as $T \to \infty$, so that

$$\bar{\mathbf{n}}(T) \sim Rx^{T+1} \sum_{r=1}^{k} \mathbf{p}_0 \mathbf{A}_r / (x - \lambda_r). \tag{3.27}$$

This result shows that the total size and the grade sizes will tend to grow geometrically. It is theoretically interesting to ask whether the *relative* grade sizes approach limits as T increases. If

$$q_j(T) = \bar{n}_j(T) \Big/ \sum_{i=1}^{k} \bar{n}_i(T)$$

then it is clear from (3.27) that

$$\mathbf{q}(T) \propto \sum_{r=1}^{k} \mathbf{p}_0 \mathbf{A}_r / (x - \lambda_r). \tag{3.28}$$

Making the same assumptions about recruitment and promotion as in the discussion leading up to (3.26), we have

$$q_j(\infty) \propto \prod_{r=1}^{j-1} p_{r,r+1} \Big/ \prod_{r=1}^{j} (x - p_{rr}). \tag{3.29a}$$

Alternative, but equivalent, expressions for $\bar{\mathbf{n}}(\infty)$ and $\mathbf{q}(\infty)$ can be obtained from the matrix expression given in (3.4). The result for $\mathbf{q}(\infty)$ is

$$\mathbf{q}(\infty) \propto \mathbf{p}_0(x\mathbf{I} - \mathbf{P})^{-1} \tag{3.29b}$$

The constants of proportionality in the above equations are, of course, determined by the fact that

$$\sum_{j=1}^{k} q_j(\infty) = 1.$$

If x is very large $(x\mathbf{I} - \mathbf{P})^{-1} \sim \mathbf{I}/x$ and

$$\mathbf{q}(\infty) \sim \mathbf{p}_0.$$

This equation expresses the obvious fact that, with a very high rate of expansion, the recruitment distribution dominates the structure. The rate at which this situation is approached when (3.29a) applies can be judged by noting that $q_j(T)$ is proportional, in the limit, to x^{-j}.

Variances and covariances of the grade sizes

The theory for closed systems given in Chapter 2 can be generalized to include open systems. For an open system (2.10) must be replaced by

$$n_j(T+1) = \sum_{i=1}^{k} n_{ij}(T) + n_{0j}(T+1) \qquad (j = 1, 2, \ldots, k) \tag{3.30}$$

where $n_{0j}(T+1)$ is the number of new entrants to grade j at time $T+1$. On taking expectations of both sides of this equation we are led back to (3.1). If we assume that the input is stochastically independent of the internal movements of the system then

$$\begin{aligned} &\operatorname{cov}\{n_j(T+1), n_l(T+1)\} \\ &\quad = \operatorname{cov}\left\{\sum_{i=1}^{k} n_{ij}(T) \sum_{i=1}^{k} n_{il}(T)\right\} + \operatorname{cov}\{n_{0j}(T+1), n_{0l}(T+1)\}. \end{aligned} \tag{3.31}$$

The first term on the right-hand side of (3.31) is essentially the same as the expression given in (2.11) of Chapter 2. It evaluation can be carried through as in that chapter without any modification. All that we have to do to make the result of (2.15) of Chapter 2 applicable in the present case is to add on the last covariance term given in (3.31) above. Thus in matrix notation we have

$$\boldsymbol{\mu}(T+1) = \boldsymbol{\mu}(T)\boldsymbol{\Pi} + \boldsymbol{\mu}_0(T+1) \tag{3.32}$$

where $\boldsymbol{\mu}(T)$ and $\boldsymbol{\Pi}$ are as defined in Chapter 2. The first k elements of the vector $\boldsymbol{\mu}_0(T+1)$ are the expected numbers of entrants to each grade at time $T+1$ listed in ascending order; the remaining k^2 elements are the covariances of these numbers listed in dictionary order. In our examples the input will be time-homogeneous so we shall suppress the argument of $\boldsymbol{\mu}_0$ and write $\boldsymbol{\mu}_0$ instead of $\boldsymbol{\mu}_0(T+1)$. In this case (3.32) leads to

$$\boldsymbol{\mu}(T) = \boldsymbol{\mu}(0)\boldsymbol{\Pi}^T + \boldsymbol{\mu}_0(\mathbf{I}-\boldsymbol{\Pi})^{-1}(\mathbf{I}-\boldsymbol{\Pi}^T) \tag{3.33}$$

The inverse of $\mathbf{I}-\boldsymbol{\Pi}$ exists because the dominant eigenvalue of $\boldsymbol{\Pi}$ is strictly less than one. If we adopt the more general error structure discussed in Chapter 2 the same equation holds with appropriate changes in $\boldsymbol{\Pi}$.

The form of $\boldsymbol{\mu}_0$ will depend upon the stochastic nature of the input and the method of allocation to grades. We shall illustrate the point using the two cases mentioned at the beginning of the section. First suppose that the input $\{R(T)\}$ is a sequence of independently and identically distributed random variables and that, at time T, a proportion p_{0j} go to grade j. In this case the first k elements of $\boldsymbol{\mu}_0$ are $\bar{R}(T)p_{0j}$ $(j = 1, 2, \ldots, k)$, where $\bar{R}(T) = ER(T)$. The covariance elements take the form

$$\operatorname{var}\{R(T)\}p_{0j}p_{0l} \qquad (j, l = 1, 2, \ldots, k).$$

Our second assumption about the allocation of recruits was that each one was to be allocated to the jth grade with probability p_{0j}. In this case the k means in the vector $\boldsymbol{\mu}_0$ are $\bar{R}(T)p_{0j}$ $(j = 1, 2, \ldots, k)$ as before, but the covariances are now given by

$$\begin{aligned} &\operatorname{cov}\{n_{0j}(T), n_{0l}(T)\} \\ &= \operatorname{var}\{R(T)\}p_{0j}p_{0l} + \bar{R}(T)(\delta_{jl}p_{0j} - p_{0j}p_{0l}) \qquad (i, j = 1, 2, \ldots, k). \end{aligned} \tag{3.34}$$

This result is obtained by first finding expectations conditional upon $R(T)$ and then averaging over the distribution of $R(T)$. Under either assumption about allocation the determination of the successive values of $\boldsymbol{\mu}(T)$ is straightforward. Some examples are discussed below but first we consider the limiting behaviour of the system.

If the input to the system is time-homogeneous, $\boldsymbol{\mu}_0$ does not depend on T. Under these circumstances we may investigate the limiting form of the vector $\boldsymbol{\mu}(T)$ as $T \to \infty$. We already know that the limit exists for the first k elements of the vector and it may be shown that the same result holds for all of the elements. This being so it is clear from (3.32) that the limiting vector must satisfy

$$\boldsymbol{\mu}(\infty) = \boldsymbol{\mu}(\infty)\boldsymbol{\Pi} + \boldsymbol{\mu}_0$$

or

$$\boldsymbol{\mu}(\infty) = \boldsymbol{\mu}_0(\mathbf{I}-\boldsymbol{\Pi})^{-1}. \tag{3.35}$$

Once the matrix $(\mathbf{I} - \mathbf{\Pi})^{-1}$ has been computed the effects of different kinds of input can easily be compared. If $R(T)$ is a fixed quantity R, then var $\{R(T)\} = 0$ and there is a substantial simplification in $\boldsymbol{\mu}_0$.

To illustrate the foregoing theory we shall present some calculations for two different kinds of input. In the first case we shall suppose that $R(T)$ is a fixed number, R, independent of T and in the second that $\{R(T)\}$ is a sequence of independent Poisson variates with constant mean equal to R. These two assumptions represent fairly extreme degrees of variability in the input. We shall assume, for the purposes of this illustration, that each new recruit is allocated to grade j with probability p_{0j}. The covariance elements of $\boldsymbol{\mu}_0$ are obtained from (3.34), where var $\{R(T)\}$ is zero for constant input and equal to $\bar{R}$ for Poisson input. The calculations in Table 3.2 relate to the example discussed earlier in this section for which the expected grade sizes were given in Table 3.1. We have further supposed that the initial size of the organization was 590 and that $\bar{R} = 60$.

The equilibrium values are approached quite rapidly for the lower grades but only very slowly for the highest grade. For both types of input there is considerable uncertainty in predictions of the size of the fifth grade in the distant future. The difference between the two kinds of input is most apparent in the lowest grades. It is not until $T = 5$ that the input has any effect on the top grade and thereafter it is only slight. In the lowest grade the effect of Poisson variability is to roughly double the variance as compared with fixed input. All of the covariances are either zero or negative, but in the Poisson case they vanish in the limit. This fact coupled with equality of the limiting means and variances suggests that the grade sizes are asymptotically distributed like independent Poisson variates. Pollard (1967) has proved this to be true in the general case when the input consists of an independent sequence of Poisson variates.

An alternative approach leading to the same result has been given by Staff and Vagholkar (1971) based on generating functions. In principle it may be used to find the joint distribution of the grade sizes as well as the moments, but in practice the manipulations required are so heavy as to make the method of very limited value. Staff and Vagholkar do give a few explicit results in very simple cases, but for most purposes the first- and second-order moments are sufficient and these can be derived more easily using Pollard's method. The one exception to this remark is when the inputs $\{R(T)\}$ are an independent sequence of Poisson random variables. In that case, as remarked above, the joint distribution of the grade sizes is asymptotically Poisson and we shall illustrate the method of generating functions by using it to derive this result.

Let $p_j(T - \tau)(j = 1, 2, \ldots, k + 1)$ be the probability that a person recruited at time τ is in grade j at time T. Those who have left are considered to

***Table 3.2** Variance–covariance matrices for the grade sizes for the example of Section 3.2 with (a) fixed input and (b) Poisson input where, in each case,* $N(0) = 9{\cdot}8333\,\bar{R}(T)$ *and* $\bar{R}(T) = 60$

T	*Fixed input*					*Poisson input*				
	64·94	−41·93	0	0	0	98·69	−30·68	0	0	0
	—	86·18	−18·59	0	0	—	89·93	−18·59	0	0
1	—	—	39·16	−9·96	0	—	—	39·16	−9·96	0
	—	—	—	18·81	−5·02	—	—	—	18·81	−5·02
	—	—	—	—	6·71	—	—	—	—	6·71
	83·82	−47·68	−4·09	0	0	131·83	−26·92	−2·99	0	0
	—	115·16	−21·51	−1·95	0	—	125·25	−20·78	−1·95	0
2	—	—	60·94	−12·81	−0·75	—	—	61·02	−12·81	−0·75
	—	—	—	31·86	−7·99	—	—	—	31·86	−7·99
	—	—	—	—	12·61	—	—	—	—	12·61
	80·20	−38·20	−7·69	−1·22	−0·07	137·86	−5·70	−2·82	−0·67	−0·05
	—	124·69	−17·15	−5·07	−0·57	—	149·79	−11·25	−4·24	−0·52
5	—	—	85·60	−11·20	−2·56	—	—	87·63	−10·85	−2·54
	—	—	—	58·36	−11·47	—	—	—	58·43	−11·46
	—	—	—	—	28·83	—	—	—	—	28·83
	71·54	−34·71	−6·73	−1·51	−0·24	129·97	−0·19	−0·25	−0·21	−0·07
	—	111·69	−11·99	−4·68	−1·16	—	142·27	−1·53	−1·57	−0·65
10	—	—	84·77	−6·72	−2·91	—	—	90·83	−4·20	−2·39
	—	—	—	79·43	−10·50	—	—	—	80·80	−10·18
	—	—	—	—	56·05	—	—	—	—	56·13
	70·13	−34·58	−6·58	−1·43	−0·24	128·57	0	0	0	0
	—	104·79	−11·10	−3·97	−1·05	—	135·71	0	0	0
∞	—	—	74·13	−4·35	−1·86	—	—	81·43	0	0
	—	—	—	76·84	−3·74	—	—	—	81·43	0
	—	—	—	—	154·50[a]	—	—	—	—	162·26
	Expected grade sizes for $T = \infty$					128·6	135·7	81·4	81·4	162·3

[a] This is the value for $T = 100$, at which point the limiting value had not been attained. The calculations in this table were carried out by

be in a grade $k + 1$. Let the number of recruits at $T - \tau$ be $R(T - \tau) = X$. Then, conditional upon X, those recruited at time $T - \tau$ will be distributed multinomially over the $k + 1$ grades at time T according to the probabilities $\{p_j(T - \tau)\}$. The joint probability generating function of these numbers may thus be written

$$g_\tau(z_{1\tau}, z_{2\tau}, \ldots, z_{k+1,\tau}|X) = \left(\sum_{j=1}^{k+1} p_j(T - \tau)z_{j\tau}\right)X. \tag{3.36}$$

The unconditional generating function is the expectation of (3.36) with respect to the distribution of X. If X has the probability generating function $l(z)$ then the generating function required is

$$g_\tau(z_{1\tau}, z_{2\tau}, \ldots, z_{k+1,\tau}) = l\left(\sum_{j=1}^{k+1} p_j(T - \tau)z_{j\tau}\right). \tag{3.37}$$

The total number in the grades is composed of the survivors from the initial establishment together with the cohorts recruited at times $1, 2, \ldots, T-1$. Since the $R(T)$'s are independent the total number of survivors from these cohorts, excluding the originals, will be obtained by taking the product of (3.37) over τ and putting $z_{j\tau} = z_j$ for all j. This gives the generating function

$$g(z_1, z_2, \ldots, z_{k+1}) = \prod_{\tau=1}^{T-1} l\left(\sum_{j=1}^{k+1} p_j(T - \tau)z_j\right) \tag{3.38}$$

for the distribution of $\mathbf{n}(T) - \mathbf{n}(0)\mathbf{P}^T$. For large T, $\mathbf{P}^T$ will be negligible and so (3.38) may then be treated as the generating function of $\mathbf{n}(T)$. This amounts to saying that T must be large enough for almost all of the original members to have left.

If the input consists of a sequence of independent Poisson variates with mean R then

$$l(s) = e^{R(s-1)}$$

Hence, in this case,

$$\begin{aligned} g(z_1, \ldots, z_{k+1}) &= \exp\left[R\left\{\sum_{\tau=0}^{T-1}\sum_{j=1}^{k+1} p_j(T \quad \tau)z_j - 1\right\}\right] \\ &= \exp\left\{\sum_{j=1}^{k+1}(z_j - 1)R\sum_{\tau=0}^{T-1} p_j(T - \tau)\right\}. \end{aligned} \tag{3.39}$$

This shows that the grade sizes, apart from original survivors, are distributed independently with expectation in the jth grade equal to

$$R\sum_{\tau=0}^{T-1} p_j(T - \tau) = R\sum_{\tau=1}^{T} p_j(\tau).$$

The probabilities $p_j(\tau)$ are obtained from

$$\mathbf{p}(\tau) = \mathbf{p}_0\mathbf{P}^{\tau-1} \tag{3.40}$$

which is the distribution of the destination of an individual recruited at time $T - \tau$. In the limit as $\tau \to \infty$ the vector of expectations tends to

$$\begin{aligned}\bar{\mathbf{n}}(\infty) &= R\mathbf{p}_0 \sum_{\tau=1}^{\infty} \mathbf{P}^{\tau-1} \\ &= R\mathbf{p}_0(\mathbf{I} - \mathbf{P})^{-1}\end{aligned} \tag{3.41}$$

which checks with the result obtained earlier for the general case. In the case of Poisson input we now see that this determines the complete distribution.

The method of derivation shows that the rate of approach to the asymptotic distribution is governed by how long it takes the original members to leave. It is easy to see that the result generalizes to the case when the mean input is a function of time, and this enables us to incorporate trends or oscillations into the input sequence.

An example with a periodic transition matrix

So far we have assumed that the transition probabilities and the recruitment distribution do not depend on T. Without this assumption (3.1b) still holds, although the subsequent analysis becomes more complicated. In particular, there may be several limiting structures or none at all. An interesting example of a non-homogeneous process arose in Gani's study of student enrolment at Michigan State University. The academic year there consisted of three 'quarters' and transitions took place at the end of each quarter. It was not considered realistic to assume that the transition and recruitment probabilities would be the same in each quarter of a given year. However, it could be assumed that these probabilities would be the same in, say, the first quarter of one year as they had been in the first quarter of previous years. This requires us to make both $\mathbf{P}$ and $\mathbf{p}_0$ functions of T satisfying

$$\mathbf{P}(T+3) = \mathbf{P}(T)$$

and

$$\mathbf{p}_0(T+3) = \mathbf{p}_0(T) \qquad (T = 0, 1, 2).$$

Suppose that $T = 0$ refers to the first quarter of the first academic year and let

$$\left.\begin{aligned}\mathbf{P}(3T+j) &= \mathbf{P}_{j+1} \\ \mathbf{p}_0(3T+j) &= \mathbf{p}_{0,j+1}\end{aligned}\right\} (j = 0, 1, 2;\, T = 0, 1, 2, \ldots).$$

Then $\mathbf{P}_j$ and $\mathbf{p}_{0j}$ relate to the jth quarter of any year. The grade structures may now be computed from the following difference equations:

$$\left.\begin{aligned}\bar{\mathbf{n}}(3T + 1) &= \bar{\mathbf{n}}(3T)\mathbf{P}_1 + R(3T + 1)\mathbf{p}_{02}\\ \bar{\mathbf{n}}(3T + 2) &= \bar{\mathbf{n}}(3T + 1)\mathbf{P}_2 + R(3T + 2)\mathbf{p}_{03}\\ \bar{\mathbf{n}}(3T + 3) &= \bar{\mathbf{n}}(3T + 2)\mathbf{P}_3 + R(3T + 3)\mathbf{p}_{01}\end{aligned}\right\} \quad (T = 0, 1, 2, \ldots). \tag{3.42}$$

These equations express the structure in a given quarter in terms of the structure in the previous quarter. For some purposes it is more convenient to relate the structures for the same quarter of succeeding years. Thus, for example, for the first quarter, (3.42) gives

$$\begin{aligned}\bar{\mathbf{n}}(3T + 3) = \bar{\mathbf{n}}(3T)\mathbf{P}_1\mathbf{P}_2\mathbf{P}_3 &+ R(3T + 1)\mathbf{p}_{02}\mathbf{P}_2\mathbf{P}_3\\ &+ R(3T + 2)\mathbf{p}_{03}\mathbf{P}_3 + R(3T + 3)\mathbf{p}_{01}.\end{aligned} \tag{3.43}$$

Similar expressions can be obtained for the second and third quarters.

The limiting behaviour of the system may be investigated by the methods of earlier sections. The case of greatest interest is when $R(T)$ is constant or tends to a limit. When this is so the limiting grade structure in the *first* quarter will be given by

$$\bar{\mathbf{n}}(\infty) = (\mathbf{I} - \mathbf{P}_1\mathbf{P}_2\mathbf{P}_3)^{-1}\{\mathbf{p}_{02}\mathbf{P}_2\mathbf{P}_3 + \mathbf{p}_{03}\mathbf{P}_3 + \mathbf{p}_{01}\}R \tag{3.44}$$

where R is the limiting value of the input. The corresponding expressions for the second and third quarters can either be obtained by substituting back in (3.42) or by a repetition of the above argument.

3.3 A MODEL FOR AN EXPANDING SYSTEM WITH GIVEN SIZE

The equations of the model

The practical context in which the present model arose was described at the beginning of the chapter. The model differs from that described in the preceding section only in that the total size of the organization rather than the input is fixed. Instead of being given the sequence of inputs $\{R(T)\}$ we now have a sequence of total sizes $\{N(T)\}$. As before, this may be a sequence of given numbers or a realization of a known stochastic process. In the latter case the symbol $N(T)$ is to be interpreted as an expected size. The distinction is of no importance until we come to consider the distributions of grade sizes. Young and Almond (1961), who proposed the model, were concerned with expanding organizations and we shall, at first, follow them in this respect. Much of the theory is also applicable to the more general case of fluctuating or decreasing sequences.

Let $M(T)$ denote the increase in size which takes place between $T - 1$ and T; thus $M(T) = N(T) - N(T-1)(T = 1, 2, \ldots)$. Equation (3.1) remains valid but cannot be used as it stands because $\{R(T)\}$ is an unknown in this version of the problem. At any time the number of recruits must be sufficient to achieve the desired expansion and to replace losses from the system. The expected number of recruits required at time $T + 1$ is thus

$$\bar{R}(T+1) = M(T+1) + \sum_{i=1}^{k} p_{i,k+1}\bar{n}_i(T) \qquad (T = 1, 2, \ldots). \qquad (3.45)$$

By substituting $\bar{R}(T + 1)$ for $R(T + 1)$ in (3.1a), the difference equations for the expected grade sizes become

$$\bar{n}_j(T+1) = \sum_{i=1}^{k} (p_{ij} + p_{i,k+1}p_{0j})\bar{n}_i(T) + M(T+1)p_{0j} \qquad (j = 1, 2, \ldots, k). \qquad (3.46a)$$

If we write $q_{ij} = p_{ij} + p_{i,k+1}p_{0j}$ this equation may be written in matrix notation as

$$\bar{\mathbf{n}}(T+1) = \bar{\mathbf{n}}(T)\mathbf{Q} + \mathbf{p}_0 M(T+1) \qquad (3.46b)$$

where $\mathbf{Q}$ is the matrix $\{q_{ij}\}$. This matrix equation has the same form as (3.1b) and leads to the result that

$$\bar{\mathbf{n}}(T) = \mathbf{n}(0)\mathbf{Q}^T + \mathbf{p}_0\left\{\sum_{\tau=0}^{T-1} M(T-\tau)\mathbf{Q}^\tau\right\}. \qquad (3.47)$$

The foregoing formulae provide all that is necessary for the numerical investigation of the expected structure. If we wish to study the form of $\bar{\mathbf{n}}(T)$ by analytical methods two approaches are possible. One of these is to use the spectral representation for the power of a matrix as in Section 3.2, but the advantages of this method are not as great here as when the input was fixed. In that case we were able to obtain many explicit results for the special but important case when the matrix $\mathbf{P}$ was triangular. This was because the eigenvalues were then equal to the diagonal elements of the matrix. It is clear from inspection of the q_{ij}'s as defined above that $\mathbf{Q}$ will not normally be triangular and hence much of the simplicity of the method is lost. An alternative approach is to use the fact that $\mathbf{Q}$ is a stochastic matrix, a conclusion which follows by observing that

$$\sum_{j=1}^{k} q_{ij} = \sum_{j=1}^{k} p_{ij} + p_{i,k+1}\sum_{j=1}^{k} p_{0j} = 1 \quad (i = 1, 2, \ldots, k).$$

This fact enables us to use known results about powers of stochastic matrices and, in particular, about their limiting behaviour. We shall use this method in the following sections.

The special case $M(T) = 0$, for all T, is particularly interesting. Our equations then relate to an open system of constant size but they are identical in form to those used in the study of closed systems in Chapter 2. They thus provide a formal justification for our earlier remark that an open system in which gains and losses were equal could be treated as closed. Each person who leaves can be paired with a new entrant and the two changes treated as one. Thus a transition from grade i to grade j can either take place within the system or by loss from grade i and replacement to grade j with total probability $p_{ij} + p_{i,k+1}p_{0j}$.

The exact solution for geometric growth rate

In order to gain insight into the behaviour of expanding organizations we shall suppose the growth rate to be geometric. That is we suppose that

$$M(T) = Mx^T \quad (T \geq 1) \tag{3.48}$$

where M is a positive constant and x is non-negative. By varying the values of M and x we can generate a considerable variety of growth patterns. The simplest case arises when there are only two strata and we shall consider this first.

When $k = 2$ the matrix $\mathbf{Q}$ can be written in the form

$$\mathbf{Q} = \begin{pmatrix} 1-\alpha & \alpha \\ \beta & 1-\beta \end{pmatrix}$$

where $1 - \alpha = p_{11} + p_{01}p_{13}$ and $1 - \beta = p_{22} + p_{02}p_{23}$. A well-known result (see, for example, Bailey, 1964, page 52) then gives

$$\mathbf{Q}^{\tau} = \frac{1}{\alpha+\beta}\left\{\begin{pmatrix} \beta & \alpha \\ \beta & \alpha \end{pmatrix} + (1-\alpha-\beta)^{\tau}\begin{pmatrix} \alpha & -\alpha \\ -\beta & \beta \end{pmatrix}\right\}. \tag{3.49}$$

Substituting for $\mathbf{Q}^{\tau}$ in (3.47), with $M(T-\tau) = Mx^{T-\tau}$, we find

$$\bar{\mathbf{n}}(T) = \frac{\mathbf{n}(0)}{\alpha+\beta}\left\{\begin{pmatrix} \beta & \alpha \\ \beta & \alpha \end{pmatrix} + (1-\alpha-\beta)^{T}\begin{pmatrix} \alpha & -\alpha \\ -\beta & \beta \end{pmatrix}\right\}$$

$$+ \frac{1}{\alpha+\beta}\sum_{\tau=0}^{T-1} Mx^{T-\tau}\mathbf{p}_0\left\{\begin{pmatrix} \beta & \alpha \\ \beta & \alpha \end{pmatrix} + (1-\alpha-\beta)^{\tau}\begin{pmatrix} \alpha & -\alpha \\ -\beta & \beta \end{pmatrix}\right\}. \tag{3.50}$$

Summing the geometric series in (3.50) and using the facts that $n_1(0) + n_2(0) = N(0)$ and $p_{01} + p_{02} = 1$, the vector of expected grade sizes becomes

$$\bar{\mathbf{n}}(T) = \frac{1}{\alpha + \beta}\left[\left\{N(0) + M\frac{(x - x^{T+1})}{1 - x}\right\} \times (\beta, \alpha) + (1 - \alpha - \beta)^T(\alpha n_1(0) - \beta n_2(0), -\alpha n_1(0) + \beta n_2(0) + M\left\{\frac{x^T - (1 - \alpha - \beta)^T}{1 - (1 - \alpha - \beta)x^{-1}}\right\}(\alpha p_{01} - \beta p_{02}, -\alpha p_{01} + \beta p_{02})\right]. \tag{3.51}$$

The total size of the organization at time T is

$$N(T) = N(0) + Mx(1 - x^T)/(1 - x). \tag{3.52}$$

The variable, T, enters the expression for $\bar{\mathbf{n}}(T)$ through the terms in x^T and $(1 - \alpha - \beta)^T$. Of these the latter exerts a diminishing influence as T increases because $1 - \alpha - \beta < 1$. If $x > 1$ the terms in x^T become dominant whereas if $x < 1$ they vanish in the limit. In the intermediate case where $x = 1$ the grade sizes increase without limit but their relative values approach a limit as $T \to \infty$. In fact, for $x \leq 1$ we have

$$\bar{\mathbf{n}}(T) \sim \frac{N(T)}{\alpha + \beta}(\beta, \alpha). \tag{3.53}$$

In order to appreciate the full implications of (3.51) it is necessary to consider special cases. One such case is considered below; the reader will find it helpful to construct others.

The following illustration is for a two-grade system in which all recruits enter grade 1. There are no demotions and the probability of withdrawal is the same for each grade. The table of transition probabilities is

$$\begin{array}{ccc} 1 & 0 & \\ \hline \frac{1}{2} & \frac{1}{4} & \frac{1}{4} \\ 0 & \frac{3}{4} & \frac{1}{4} \end{array}$$

For this example

$$1 - \alpha = 1 - \beta = \tfrac{3}{4}.$$

Substitution in (3.51) gives

$$\bar{\mathbf{n}}(T) = N(T)(\tfrac{1}{2}, \tfrac{1}{2}) + (\tfrac{1}{2})^{T+2}(n_1(0) - n_2(0), n_2(0) - n_1(0)) + Mx\left\{\frac{x^T - (\frac{1}{2})^T}{x - \frac{1}{2}}\right\}(\tfrac{1}{2}, -\tfrac{1}{2}). \tag{3.54}$$

If x is fairly small the approach to the limit will be rapid. For example, if $x = \frac{1}{2}$

$$n_1(T) = \tfrac{1}{2}\{N(0) + M(1 - (\tfrac{1}{2})^T)\} + (\tfrac{1}{2})^{T+1}\{n_1(0) - n_2(0)\} + (\tfrac{1}{2})^{T+2}TM. \quad (3.55)$$

The difference between the initial and final structures and the size of M both affect the rate of approach to the limit but their effect is not great. If x is near to 1 the approach to the limit may be very slow, as we shall see below.

When $k > 2$ there is no simple expression for $\mathbf{Q}^T$ but similar conclusions can be drawn. If the growth rate is geometric the matrix series in (3.47) can almost always be summed. It is easy to verify that

$$(x\mathbf{I} - \mathbf{Q})\left(\sum_{\tau=0}^{T-1} \mathbf{Q}x^{T-\tau}\right) = x(x^T\mathbf{I} - \mathbf{Q}^T)$$

and hence, if $x\mathbf{I} - \mathbf{Q}$ possesses an inverse

$$\sum_{\tau=0}^{T-1} \mathbf{Q}^\tau x^{T-\tau} = x(x\mathbf{I} - \mathbf{Q})^{-1}(x^T\mathbf{I} - \mathbf{Q}^T).$$

Thus we have

$$\bar{\mathbf{n}}(T) = \mathbf{n}(0)\mathbf{Q}^T + Mx\mathbf{p}_0(x\mathbf{I} - \mathbf{Q})^{-1}(x^T\mathbf{I} - \mathbf{Q}^T). \quad (3.56)$$

The inverse of $x\mathbf{I} - \mathbf{Q}$ does not exist if the determinant of the matrix vanishes. This will happen whenever x is equal to an eigenvalue of the matrix $\mathbf{Q}$. For a stochastic matrix it is known that at least one eigenvalue is equal to unity and that the remainder lie between zero and one. Hence (3.56) holds for all $x > 1$ and when $x < 1$ for all but at most $k - 1$ values of x. For numerical work with $x < 1$ it is preferable to use the recursive formula of (3.46b). The usefulness of the representation of (3.56) is that it provides direct information about the transient behaviour of the system. This behaviour depends on the terms x^T and $\mathbf{Q}^T$. It is known from the general theory of Markov chains that, if $\mathbf{Q}$ is regular, $\mathbf{Q}^T$ tends to a limit as T increases. As in the case $k = 2$ the long-term behaviour of the system depends crucially on the magnitude of x.

In order to illustrate the theory we shall take $k = 5$ and use the same transition probabilities as in the example of Section 3.2. Let the rate of growth be

$$M(T) = \frac{1}{10}N(0)x^T$$

with $x = 0, \frac{1}{2}, 1$ and 2. Because the total size of the organization is changing we have tabulated (Table 3.3) the relative expected grade sizes given by

$$q_j(T) = \bar{n}_j(T)/N(T) \qquad (j = 1, 2, \ldots, k).$$

Table 3.3 Percentage values of the expected grade sizes for an organization with geometric growth rate and transition probabilities as in Section 3.2

		$100q_i(T)$				
		i				
T	x	1	2	3	4	5
0	0	40	30	15	10	5
	$\frac{1}{2}$	40	30	15	10	5
	1	40	30	15	10	5
	2	40	30	15	10	5
1	0	35·6	32·2	15·8	10·8	5·8
	$\frac{1}{2}$	37·4	31·8	15·0	10·2	5·5
	1	39·1	31·5	14·3	9·8	5·2
	2	42·1	31·0	13·1	9·0	4·8
2	0	32·5	32·8	16·6	11·5	6·5
	$\frac{1}{2}$	34·8	32·8	15·7	10·7	6·1
	1	38·4	32·4	14·2	9·6	5·4
	2	46·6	31·3	10·9	7·2	4·1
5	0	27·8	31·1	18·2	13·8	9·1
	$\frac{1}{2}$	29·2	32·0	17·7	12·8	8·3
	1	36·3	33·2	14·8	9·6	6·1
	2	59·6	31·6	5·4	2·2	1·3
10	0	24·8	27·5	17·7	16·2	13·8
	$\frac{1}{2}$	25·3	28·2	17·9	15·7	12·8
	1	33·8	32·4	15·8	10·6	7·4
	2	63·6	31·9	3·9	0·6	0·1
25	0	22·3	23·8	14·6	15·5	23·8
	$\frac{1}{2}$	22·4	23·9	14·7	15·6	23·3
	1	29·3	29·4	15·9	13·0	13·0
	2	63·8	31·9	3·8	0·5	0·0
50	0	21·8	23·0	13·8	13·9	27·3
	$\frac{1}{2}$	21·8	23·1	13·9	14·0	27·3
	1	26·2	26·8	15·2	13·6	18·1
	2	63·8	31·9	3·8	0·5	0·0

When $x = 0$ there is no expansion and the limit is approached rather slowly, especially in the highest grade. This is due mainly to the big difference between the initial and limiting structures. When $x = \frac{1}{2}$ the ultimate increase in total size is 10 per cent but the rate of approach to the limiting structure is hardly affected. The case $x = 1$ shows a very slow approach to the limit though here again the lower grades attain their limiting values more quickly than the higher ones. In this example grade 5 is little more than half its limiting value after 50 years. Under this kind of expansion the structure would not achieve equilibrium in any period likely to be of practical interest. By contrast, when $x = 2$, the limit is reached quite rapidly although in this case it is a different limit. Few organizations would be able to maintain such a rapid rate of growth long enough for this final structure to be of real importance.

The limiting structure

We have already investigated the limiting behaviour of expanding organizations when the growth rate is geometric. In the special cases considered we observed that the quantities $\{q_j(T)\}$ approached the same limiting values for $x \leq 1$ but had different limits when $x > 1$. We shall now give a general method of finding the limiting structure and use it to show how this structure depends on the growth rate.

Let us consider $\bar{\mathbf{n}}(T)$ in the form given by (3.47). In Chapter 2 we made use of the fact that if $\mathbf{Q}$ is a regular stochastic matrix then

$$\lim_{T\to\infty} \mathbf{Q}^T = \mathbf{Q}^\infty$$

exists and is a matrix with identical rows. We introduce the vector $\mathbf{q} = (q_1, q_2, \ldots, q_k)$ to denote any row of this matrix. This result enables us to deduce that the limit of the first term on the right-hand side of (3.47) is $N(0)\mathbf{q}$. To find the limit of the second term consider the matrix

$$\mathbf{Q}_T^* = \left(\sum_{\tau=0}^{T-1} M(T-\tau)\mathbf{Q}^\tau\right) \Bigg/ \sum_{\tau=0}^{T-1} M(T-\tau). \tag{3.57}$$

The (i, j)th element in this matrix may be written in the form

$$S_{ij}(T) = \sum_{\tau=0}^{T-1} q_{ij}^{(\tau)} d(T-\tau)$$

where $q_{ij}^{(\tau)}$ is the τ-step transition probability for the Markov chain with transition matrix $\mathbf{Q}$: we define $q_{ii}^{(0)} = 1$. The weights $d(T-\tau)$ obviously satisfy the condition

$$\sum_{\tau=0}^{T-1} d(T-\tau) = 1.$$

Further, since $0 \leq q_{ji}^{(\tau)} \leq 1$ and $d(T-\tau) \geq 0$ for all τ, it follows that the elements of $\mathbf{Q}_T^*$ all lie in the interval (0, 1) for all T. Using this notation (3.47) may be written as

$$\bar{\mathbf{n}}(T) = \mathbf{n}(0)\mathbf{Q}^T + \mathbf{p}_0\mathbf{Q}_T^*\{N(T) - N(0)\}. \tag{3.58}$$

Two cases must now be distinguished. If $M(T)$ is such that $N(T)$ approaches a limit then

$$\bar{\mathbf{n}}(\infty) = N(0)\mathbf{q} + \mathbf{p}_0\mathbf{Q}_\infty^*(N(\infty) - N(0)) \tag{3.59}$$

assuming that

$$\lim_{T\to\infty} \mathbf{Q}_T^* = \mathbf{Q}_\infty^*$$

exists. If $N(T)$ increases without limit the expected grade sizes also tend to infinity but we can still calculate their relative sizes. We then find that

$$\mathbf{q}(\infty) = \mathbf{p}_0\mathbf{Q}_\infty^* \, . \tag{3.60}$$

In either case the solution to our problem reduces to finding

$$\lim_{T\to\infty} \mathbf{Q}_T^*.$$

Provided $\mathbf{Q}$ is regular we know that

$$\lim_{T\to\infty} q_{ij}^{(T)} = q_i \qquad (i = 1, 2, \ldots, k).$$

Thus for any positive ε, however small, there is a value T_0 of T such that for $T \geq T_0$, $|q_{ij}^{(T)} - q_i| < \varepsilon$. The element $S_{ij}(T)$ of $\mathbf{Q}_T^*$ may be written

$$S_{ij}(T) = \sum_{\tau=0}^{T_0-1} q_{ij}^{(\tau)} d(T-\tau) + \sum_{\tau=T_0}^{T-1} q_i d(T-\tau) + \sum_{\tau=T_0}^{T-1} (q_{ij}^{(\tau)} - q_i) d(T-\tau). \tag{3.61}$$

The modulus of the last term is certainly less than ε and can thus be made arbitrarily small. To find the limit of $S_{ij}(T)$ we therefore need to consider only the first two terms. There are two cases:

(a) If

$$\lim_{T\to\infty} \sum_{\tau=T_0}^{T-1} d(T-\tau) = 1$$

the second term in (3.61) tends to q_i and the first term vanishes. In this case, therefore, the limit exists and is equal to q_i. We shall refer to this condition as condition A. It is a condition imposed on the rate of growth of the organization. If it is satisfied we find from both (3.59) and (3.60) that

$$\mathbf{q}(\infty) = \mathbf{q}. \tag{3.62}$$

This means that *the same limiting structure is reached for all rates of growth satisfying condition A*. In particular, the limit is that reached by a system in which there is no expansion at all.

(b) If condition A does not hold $\mathbf{Q}_T^*$ may or may not tend to a limit, depending on the nature of the sequence $\{M(T)\}$. An example is considered below and, in general, each problem must be considered separately.

Condition A may be expressed in various forms. An equivalent version is that

$$\lim_{T \to \infty} d(T - C) = 0$$

for any positive integer C. Perhaps the form which shows most clearly the nature of the restraint which it imposes on growth is

$$\lim_{T \to \infty} M(T)/N(T) = 0.$$

Expressed in words this means that the *proportionate* growth rate must vanish in the limit.

For the geometric growth rate, condition A holds whenever $x \leq 1$. The case $x = 1$ is particularly interesting because, although the organization grows without limit, the limiting grade structure is the same as when there is no growth at all. However, as we saw in Table 3.3 the rate of approach to the limit is much slower. If $x > 1$

$$\lim_{T \to \infty} \sum_{\tau = T_0}^{T-1} d(T - \tau) = 0$$

and hence condition A does not hold. The limiting grade structure may be deduced directly in this case from (3.56). If T is large

$$\bar{\mathbf{n}}(T) \sim M x \mathbf{p}_0 (x\mathbf{I} - \mathbf{Q})^{-1} \tag{3.63}$$

and

$$\mathbf{q}(\infty) \propto \mathbf{p}_0 (x\mathbf{I} - \mathbf{Q})^{-1}. \tag{3.64}$$

The inverse matrix may be formally expanded in a power series to give

$$(x\mathbf{I} - \mathbf{Q})^{-1} = \frac{1}{x}\left(\mathbf{I} + \frac{\mathbf{Q}}{x} + \frac{\mathbf{Q}^2}{x^2} + \dots\right). \tag{3.65}$$

The elements of this matrix are thus

$$S_{ij}(\infty) = \frac{1}{x} \sum_{\tau=0}^{\infty} q_{ij}^{(\tau)} x^{-\tau}.$$

If x is very large the first term is dominant and

$$\mathbf{q}(\infty) \doteqdot \mathbf{p}_0. \tag{3.66}$$

This result could have been anticipated by observing that with such a rapid rate of growth most members of the system would be new recruits.

Contracting systems

The argument leading to (3.46a) and (3.46b) may break down if $M(T+1)$ is negative for any T. If this happens it may be that the expected number of losses, given by

$$\sum_{i=1}^{k} p_{i,k+1}\bar{n}_i(T),$$

may not be sufficient to achieve the reduction in size implied by $M(T+1)$. In such circumstances $\bar{R}(T+1)$ as given by (3.45) will be negative. Since the model as described so far makes no provision for dealing with redundancies such an occurrence would bring the calculations to a halt.

If we wish to allow for redundancies the model must be extended. Two possibilities are as follows. One is to specify a redundancy vector $\mathbf{s} = (s_1, s_2, \ldots, s_k)$ whose elements are the proportions of $|R(T)|$ to be declared redundant in each grade. Instead of adding the vector $R(T)\mathbf{p}_0$ to $\mathbf{n}(T)\mathbf{P}$ we must then subtract $|R(T)|\mathbf{s}$. The basic difference equation for the expected grade sizes now takes one of two forms according to the sign of $R(T)$ as follows:

$$\bar{n}_j(T+1) = \sum_{i=1}^{k} (p_{ij} + p_{i,k+1}p_{0j})\bar{n}_i(T) + M(T+1)p_{0j} \qquad (j = 1, 2, \ldots, k) \tag{3.67a}$$

if $\bar{R}(T+1) \geq 0$ or

$$\bar{n}_j(T+1) = \sum_{i=1}^{k} (p_{ij} + p_{i,k+1}s_j)\bar{n}_i(T) + M(T+1)s_j, \qquad (j = 1, 2, \ldots, k) \tag{3.67b}$$

if $\bar{R}(T+1) < 0$. Note that $M(T+1)$ will certainly be negative in (3.67b) and may be negative in (3.67a). We must therefore alternate between (3.67a) and (3.67b) according to whether the current situation requires recruitment or redundancy. If the vector $\mathbf{s}$ is chosen to be the same as $\mathbf{p}_0$ it is clear that the two equations become identical and we can use the original equation (3.46b) as if it were valid for any $M(T+1)$. In doing this, however, we must check at each step that none of the elements of $\bar{\mathbf{n}}(T+1)$ has become negative. If negative elements do occur further progress can only be made by choosing a new $\mathbf{s}$.

An alternative method, which avoids the possibility of negative stock sizes, is to compute the number of redundancies as a proportion of the stocks after wastage and promotion have taken place. In the absence of recruitment or redundancy the expected size of the jth grade at $T+1$ is

$$\bar{n}'_j(T+1) = \sum_{i=1}^{k} p_{ij}\bar{n}_i(T) \qquad (j = 1, 2, \ldots, k).$$

Redundancies will be necessary if

$$\sum_{j=1}^{k} \bar{n}'_j(T+1) > N(T+1)$$

and the rule proposed is to reduce each $\bar{n}'_j(T+1)$ by the same proportion. If this proportion is denoted by x then it must satisfy

$$x \sum_{j=1}^{k} \bar{n}'_j(T+1) = N(T+1)$$

or

$$x = N(T+1)\Big/\sum_{i=1}^{k} (1 - p_{i,k+1})\bar{n}_i(T). \tag{3.68}$$

If this fraction is applied to each grade the resulting stock numbers will be

$$\bar{n}_j(T+1) = N(T+1) \sum_{i=1}^{k} p_{ij}\bar{n}_i(T)\Big/\left\{N(T) - \sum_{i=1}^{k} p_{i,k+1}\bar{n}_i(T)\right\} \qquad (j = 1, 2, \ldots, k) \tag{3.69a}$$

or

$$\mathbf{n}(T+1) = \mathbf{n}(T)\mathbf{P}\{1 + M(T)/N(T)\}/\{1 - W(T)\} \tag{3.69b}$$

where $W(T) = \bar{\mathbf{n}}(T)\mathbf{p}'_{k+1}$. In this version of the model the last equation would be used whenever $\bar{R}(T+1)$ as given by (3.45) turned out to be negative.

Variances and covariances of the grade sizes

The method used to obtain the variances and covariances for the model with given input can be applied here with only minor modifications. If we start from (3.30) we can no longer assume that the input is independent of the internal movements, because the latter influence the wastage which, in turn, determines the input. To the two terms on the right-hand side of (3.31) we must therefore add terms involving the covariances between the $n_{0j}(T+1)$'s and the internal flows. However, the fairly heavy algebra can be circumvented by arguing as follows. Each loss is directly responsible for a recruitment; if the loss occurs from grade i the probability that the consequential recruitment is in grade j is $p_{i,k+1}p_{0j}$. The replacement of a leaver can thus be treated as if it were an internal transfer. The total probability of a move (of any kind) out of grade i which results in an addition to grade j is thus

$$q_{ij} = p_{ij} + p_{i,k+1}p_{0j}. \tag{3.70}$$

This is the same point as we made in another connection on page 81.

If $M(T) = 0$ for all T we can therefore use the theory developed in Chapter 2 for the case of a closed system. The vector of expectations, variances and covariances will thus be given by

$$\boldsymbol{\mu}(T+1) = \boldsymbol{\mu}(T)\boldsymbol{\Pi}' \tag{3.71}$$

where the elements of $\boldsymbol{\Pi}'$ are the same functions of the q_{ij}'s as $\boldsymbol{\Pi}$ in (2.15) is of the p_{ij}'s.

When the system is expanding there are additional recruits who fill the vacancies created by expansion. If these are allocated to grades independently of the replacements and with the same probabilities $\{p_{0j}\}$, then these flows will be independent of all other flows and so their contribution to the variances and covariances can simply be added to (3.71). Let the vector of expectations, variances and covariances attributable to the expansion be $\boldsymbol{\mu}_0'(T+1)$. Then

$$\boldsymbol{\mu}_0'(T+1) = M(T+1)\{p_{01}, p_{02}, \ldots, p_{0k}, p_{01}(1-p_{01}), -p_{01}p_{02} \cdots \\ \cdots p_{0k}(1-p_{0k}\}, \tag{3.72}$$

the covariance of the flows into grades i and j being

$$\delta_{ij}p_{0j} - p_{0j}p_{0l} \qquad (i, j = 1, 2, \ldots, k).$$

The total expression for the difference equation is thus

$$\boldsymbol{\mu}(T+1) = \boldsymbol{\mu}(T)\boldsymbol{\Pi}' + \boldsymbol{\mu}_0'(T+1). \tag{3.73}$$

The variances and covariances can therefore be computed using the same computer program as for the fixed input, but with $\mathbf{\Pi}$ replaced by $\mathbf{\Pi}'$ and $R(T+1)$ by $M(T+1)$.

Generalizations and extensions

When dealing with closed Markov models in Chapter 2 we discussed a number of possible generalizations designed to give the models greater realism. The same kind of approach can be made to the open Markov models of the present chapter. In practice, the simple assumption of time-homogeneous transition rates, applicable to all individuals, has often been successful in describing the movement of individuals through an organization even when general knowledge would suggest otherwise. However, there are many instances where the simple model is clearly not applicable and we now briefly describe some extensions.

Time-dependent transition matrices present no problems and we have already illustrated the use of a periodic matrix earlier in this section. Heterogeneity of the kind where the transition matrix differs from one individual to another can be handled in the same way as for the closed case. In manpower and educational applications it is doubtful whether the mover-stayer' type of model is applicable—at least in the long term—but less extreme forms of variability doubtless occur. Where it is possible to identify sources of variability different transition matrices can be used for different subgroups. For example, in a firm it is likely that the transition rates for men and women will be different and so the model would be applied to each separately.

The idea of 'cumulative inertia' can be easily incorporated into the model by subdividing grades according to length of service as in Chapter 2. In fact, the freedom to choose the grades of our system in any way we please makes the Markov model much more flexible than appears at first sight. Much of the art in the successful application of the Markov model lies in so defining the 'grades' that the assumptions of the model are most nearly satisfied. For example, since propensity to leave an organization depends strongly on length of service (see Chapter 6) it is usually desirable to define the grades so that they are reasonably homogeneous with respect to length of service. Other relevant variables can be treated in the same way, but repeated subdivision in the interests of homogeneity creates problems of small numbers when it comes to estimating the parameters of the model. This is an important statistical question which merits further attention, but it is outside our present scope.

Marshall (1973) has devised an ingenious method of investigating the adequacy of the simple Markov model under very general assumptions about the stochastic nature of the flows. The basic idea is quite simple and we shall describe it as applied to a system with fixed input. Consider the joint distribution of the vectors $\mathbf{n}(T+1)$ and $\mathbf{n}(T)$. Let h index all original members of the system, and those who join prior to T, in some convenient order. We shall introduce random variables X_{hij} such that

$$\begin{aligned} X_{hij} &= 1 \text{ if the } h\text{th individual is in grade } i \text{ at} \\ &\quad\ \text{time } T \text{ and grade } j \text{ at time } T+1 \\ &= 0 \text{ otherwise.} \end{aligned}$$

The probability that $X_{hij} = 1$ will depend on the stochastic laws of movement through the system, but all we need to assume for the present analysis is that individuals behave independently of one another. From the definition it follows that

$$\left.\begin{aligned} n_i(T) &= \sum_{j=1}^{k} \sum_{h} X_{hij} \qquad (i = 1, 2, \ldots, k) \\ \text{and} \qquad & \\ n_j(T) &= \sum_{i=1}^{k} \sum_{h} X_{hij} \qquad (j = 1, 2, \ldots, k) \end{aligned}\right\} \tag{3.74}$$

An appeal to the central limit theorem now establishes, under very general conditions on the distribution of the X_{hij}'s, that $(\mathbf{n}(T), \mathbf{n}(T+1))$ has, approximately, a $2k$-variate normal distribution.

The basic Markov model predicts $\mathbf{n}(T+1)$ given $\mathbf{n}(T)$ and so it is of interest to consider the distribution of $\mathbf{n}(T+1)$ conditional on $\mathbf{n}(T)$. Assuming multivariate normality it follows that $\mathbf{n}(T+1)$ given $\mathbf{n}(T)$ has a k-variate normal distribution and, in particular, that the mean of $\mathbf{n}(T+1)$ is a linear function of $\mathbf{n}(T)$. Thus, under very general circumstances the expected grade sizes at successive points in time will be related by an equation of the form (3.1). However, in general, $\mathbf{P}$ will depend on time and on the recruitment pattern as well as flow laws. The process will not, in general, be Markovian. Marshall (1973) investigates a number of special cases of the general model and compares it with the Markov case. One useful by-product of this analysis is that it provides an approximation to the distributions of the grade sizes. Assuming their normality we require only a knowledge of first and second moments for a complete specification of the distribution. The Wynn and Sales (1973a) extension of Pollard's matrix method provides this information when the Markov property holds for a general error structure of the flows.

Many extensions of the Markov model are possible which will accommodate restrictions on the stock sizes. An important feature of the transient behaviour of the Markov chain model is that the expected grade sizes may vary with the passage of time. In practice, the amount of variation may be limited. For example, the number of school leavers desiring university places may well be a fixed proportion of the age group, but the number who can actually take up a university place will be limited by the number of places available. If there is an upper limit on the size of any stock the model must be extended by the addition of a rule saying how the excess flow is to be reallocated among the grades. This feature is the distinguishing characteristic of the bottleneck models introduced by Armitage, Smith and Alper (1969) and mentioned earlier. In manpower planning in a firm there may also be lower limits below which the stock must not be allowed to fall. For this case the model must include rules specifying how deficits are to be made good.

In some applications the upper and lower limits on the stocks will coincide if there is a fixed establishment in each grade. Models appropriate for this situation, based on renewal theory, will be described in Chapter 8. Forbes, in unpublished work, has generalized the Markov model in such a way that the simple Markov and discrete time renewal models can be treated within the same mathematical framework. This development opens up the prospect of a model which covers not only these special cases but a range of intermediate ones also.

White (1970a) applies Markovian methods to systems with fixed grade sizes by modelling the flows of vacancies rather than those of people. Thus a loss corresponds to the 'recruitment' of a new vacancy; when this vacancy is filled the vacancy moves to the place from which the new incumbent came, and so on. Ultimately, the vacancy leaves the system at the point where a recruitment is made from outside the system. White's main interest is in vacancy chains, which is the name given to the path of a vacancy through the system. Such a chain is similar to (but not equivalent to) the career path of an individual from the viewpoint of the present chapter. White discusses the relationship between his approach and the more usual one in Chapter 9 of his book.

The basic Markov model described in this chapter and the various extensions outlined above all have an important defect—at least when applied to the flows of manpower in a firm. They all assume that individual behaviour is unaffected by how individuals perceive their environment and, in particular, their promotion chances. It seems plausible to suppose that an individual's assessment of his promotion chances will affect the likelihood of his leaving. If this is so our model ought to include a statistical relationship (probably lagged) between the wastage rates and the promotion rates. In general this will produce a non-linear model which would be less susceptible to mathematical analysis than the linear models discussed here. Young (1971) proposed such a model in which the 'normal' number of leavers was increased by a number of 'frustrated' leavers, proportional to the difference between an 'expected' number of promotees and the number actually arising. He stated that such a model did succeed in reproducing certain features of real systems which cannot be accounted for by the linear model.

CHAPTER 4

Control Theory for Markov Models

4.1 INTRODUCTION

In the last chapter we formulated various versions of a Markov chain model and applied it to the study of graded systems. These models enabled us to study the effect of constant flow rates on the changing grade structure. In a manpower system some of the flows are under the control of management and the problem then becomes one of knowing how to control these flows in such a way as to achieve and maintain a desired structure. We observed in the last chapter how a system with reasonable looking promotion rates exhibited a tendency for the higher grades to grow at the expense of the lower. The same characteristic has often been observed in practice—see, for example, Young (1971). It commonly occurs when a period of expansion comes to an end: promotion rates then become inflated during expansion and the attempt to maintain them through a period of reduced growth creates pressure for growth at the top. Universities in both the United Kingdom and the United States have passed through such an experience in recent years, but these are not isolated examples. Indeed, the phenomenon seems to be a normal characteristic of bureaucracy having been noted as far back as the Roman Empire (in this connexion see the Cambridge *Medieval History*, Vol. IV, Part I, pp. 22 ff.).

The situation we have described can easily be cured if sacking and demotion on any scale are acceptable. In practice, such action would almost always be regarded as highly undesirable, if not impossible. Measures designed to control a system's structure affect the lives of people whose aspirations must be taken seriously if the system is to function harmoniously and efficiently. The essence of the problem is thus to maintain an effective degree of control within the restraints imposed by the requirements of good personnel management.

The problem takes on a slightly different form when the 'grades' of the system represent age groups. The control of age structures is important in avoiding promotion bottlenecks and recruitment difficulties. For historical reasons many organizations find themselves with 'bulges' in their age distribution stemming from abnormally high recruitment at some point in the past. For example, at the end of World War II many British firms, including the Civil Service, recruited large numbers of people coming out of the forces of similar age. When such a group reaches the age of retirement obvious problems arise for recruitment. It might be hoped that a theory of control for such systems would provide some guidance on the ages at which people should be recruited to smooth out unwanted irregularities. In this application we cannot, of course, control 'promotion', which here corresponds to the inevitable process of ageing, but recruitment and, to some extent leaving, do provide the opportunity to change the age structure.

It will be helpful to specify clearly what it is that we wish to control and the means which are available to exercise control. We shall aim to control the numbers of people (the stocks) in the grades of a system over a period of time. This has two aspects; the control of the total stock of people in the whole organization and of the way in which they are distributed among the grades. We may have the right total without the right distribution, and vice versa. The problem of controlling the total stock has been dealt with, in part at least, in Section 3.3 of Chapter 3, where we showed how to compute the number of recruits to achieve a required total size. In this chapter the main emphasis will be on controlling the stocks in the individual grades.

As to the means of control, this must be exercised through the flows of people which it is convenient to classify into three categories:

(a) Wastage (the vector $\mathbf{w}$ which is identical to the loss vector $\mathbf{p}_{k+1}$ of the previous chapter);

(b) Promotion, taken here to include transfers and demotion (the matrix $\mathbf{P}$);

(c) Recruitment. This has two parts, the total number $\{R(T)\}$ and their allocation to the grades $\mathbf{r} \equiv \mathbf{p}_0$. It is with the latter that we shall be primarily concerned in this chapter.

In the model wastage and promotion are intimately connected. Because

$$\sum_{j=1}^{k} p_{ij} + w_i = 1 \quad (i = 1, 2, \ldots, k)$$

any change in either w_i or a p_{ij} requires compensating changes in one or more of the other parameters to maintain the equality. This will not cause any particular problems of interpretation in the applications we shall discuss and we prefer to keep the distinction.

The wastage flow can be controlled to some extent. It can be increased by sacking people or by offering them financial or other inducements to leave, and decreased by improved conditions or inducements to stay. These methods of control are, however, somewhat uncertain in their operation and, in the case of sacking, certainly undesirable.

More precise control can be exercised over the promotion flows in that these result from direct management decisions. Even here, however, there are often compelling reasons for varying the promotion rates as little as possible. For example, an increase in the promotion rate may involve the promotion of inadequately qualified people and make it difficult to return to the original standards should this prove necessary later. Equally, a decrease in the promotion rates is likely to create problems among people who see what they regarded as their expectations for advancement being eroded.

The recruitment flows offer the most attractive means of control since decisions to recruit more or fewer people at a given level can be taken without the same immediate impact on those already serving. There may be practical difficulties in finding enough recruits at the required levels and existing members may express concern about the effects, for example, of high recruitment near the top on their own promotion prospects. Nevertheless, of the three main methods of control the adjustment of the recruitment vector seems to offer the least painful means of control. We have already noted that this is the only means, other than wastage, which is available when controlling age distributions. It may well happen, of course, that no one of these sets of control parameters will be sufficient in itself and that two or more will have to be used in conjunction.

We shall treat the control problem in two stages. The term *steady-state control* will be used to refer to the problem of holding the grade structure at some specified value. *Sequential* (or *adaptive*) *control* will refer to the problem of changing a given structure to some desired structure by a sequence of adjustments to the control parameters. Although in many applications the problem of reaching a target will arise before that of maintaining the structure once achieved, we shall treat them in the reverse order because of the greater simplicity of the steady-state problem.

The two problems may be formally stated, as follows:

(a) *Steady state control.* Let $\mathbf{n}^*$ denote the structure which is to be maintained. Then we have to find values of the control parameters satisfying

$$\mathbf{n}^* = \mathbf{n}^*\mathbf{P} + \mathbf{n}^*\mathbf{w}'\mathbf{r}.$$

Notice that if $\mathbf{n}^*$ is to be maintained the total size must be constant and so the number recruited must be $\mathbf{n}^*\mathbf{w}'$. If we wish to hold the relative sizes of the grades fixed while the total size increases or decreases then the equation to be satisfied is

$$(1 + \alpha)\mathbf{q}^* = \mathbf{q}^*\mathbf{P} + \mathbf{q}^*\mathbf{w}'\mathbf{r} + \alpha\mathbf{r} \tag{4.1}$$

where $\alpha = \{N(T + 1) - N(T)\}/N(T)$ and $\mathbf{q}^* = \mathbf{n}^*(T)/N(T)$. This equation will be derived and discussed in Section 4.2. If $\mathbf{P}$ is given and $\mathbf{r}$ has to be found we shall speak of *recruitment control*. When $\mathbf{r}$ is fixed and $\mathbf{P}$ is at choice the term used will be *promotion control*. The set of vectors $\mathbf{n}^*$ or $\mathbf{q}^*$ which it is possible to maintain, that is for which a solution to (4.1) exists, will be called the *maintainable set*. Later on, when the problem is viewed geometrically, we shall talk about the *maintainable region*. In either case the set will be denoted by $\mathscr{M}$; it will depend, of course, on whether control is by promotion, recruitment or a mixture of the two, but this will not be made explicit in the notation.

(b) *Sequential control*. The problem here is to find a sequence of control parameters $\{\mathbf{r}(T + 1)\}$ or $\{\mathbf{P}(T)\}$ or both such that the system moves from its present structure $\mathbf{n}^*(0)$ to a desired $\mathbf{n}^*$ in an optimal manner. Optimality may be defined in terms of costs of one kind or another, but here we shall concentrate on achieving the goal as quickly as possible. Observe that in sequential control we express the control parameters as functions of the argument T because it is of the essence of the problem that these can be adjusted at each time point. We shall see that just as all structures need not be maintainable so all structures are not necessarily attainable. Hence the concept of *attainability* plays a fundamental role in the theory of sequential control; Section 4.3 of the present chapter is devoted to it. It is useless to search for a control strategy to achieve a certain goal if no such strategy exists; the theory of attainability aims to identify attainable structures.

The problem of the control of a Markov system in discrete time was posed in the first edition and developed by Bartholomew (1969b), Forbes (1970) and Davies (1973). The original work in Chapter 3 of the first edition of this book was concerned only with steady-state control in a fixed size organization, though it foreshadowed the programming formulation developed later in this chapter. Forbes was also concerned with steady-state control but he extended the theory to expanding systems. Davies work was a direct development of Bartholomew (1969b) and gives a fuller analysis of the concepts of attainability and maintainability for fixed size and expanding or contracting organizations.

Rowe, Wagner and Weathersby (1970) have formulated the control problem in terms which require the minimization of a discounted cost function over a finite planning horizon. They confine themselves to the question of control by recruitment. Charnes, Cooper, Niehaus and others (see, for example, Charnes and coworkers, 1970) have developed techniques of goal programming based on the Markov model. Like Rowe and coworkers they seek to find a recruitment strategy which minimizes a weighted function of discrepancies between the actual and desired structures over a fixed period of time. They also introduce budgetary constraints which restrict the total salary bill in any period and hence place restrictions on the recruitment pattern. This formulation is more complex but essentially equivalent to the approach we shall discuss in Section 4.4. The two approaches are, however, best viewed in a complementary sense. Charnes and coworkers aim to provide efficient numerical procedures for large manpower planning models incorporating a high degree of realism. The approach of the present book, and this chapter in particular, is to develop the analysis of relatively simple models in a way which gives insight into and understanding of the way the system works.

Morgan (1970) also formulated a control problem as a linear programme in which the variables are recruitment and promotion numbers and costs are attached to recruitment, redundancy and overmanning. In a subsequent paper (Morgan, 1971) the author considers the control of an age structure in which age groups take the place of grades, and he proposed a strategy which bears some similarity to those which we shall discuss in Section 4.4. The framework of theory in which our discussion is set may form a useful starting point for further development of Morgan's work.

Control theory has a lengthy history and a rapidly growing literature. Most of the existing work has been done in an engineering context but recently there has been much discussion about the possibility of applying it to the regulation of the economy. Very little of the theory appears to be directly applicable to the control problem formulated in this chapter. However, in a recent book Canon, Cullum and Polack (1970) have expounded the theory in terms very similar to those which form the basis of our discussion. In fact, much of this chapter could be regarded as an application of their approach to the problems of manpower planning, though most of the work described here pre-dates the publication of that book. We have, however, adapted some of their terminology and a further study of their work should prove fruitful for further developments.

4.2 STEADY-STATE CONTROL

The object of this section is to consider how to maintain a given grade structure over a period of time. There are various reasons why we may wish to do this. For example, the question may arise as part of a planning exercise when a grade structure has been decided upon and it is required to know what promotion and recruitment policies are compatible with this structure. Alternatively, an existing system may have a structure which is becoming too top-heavy and then the question is how to prevent matters becoming any worse. Not every structure can be maintained and an important part of the investigation will be to delineate those structures which can be maintained from those which cannot.

Control of an organization of constant size by recruitment policy

As our prime interest is in the relative sizes of the grades it will be convenient throughout this section to work in terms of the variables

$$q_i(T) = \bar{n}_i(T)/N \qquad (i = 1, 2, \ldots, k)$$

where N is the total size of the system. The basic difference equation for the Markov model is then

$$\mathbf{q}(T+1) = \mathbf{q}(T)\mathbf{P} + \mathbf{q}(T)\mathbf{w}'\mathbf{r}. \tag{4.2}$$

A structure $\mathbf{q}(T)$ can be maintained if we can find values of the control parameters such that $\mathbf{q}(T+1) = \mathbf{q}(T)$. In other words, we are interested in values of the parameters which satisfy

$$\mathbf{q} = \mathbf{q}\mathbf{P} + \mathbf{q}\mathbf{w}'\mathbf{r}. \tag{4.3}$$

If the recruitment vector $\mathbf{r}$ is the only set of parameters amenable to control we have to find an $\mathbf{r}$ satisfying (4.3). The possibility of doing this is easily investigated by solving (4.3) for $\mathbf{r}$ to give

$$\mathbf{r} = \mathbf{q}(\mathbf{I} - \mathbf{P})/\mathbf{q}\mathbf{w}'. \tag{4.4}$$

It is easy to check that the elements of $\mathbf{r}$ so obtained add up to one, but they may not all be positive. If they are, (4.4) gives the unique policy meeting the requirements; if not, the structure is not maintainable.

It would be useful to be able to characterize the set of structures which can be maintained with a given $\mathbf{P}$ and $\mathbf{w}$. A simple characterization of the maintainable region which follows directly from (4.4) is that it is the set of $\mathbf{q}$'s for which

$$\mathbf{q} \geq \mathbf{q}\mathbf{P}.$$

To obtain a more illuminating characterization we view the problem geometrically. Any structure $\mathbf{q}$ can be represented by a point in k-dimensional Euclidean space with the elements of $\mathbf{q}$ as its coordinates. Since all the $\mathbf{q}$'s must be non-negative and add up to one, all allowable structures lie on the hyper-plane $\sum_{i=1}^{k} q_i = 1$ in the positive orthant. When $k = 2$ this is the hypotenuse of the right-angled triangle with vertices (0, 0), (0, 1), (1, 0). When $k = 3$ it is the equilateral triangle illustrated in Figure 4.1. In higher dimensions the position is difficult to visualize but the geometrical terminology is still useful. The set of possible structures will be denoted by $\mathscr{X}$. The maintainable region $\mathscr{M}$ will be a subset of $\mathscr{X}$ whose boundaries we wish to determine.

The boundary of $\mathscr{M}$ may be found as follows. From (4.3)

$$\mathbf{q}(\mathbf{I} - \mathbf{P}) = \mathbf{q}\mathbf{w}'\mathbf{r}.$$

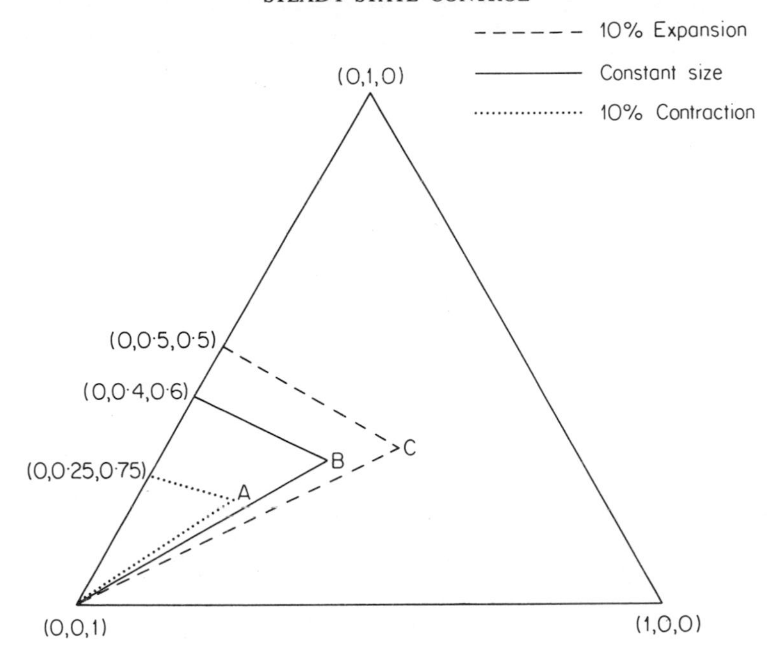

A=(0·158, 0·210, 0·632), B=(0·286, 0·286, 0·428), C=(0·386, 0·307, 0·307)

Figure 4.1 The maintable region for recruitement control with $k = 3$ and

$$\mathbf{P} = \begin{pmatrix} 0{\cdot}5 & 0{\cdot}4 & 0 \\ 0 & 0{\cdot}6 & 0{\cdot}3 \\ 0 & 0 & 0{\cdot}8 \end{pmatrix}$$

Hence

$$\mathbf{q} = \mathbf{q}\mathbf{w}'\mathbf{r}(\mathbf{I} - \mathbf{P})^{-1}; \tag{4.5}$$

the inverse always exists, as we noted earlier in (3.4). The vector $\mathbf{r}$ may be written $\mathbf{r} = \sum_{i=1}^{k} r_i\mathbf{e}_i$ where $\mathbf{e}_i$ is the vector with a one in the ith position and zeros elsewhere. Substituting in (4.5) then gives

$$\mathbf{q} = \mathbf{q}\mathbf{w}' \sum_{i=1}^{k} r_i\{\mathbf{e}_i(\mathbf{I} - \mathbf{P})^{-1}\}. \tag{4.6}$$

Post-multiplying both sides of (4.6) by a column vector of ones gives

$$1 = \mathbf{q}\mathbf{w}' \sum_{i=1}^{k} r_i d_i$$

where d_i is the sum of the elements in the ith row of $(\mathbf{I} - \mathbf{P})^{-1}$. Substituting for $\mathbf{qw}'$ in (4.6) then yields

$$\mathbf{q} = \sum_{i=1}^{k} \left(\frac{r_i d_i}{\sum_{j=1}^{k} r_j d_j} \right) \frac{1}{d_i} \mathbf{e}_i (\mathbf{I} - \mathbf{P})^{-1}. \tag{4.7}$$

The vector $\mathbf{q}$ has thus been represented as a convex combination (that is a weighted average with non-negative weights) of the points with coordinates $d_i^{-1}\mathbf{e}_i(\mathbf{I} - \mathbf{P})^{-1} (i = 1, 2, \ldots, k)$. For each $\mathbf{r}$ there is exactly one set of weights and hence one $\mathbf{q}$, and as $\mathbf{r}$ ranges over its possible values $\mathbf{q}$ ranges over all points in the convex hull formed by the points $d_i^{-1}\mathbf{e}_i(\mathbf{I} - \mathbf{P})^{-1} (i = 1, 2, \ldots, k)$. The maintainable region is thus the convex region with these points as vertices. These vertices are easily computed by taking the rows of $(\mathbf{I} - \mathbf{P})^{-1}$ in turn and scaling their elements so that the row sums are one. If the w's happen to be equal then $\mathbf{qw}'$ is a constant and the result follows at once from (4.6).

We now have a method of finding the vertices of $\mathscr{M}$ by scaling the elements in the rows of $(\mathbf{I} - \mathbf{P})^{-1}$. When $k = 3$ the region can be plotted in two dimensions; this has been done for a system with matrix

$$\mathbf{P} = \begin{pmatrix} 0{\cdot}5 & 0{\cdot}4 & 0 \\ 0 & 0{\cdot}6 & 0{\cdot}3 \\ 0 & 0 & 0{\cdot}8 \end{pmatrix}$$

in Figure 4.1. The region is the one with vertices (0, 0, 1), (0, 0·4, 0·6), (0·286, 0·286, 0·428). We note that all of the structures which can be maintained by recruitment are top-heavy. In this example the least top-heavy structure is represented by the structure at the point **B** on the diagram, which has a large top grade. In higher dimensions it is less easy to represent the region diagrammatically but inspection of the vertices gives a good idea of the structures which are maintainable.

The numerical example is a special case of an important class of transition matrices, which we shall describe as super-diagonal, having the form

$$\mathbf{P} = \begin{pmatrix} p_{11} & p_{12} & 0 & 0 & 0 \\ 0 & p_{22} & p_{23} & 0 & 0 \\ \cdot & & \cdot & \cdot & \cdot \\ \cdot & & & \cdot & \cdot \\ \cdot & & & \cdot & p_{k-1,k} \\ 0 & \cdot & \cdot & \cdot & p_{kk} \end{pmatrix}.$$

This means that promotion is possible only into the next higher grade; it also arises if the grades are defined by age groups. The matrix $(\mathbf{I} - \mathbf{P})$ can easily be inverted; its ith row is

$$\mathbf{e}_i(\mathbf{I} - \mathbf{P})^{-1} = \left(0, \ldots, 0, \frac{1}{1 - p_{ii}}, \frac{p_{i,i+1}}{(1 - p_{ii})(1 - p_{i+1,i+1})}, \ldots, \frac{1}{1 - p_{ii}} \prod_{r=i}^{k-1} \frac{p_{r,r+1}}{(1 - p_{r+1,r+1})}\right) \quad (i = 1, 2, \ldots, k-1) \tag{4.8}$$

and

$$\mathbf{e}_k(\mathbf{I} - \mathbf{P})^{-1} = \left(0, 0, \ldots, 0, \frac{1}{1 - p_{kk}}\right).$$

There is no simple expression for the row sums of $(\mathbf{I} - \mathbf{P})^{-1}$ but the vertices of $\mathscr{M}$ are easily calculated numerically. We shall see later that the structures corresponding to the vertices are maintained by recruitment into a single grade. The first row of $(\mathbf{I} - \mathbf{P})^{-1}$ is of particular interest because it gives the structure which can be maintained by recruitment into the bottom grade of a hierarchy. Under these circumstances the jth grade has size proportional to

$$\frac{1}{(1 - p_{jj})} \prod_{r=1}^{j-1} \frac{p_{r,r+1}}{(1 - p_{r+1,r+1})} \tag{4.9}$$

where the product is defined to be one if $j = 1$. This should be compared with (3.26) which gives the structure which will arise in equilibrium with a constant rate of recruitment at the lowest level. The two situations are essentially the same and we note again the crucial importance of $(1 - p_{jj})$ in determining what size of grade can be maintained. Other maintainable structures are weighted averages of the rows, suitably scaled, of $(\mathbf{I} - \mathbf{P})^{-1}$. Since the first row is the only one with a non-zero entry in the lowest grade it is obvious that no structure requiring a bottom grade larger than $d_1/(1 - p_{11})$ can be maintained.

These expressions simplify even further if the diagonal elements in $\mathbf{P}$ are also zero. This occurs if the k grades are based on age (or seniority) groups of the same length as the discrete time interval between events. Only the numerators in (4.8) remain and the non-zero elements in each row then form a decreasing sequence as intuition suggests they must. It does not follow that any maintainable age structure must be a decreasing function of age because the scaling and weighting of the rows may produce 'bumps'. For example, a heavy recruitment in the middle age range will obviously inflate the size of that and, to a lesser extent, the following age groups. If 'age' refers to length of service in the system then recruits must, by definition, enter at the lowest level; only the first row applies and any maintainable length of service structure must be monotonic non-increasing. This way of defining the grades provides an important link between Markov chain models and the renewal theory models to be discussed in later chapters.

The concept of maintainability can be extended. In practice, it may not matter that a structure cannot be precisely maintained if it is possible to remain sufficiently near to it. One way of formalizing this is to introduce the idea of *n-step maintainability*. A structure is n-step maintainable if a sequence of control parameters can be found such that the system returns to the goal at every nth step. Maintainability as discussed in this section is thus one-step maintainability. Since a one-step maintainable structure is obviously n-step maintainable the n-step maintainable region will include $\mathscr{M}$. By this means a wider range of structures can be maintained, in the weaker sense, without resort to anything other than recruitment policy. A fuller discussion of n-step maintainability is given by Davies (1973).

Control by recruitment under growth and contraction

The set of maintainable structures turns out to be surprisingly small for plausible looking $\mathbf{P}$ matrices. In particular, if $\mathbf{P}$ is upper-triangular, $\mathscr{M}$ usually excludes the typical 'bottom-heavy' structure of the traditional staff 'pyramid'. In this section we shall examine how the position is affected if the system is growing or contracting. This investigation may be viewed from two points of view. Expansion or contraction may be looked upon as a further control parameter by means of which the set of maintainable structures can be varied or as something imposed from outside over which we have no control. In one sense we cannot maintain a structure if the system is changing size since some grade sizes must inevitably change. We can, however, consider what Forbes (1970) has termed a quasi-stationary state in which the relative sizes of the grades are to be maintained. This is likely to be the most relevant case in practice if the relative grade sizes are determined by the *kind* of work the organization does and the total size by the *amount* of work it does.

Let $N(T)$ be the total size of the system at time T and let $N(T + 1) = (1 + \alpha)N(T)$. The basic equation for an organization of fixed size (3.46a) is then

$$\bar{\mathbf{n}}(T + 1) = \bar{\mathbf{n}}(T)\mathbf{P} + \bar{\mathbf{n}}(T)\mathbf{w}'\mathbf{r} + \alpha N(T)\mathbf{r}. \tag{4.10}$$

Introducing $\mathbf{q}(T) = \bar{\mathbf{n}}(T)/N(T)$ the equation becomes

$$(1 + \alpha)\mathbf{q}(T + 1) = \mathbf{q}(T)\mathbf{P} + (\mathbf{q}(T)\mathbf{w}' + \alpha)\mathbf{r}. \tag{4.11}$$

The condition for a quasi-stationary structure **q** to be maintainable is thus that

$$(1 + \alpha)\mathbf{q} = \mathbf{qP} + (\mathbf{qw}' + \alpha)\mathbf{r} \tag{4.12}$$

for some $\mathbf{r} \geq \mathbf{0}$ with $\sum_{i=1}^{k} r_i = 1$. If such an **r** exists it is given by

$$\mathbf{r} = \{\mathbf{q}(\mathbf{I} - \mathbf{P}) + \alpha\mathbf{q}\}/(\mathbf{qw}' + \alpha). \tag{4.13}$$

This should be compared with (4.4). As before, the elements of **r** always sum to one, but we are interested in those **q**'s for which they are all non-negative. It is clear that any **q** for which **r** is given by (4.4) is non-negative will also yield a non-negative **r** in (4.13) if $\alpha > 0$. Hence, if the system is expanding the maintainable region will be larger than when it is constant in size. The converse holds if $\alpha < 0$. Thus expansion increases the range of maintainable structures and contraction decreases it.

The argument leading to the determination of the vertices of $\mathscr{M}$ goes through in the present case, with obvious modifications to give the vertices of $\mathscr{M}$ with coordinates proportional to

$$\mathbf{e}_i(\mathbf{I}(1 + \alpha) - \mathbf{P})^{-1} \qquad (i = 1, 2, \ldots, k). \tag{4.14}$$

The only change required in (4.8) is the replacement of $1 - p_{ii}$ by $1 + \alpha - p_{ii}$ for all i in the denominators. If α is positive this clearly increases the amount by which the vertex structures taper towards the top. If α is negative a complication arises, as we see from the fact that $1 + \alpha - p_{ii}$ will be negative if $|\alpha| > 1 - p_{ii}$. Suppose that the present size of the ith grade is $n_i(T)$; then at $T + 1$ the size of that grade will have to be $(1 + \alpha)n_i(T)$ if the structure is to be maintained. However, the expected number of losses is $w_i n_i(T)$ so that if $|\alpha| < w_i$ the structure certainly cannot be maintained. The region $\mathscr{M}$ will thus only be non-empty if $|\alpha| \geq w_i$ for all i, and the maximum rate of contraction for which at least one maintainable structure exists is $\alpha = -\min_i w_i$.

To illustrate these results we return to the numerical example with $k = 3$, shown in Figure 4.1. The new vertices of the region for 10 per cent expansion and 10 per cent contraction are as follows:

10 per cent expansion	(0, 0, 1), (0, 0·5, 0·5), (0·386, 0·307, 0·307)
10 per cent contraction	(0, 0, 1), (0, 0·25, 0·75), (0·158, 0·210, 0·632).

These regions are also plotted on the figure, from which it is clear that a change in the total size of this amount has a big effect on the size of the region, especially around the vertex represented by the last points listed above.

The foregoing calculations can be given a different interpretation. Suppose that we had set out to investigate the effect of changing the wastage rate on the size of the maintainable region. If we had done this by adding an amount α on to each wastage rate and subtracting the same amount from each p_{ii} the recruitment vector required to maintain $\mathbf{q}$ would be

$$\mathbf{r} = \mathbf{q}\{\mathbf{I} - (\mathbf{P} - \alpha\mathbf{I})\}/(\mathbf{q}\mathbf{w}' + \alpha). \tag{4.15}$$

Equation (4.15) is identical with (4.13) and the effects of a change in size and of a fixed change in the wastage rate are equivalent. This interpretation can only be made when $\alpha > 0$ if $\alpha < \min_i p_{ii}$.

Control by promotion with constant size

There are circumstances in which the recruitment vector is not amenable to control, as, for example, when it is linked to the supply at different levels or dictated by management policy. In such a case we wish to know what degree of control can be exercised by the promotion (or demotion) rates. For a fixed size organization the problem is that of finding a matrix $\mathbf{P}$ satisfying

$$\mathbf{q} = \mathbf{q}\mathbf{P} + \mathbf{q}\mathbf{w}'\mathbf{r}. \tag{4.16}$$

To delineate the maintainable region we must determine the set of $\mathbf{q}$'s for which there exists an admissible $\mathbf{P}$. Such a $\mathbf{P}$ must have non-negative elements with ith row summing to $1 - w_i (i = 1, 2, \ldots, k)$. We cannot proceed to solve (4.16) for $\mathbf{P}$ in the way we solved it for $\mathbf{r}$ to see whether the solution has the desired property because there is not, in general, a unique solution. However, it is easy to determine the maintainable region if no other restrictions are placed upon $\mathbf{P}$. Writing (4.16) in the form

$$\mathbf{q}\mathbf{P} = \mathbf{q} - \mathbf{q}\mathbf{w}'\mathbf{r} \tag{4.17}$$

we see that the condition

$$\mathbf{q} \geq \mathbf{q}\mathbf{w}'\mathbf{r} \tag{4.18}$$

is necessary for $\mathbf{q}$ to be maintainable since all the elements of the vector $\mathbf{q}\mathbf{P}$ must be non-negative. It is also sufficient because it is obviously possible to choose the p_{ij}'s so as to distribute those who do not leave in any way whatsoever.

The case of a general $\mathbf{P}$ is of rather limited practical interest because it is rare for all types of transition to be possible. In a hierarchical organization, $\mathbf{P}$ is usually upper-triangular or super-diagonal (that is non-zero elements in the main diagonal and the diagonal above it) in form. The latter case also lends itself to simple analysis, so we shall concentrate on this case in the remainder of this section.

When **P** has the super-diagonal form the equation (4.17) admits a unique solution which may then be tested to see whether its elements are non-negative. It is more convenient to abandon the matrix notation temporarily and to write (4.17) as

$$\left.\begin{aligned} q_1p_{11} &= q_1 - r_1 \sum_{i=1}^{k} q_iw_i \\ q_1p_{12} + q_2p_{22} &= q_2 - r_2 \sum_{i=1}^{k} q_iw_i \\ &\cdots\cdots\cdots\cdots \\ q_{k-1}p_{k-1,k} + q_kp_{kk} &= q_k - r_k \sum_{i=1}^{k} q_iw_i \end{aligned}\right\}. \tag{4.19}$$

Eliminating p_{ii}, using $p_{ii} = 1 - w_i - p_{i,i+1}(i = 1, 2, \ldots, k - 1)$, and solving for $p_{i,i+1}$ we obtain

$$p_{i,i+1} = \sum_{j=1}^{i} r_j\left\{\sum_{j=1}^{k} \frac{q_jw_j}{q_i}\right\} - \sum_{j=1}^{i} \frac{q_jw_j}{q_i} \qquad (i = 1, 2, \ldots, k - 1) \tag{4.20}$$

where the last term is defined as zero if $i = 1$. A structure **q** is thus maintainable if

$$0 \leq p_{i,i+1} \leq 1 - w_i \qquad (i = 1, 2, \ldots, k)$$

and so the maintainable region is that set of **q**'s such that

$$0 \leq \sum_{j=1}^{i} r_j\left\{\sum_{j=1}^{k} q_jw_j\right\} - \sum_{j=1}^{i} q_jw_j \leq q_i(1 - w_i) \qquad (i = 1, 2, \ldots, k - 1). \tag{4.21}$$

Equation (4.20) has a simple intuitive derivation most easily seen when $r_1 = 1$, $r_i = 0$ $(i > 1)$, in which case

$$p_{i,i+1} = \sum_{j=i+1}^{k} q_jw_j/q_i.$$

In words, this says that the proportions requiring to be promoted from grade i must be equal to the number leaving from grades $i + 1$ to k divided by the size of grade i. This result is, in effect, a discrete time version of a result we shall meet later in the discussion of renewal models in Chapter 8.

The shape of the maintainable region given by (4.21) is interesting, especially when it is compared with the corresponding regions for control by recruitment. Figure 4.2 shows regions plotted for a three-grade system having the same wastage rates as the system illustrated in Figure 4.1 (viz. 0·1, 0·1, 0·2) and two recruitment vectors (1, 0, 0) and (0·5, 0·5, 0). In contrast to the situation in Figure 4.1 the very top-heavy structures are now excluded from $\mathscr{M}$ and many of the bottom-heavy structures are included. The precise situation depends, of course, on **r**; **r** = (1, 0, 0) is more favourable to the

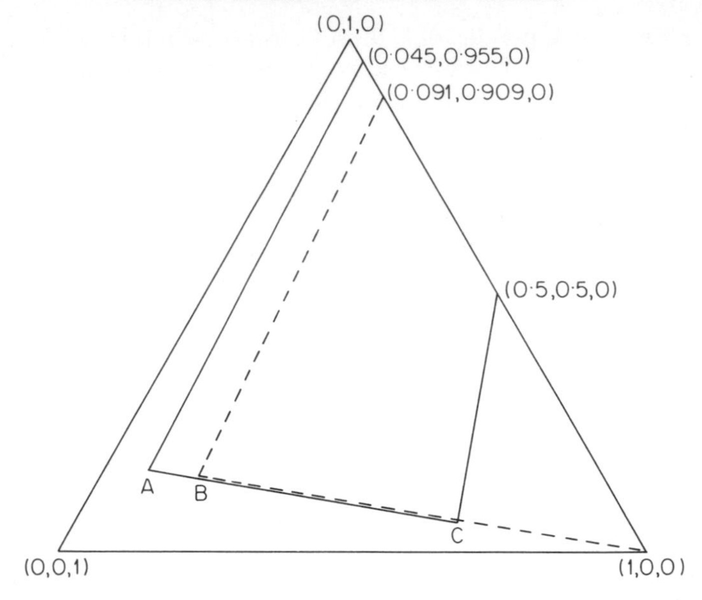

Figure 4.2 The maintainable regions for promotion control with $k = 3$, wastage vector (0·1, 0·1, 0·2) and recruitement vectors (a) (1, 0, 0) and (b) (0·5, 0·5, 0). The dashed boundary gives the region using (a) and the continuous region using (b).

maintainence of bottom-heavy structures than any other **r**. If the problem is to arrest a tendency to grow at the top it is clear that promotion control is likely to be more effective than recruitment control.

If both **P** and **r** are subject to control it is easy to see that any structure is maintainable. One way of maintaining a given structure is to make all the promotion rates zero and to recruit to each grade the same number of people as leave. There will be other, more reasonable, ways.

Control by promotion under expansion and contraction

There is no difficulty about extending the argument of the last section to cover the case of an expanding or contracting system. We shall state the main results and the reader should have no difficulty in justifying them.

The maintainable region is defined by

$$\mathbf{q} \geq (\mathbf{q}\mathbf{w}' + \alpha)\mathbf{r}/(1 - \alpha) \tag{4.22}$$

for general **P**. For super-diagonal **P** the elements of **P** must satisfy

$$q_1 p_{11} = q_1(1 + \alpha) - r_1 \left\{ \sum_{i=1}^{k} q_i w_i + \alpha \right\}$$

and

$$q_i p_{i,i+1} + q_{i+1} p_{i+1,i+1} = q_i(1 + \alpha) - r_{i+1} \left\{ \sum_{i=1}^{k} q_i w_i + \alpha \right\}$$

$$(i = 1, 2, \ldots, k - 1). \quad (4.23)$$

Solving, as before, for $p_{i,i+1}$ and making use of the inequalities $0 \leq p_{i,i+1} \leq 1 - w_i$ we arrive at the following specification of the maintainable region:

$$0 \leq \sum_{j=1}^{i} r_j \sum_{j=1}^{k} q_j(w_j + \alpha) - \sum_{j=1}^{i} q_i(w_i + \alpha) \leq q_i \ (i = 1, 2, \ldots, k). \quad (4.24)$$

These inequalities are identical with those of (4.21) except that where before we had w_j we now have $w_j + \alpha$. The effect of the change in size on the maintainable region is thus the same as an equivalent change in the wastage rates subject to the changed rates being between zero and one. As with control by recruitment, the effect of a positive α is to enlarge $\mathscr{M}$ and of a negative α to reduce it.

4.3 ATTAINABILITY

Introduction

We have been concerned in the last section with the possibility of exercising control through the recruitment and promotion flows to maintain a given structure. Often we shall wish to do more than this since the present structure may be far away from what we would wish. Let us suppose that there is some desired structure $\mathbf{q}^*$ and that the present structure is $\mathbf{q}(0)$. Two questions now arise: is it possible to reach $\mathbf{q}^*$ from $\mathbf{q}(0)$ and, if so, how. In this section we shall investigate the first of these questions, leaving the second for a later section.

A partial answer to the question follows almost at once from the work on maintainability. In fact, any $\mathbf{q}^*$ in $\mathscr{M}$ can be attained or approached arbitrarily closely no matter what the starting point $\mathbf{q}(0)$. To see this, suppose that $\mathbf{P}^*$ and $\mathbf{r}^*$ are such that

$$\mathbf{q}^* = \mathbf{q}^*\mathbf{P}^* + \mathbf{q}^*\mathbf{w}'\mathbf{r}^*. \quad (4.25)$$

Then if a system starts from $\mathbf{q}(0)$ with these parameters it will have $\mathbf{q}^*$ as its limiting structure since this is the equation which the limiting structure is known to satisfy. (Note that in using the terms such as 'arbitrarily closely' we are treating the space $\mathscr{X}$ of structures as continuous. This is legitimate since the predicted structures are expected values but the space of $\mathbf{q}$'s will be discrete because the numbers in the grades must be integers and so the elements of $\mathbf{q}^*$ must be rational.) The maintainability criteria thus have a further use in that they serve to provide a condition which will ensure the attainability of any target structure which satisfies them.

When we come to consider points outside $\mathscr{M}$ the position is less clear. Bartholomew (1969b) and Davies (1973) have investigated the question in some detail. There are two methods of approach. One is to start with $\mathbf{q}(0)$ and determine the set of structures which can be reached from it in 1, 2, 3, ... steps. The other is to work in the reverse direction starting with $\mathbf{q}^*$ and finding the set of structures from which it can be reached in 1, 2, 3, ... steps. To establish attainability we must then show that either sequence of sets ultimately includes the other end-point. A hybrid approach may also be used by working from both ends until the backward and forward sequences overlap. Since we shall be describing algorithms in the next section which automatically determine whether a point is attainable, there is little to be gained by going into a full attempt to delineate the various classes of attainable sets. Instead we shall give some elementary results, of a negative kind, which serve to identify some goals as unattainable. In particular, we shall determine a set of structures $\mathscr{A}$ having the property that any member of the set can be attained from at least one other point in $\mathscr{X}$. The interest of this set lies in its complement, $\bar{\mathscr{A}}$, which consists of all those structures which cannot be reached from *any* other structure. If $\mathbf{q}^*$ belongs to $\bar{\mathscr{A}}$ we can say at once that it cannot be reached from anywhere else and in particular from $\mathbf{q}(0)$.

Attainability using control by recruitment

If a structure is attainable from at least one other point then it must be reachable from at least one other point in one step. That is, there must exist a $\mathbf{y}$ such that

$$\mathbf{q}^* = \mathbf{y}(\mathbf{P} + \mathbf{w}'\mathbf{r}) \tag{4.26}$$

for some admissible recruitment vector $\mathbf{r}$. This is equivalent to saying that there is a $\mathbf{y}$ such that

$$\mathbf{q}^* \geq \mathbf{y}\mathbf{P}, \tag{4.27}$$

because if this were not so $\mathbf{r}$ would have at least one negative element. Equation (4.27) thus characterizes the attainable region, $\mathscr{A}$, but it is not particularly convenient for finding the boundary of the region. A more useful result may be obtained which enables us to determine the co-ordinates of the vertices of the region. The result may be stated as follows: $\mathscr{A}$ is the convex hull of the points with coordinates

$$\mathbf{e}_i\mathbf{P} + \mathbf{w}_i\mathbf{e}_j \qquad (i, j = 1, 2, \ldots, k) \tag{4.28}$$

where $\mathbf{e}_i$ denotes a k-dimensional vector with one in the ith position and zeros elsewhere. (The convex hull of a set of points is all those points whose coordinates can be expressed as weighted averages, with non-negative weights, of the coordinates of the given set.)

Not all the points in (4.28) will be vertices—some will be interior points. For small k it is fairly easy to identify the vertices from geometrical considerations, but for large k special routines of the kind used in finding feasible solutions for linear programmes are required. The proof of this result depends on standard arguments, which will be used later, relating to convex sets.

Consider first the set of structures which can be reached in one step starting from the vertex $\mathbf{e}_i$ of $\mathscr{X}$. Clearly, all these points belong to $\mathscr{A}$. This set is determined by

$$\mathbf{q}(1) = \mathbf{e}_i\mathbf{P} + \mathbf{e}_i\mathbf{w}'\mathbf{r}, \tag{4.29}$$

as $\mathbf{r}$ varies over the set $\mathbf{r} \geq 0, \sum_{i=1}^{k} r_i = 1$. The right-hand side of (4.29) may be written

$$\sum_{j=1}^{k} r_j\{\mathbf{e}_i\mathbf{P} + \mathbf{w}_i\mathbf{e}_j\}, \tag{4.30}$$

from which it follows that the set (4.29) in the convex hull of the points with coordinates

$$\mathbf{e}_i\mathbf{P} + \mathbf{w}_i\mathbf{e}_j \qquad (j = 1, 2, \ldots, k). \tag{4.31}$$

There is one such set for each i and the union of their one-step attainable sets must be in $\mathscr{A}$.

Consider next the set of points which can be reached in one step from an arbitrary starting point $\mathbf{y}$. Writing

$$\mathbf{y} = \sum_{i=1}^{k} \mathbf{e}_i y_i$$

$\mathbf{q}(1)$ is given by

$$\begin{aligned} \mathbf{q}(1) &= \sum_{i=1}^{k} \mathbf{y}_i\mathbf{e}_i\mathbf{P} + \sum_{i=1}^{k} y_i\mathbf{e}_i\mathbf{w}'\mathbf{r} \\ &= \sum_{i=1}^{k} y_i\{\mathbf{e}_i\mathbf{P} + \mathbf{w}_i\mathbf{r}\}. \end{aligned} \tag{4.32}$$

Thus $\mathbf{q}(1)$ has been expressed as a convex combination of points in the regions which can be reached from the vertices of $\mathscr{X}$. Any point which can be expressed as a convex combination of points in these regions can obviously be expressed as a convex combination of their vertices, because a weighted average of weighted averages is itself a weighted average. The total set of points attainable from any starting point is thus the convex hull of the points with coordinates given by (4.28).

For the three-grade system with parameters

$$\mathbf{P} = \begin{pmatrix} 0{\cdot}5 & 0{\cdot}4 & 0 \\ 0 & 0{\cdot}6 & 0{\cdot}3 \\ 0 & 0 & 8 \end{pmatrix}, \qquad \mathbf{w} = (0{\cdot}1, 0{\cdot}1, 0{\cdot}2)$$

the set of coordinates given by (4.28) is easily found to be

$$\{(0{\cdot}5 + 0{\cdot}1), 0{\cdot}4, 0\}^*, \{(0 + 0{\cdot}1), 0{\cdot}6, 0{\cdot}3\}, \{(0 + 0{\cdot}2), 0, 0{\cdot}8\}^*$$

$$\{0{\cdot}5, (0{\cdot}4 + 0{\cdot}1), 0\}^*, \{0, (0{\cdot}6 + 0{\cdot}1), 0{\cdot}3\}^*, \{0, (0 + 0{\cdot}2), 0{\cdot}8\}$$

$$\{0{\cdot}5, 0{\cdot}4, (0 + 0{\cdot}1)\}, \{0, 0{\cdot}6, (0{\cdot}3 + 0{\cdot}1)\}, \{0, 0, (0{\cdot}8 + 0{\cdot}2)\}^*.$$

The vertices are marked by an asterisk and the region which they determine is plotted on Figure 4.3. Looking at the complement of $\mathscr{A}$ we see that the

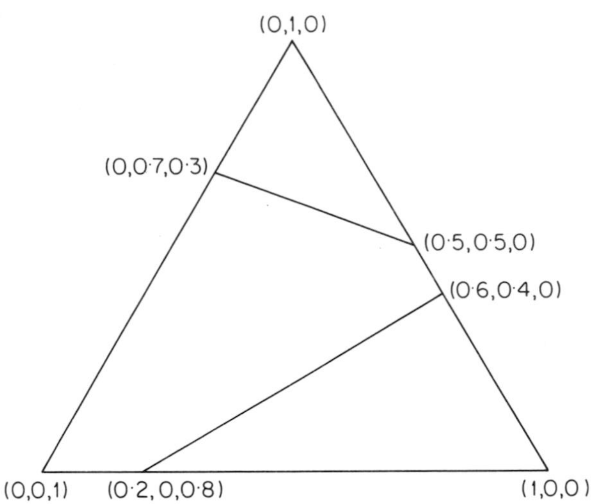

Figure 4.3 The attainable region for recruitment control with $k = 3$ and

$$\mathbf{P} = \begin{pmatrix} 0{\cdot}5 & 0{\cdot}4 & 0 \\ 0 & 0{\cdot}6 & 0{\cdot}3 \\ 0 & 0 & 0{\cdot}8 \end{pmatrix}$$

structures which cannot be reached from any starting structure lie in two triangular regions near to the vertices (0, 1, 0) and (1, 0, 0).

Attainability using control by promotion

The same kind of argument used for control by recruitment works also for control by promotion. Thus if a point $\mathbf{q}^*$ is attainable at all it must be attainable in one step from some point. Hence there must exist a $\mathbf{y}$ and a $\mathbf{P}$ such that

$$\mathbf{q}^* = \mathbf{y}(\mathbf{P} + \mathbf{w}'\mathbf{r})$$

for given $\mathbf{w}$ and $\mathbf{r}$. There is a duality between the matrices $\mathbf{P}$ and $\mathbf{w}'\mathbf{r}$ which enables us to repeat the argument based on (4.26). Both are non-negative matrices with row sums less than one, so $\mathbf{q}^*$ is attainable if there is a $\mathbf{y}$ such that

$$\mathbf{q}^* \geq \mathbf{y}\mathbf{w}'\mathbf{r}. \tag{4.33}$$

This argument only works if $\mathbf{P}$ is allowed to have non-zero elements anywhere since, if the inequality holds, it must be possible to choose $\mathbf{P}$ such that

$$\mathbf{yP} = \mathbf{q}^* - \mathbf{y}\mathbf{w}'\mathbf{r} \quad \text{with} \quad \mathbf{P1}' = \mathbf{1}' - \mathbf{w}'. \tag{4.34}$$

If, for example, $\mathbf{P}$ is upper-triangular then (4.34) would require

$$y_1 p_{11} = q_1^* - r_1 \sum_{i=1}^{k} y_i w_i,$$

which need not give a p_{11} less than one.

If $\mathbf{P}$ has a special structure, $\mathscr{A}$ will be smaller than the region given by (4.33). When $\mathbf{P}$ has the super-diagonal form (4.26) may be solved for $\mathbf{P}$ and the condition for attainability is obtained by requiring $0 \leq p_{i,i+1} \leq 1 - w_i$. These conditions become

$$0 \leq \sum_{j=1}^{i} r_j \left\{ \sum_{j=1}^{h} y_j w_j \right\} - \sum_{j=1}^{i} w_j y_j + \sum_{j=1}^{i} (y_j - q_j) \leq y_j(1 - w_j)$$

$$(j = 1, 2, \ldots, k - 1), \tag{4.35}$$

which should be compared with the corresponding maintainability condition of (4.21). The attainable region $\mathscr{A}$ is thus the set of points $\mathbf{q}$ such that there exists a $\mathbf{y}$ satisfying (4.35).

As with the corresponding results on maintainability these inequalities are not particularly helpful for plotting the boundaries of $\mathscr{A}$. However, it is again possible to express $\mathscr{A}$ as a convex hull and in some circumstances to identify its vertices. It turns out that the derivation is much simpler for the case of a general $\mathbf{P}$, so we shall solve the problem in this case and then specialize later. As before, we start by considering the set of points which can be attained, starting from the vertex $\mathbf{e}_i$. These are given by

$$\mathbf{q}_i(1) = \mathbf{e}_i\mathbf{P} + \mathbf{e}_i\mathbf{w}'\mathbf{r}$$

where $\mathbf{r}$ and $\mathbf{w}'$ are now fixed and the elements of $\mathbf{P}$ vary over the set

$$p_{ij} \geq 0, \sum_{j=1}^{k} p_{ij} = 1 - w_i \qquad (i, j = 1, 2, \ldots, k).$$

Let us introduce the conditional transition probabilities $p'_{ij} = p_{ij}/(1 - w_i)$. Then

$$\begin{aligned}\mathbf{q}_i(1) &= (1 - w_i)(p'_{i1}, p'_{i2}, \ldots, p'_{ik}) + w_i(r_1, r_2, \ldots, r_k) \\ &= (1 - w_i)\sum_{j=1}^{k} p'_{ij}\mathbf{e}_j + w_i\mathbf{r} \\ &= \sum_{j=1}^{k} p'_{ij}\{(1 - w_i)\mathbf{e}_j + w_i\mathbf{r}\}. \qquad (4.36)\end{aligned}$$

Since $\sum_{j=1}^{k} p'_{ij} = 1$ and $p'_{ij} \geq 0$ for all i and j the points which can be reached from the vertex $\mathbf{e}_i$ have been expressed as the convex hull of the points with vertices

$$(1 - w_i)\mathbf{e}_j + w_i\mathbf{r} \qquad (j = 1, 2, \ldots, k).$$

Repeating the line of argument used in justifying (4.32) we deduce that the set of points reachable from at least one point is the convex hull of the points with coordinates

$$(1 - w_i)\mathbf{e}_j + w_i\mathbf{r} \qquad (i, j = 1, 2, \ldots, k). \qquad (4.37)$$

These points are weighted averages of each vertex, in turn, and the point $\mathbf{r}$. In geometrical terms these points lie on the line joining $\mathbf{e}_h$ and $\mathbf{r}$, dividing the line in the ratio $w_i : 1 - w_i$. The set of points (4.37) is shown in Figure 4.4 for $k = 3$. It is clear from the diagram that the vertices of the convex hull are the outermost points on the lines joining $\mathbf{r}$ to the vertices. In the general case these points have coordinates

$$(1 - \min_i w_i)\mathbf{e}_j + \min_i w_i\mathbf{r} \qquad (j = 1, 2, \ldots, k). \qquad (4.38)$$

The effect of this result is to exclude structures around the edges of $\mathscr{X}$ (that is having small numbers in any grade). The reason for this is clear intuitively if we observe that, when $\mathbf{r}$ is fixed, we can never have fewer people in a grade than recruitment provides. Had we chosen to illustrate the region for a recruitment vector with one or more zero elements, $\mathbf{r}$ would have been at an edge or corner of $\mathscr{X}$ and the above observation would have to be modified accordingly (see Figure 4.5).

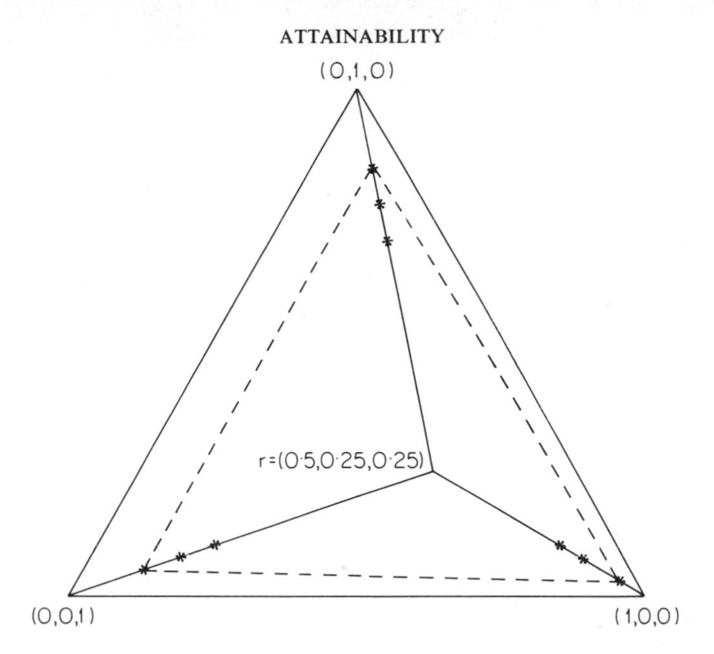

Figure 4.4 The attainable region for promotion control with $k = 3$, $\mathbf{w} = (0{\cdot}4, 0{\cdot}3, 0{\cdot}2)$, $\mathbf{r} = (0{\cdot}5, 0{\cdot}25, 0{\cdot}25)$. The asterisks are the points $(1 - w_i)\mathbf{e}_j + w_i\mathbf{r}$ $(i, j = 1, 2, \ldots, k)$ and the dashed line is the boundary of the region

The simplicity of this result depends on our having assumed a general form for **P**. If some transitions are impossible the representation of (4.36) still applies, but some of the p_{ij}'s are now zero. To obtain the vertices of $\mathscr{A}$ we must delete from the list in (4.37) all those points having a j for which $p_{ij} = 0$. Thus if **P** is upper-triangular then $\mathscr{A}$ is the convex hull of the points

$$(1 - w_i)\mathbf{e}_j + w_i\mathbf{r} \qquad (i = 1, 2, \ldots, k, j = i, i+1, \ldots, k). \qquad (4.39)$$

Similarly, if **P** is super-diagonal in form j takes only the values i and $i + 1$ for each i. The region $\mathscr{A}$ is plotted on Figure 4.5 for this case when $k = 3$ and $w = (0{\cdot}1, 0{\cdot}1, 0{\cdot}2)$ and for the two recruitment vectors $\mathbf{r} = (1, 0, 0)$ and $\mathbf{r} = (0{\cdot}5, 0{\cdot}5, 0)$. In each case the excluded structures are confined to a narrow strip along the boundaries. The size of $\mathscr{A}$ depends on the w_i's; the smaller they are the larger is $\mathscr{A}$ because more people remain to be moved around by the promotion rates.

It follows at once from the foregoing results that if both **P** and **r** are subject to control then $\mathscr{A}$ coincides with $\mathscr{X}$. Indeed, it would be sufficient to achieve this if **r** was restricted to taking the three values (1, 0, 0), (0, 1, 0), (0, 0, 1).

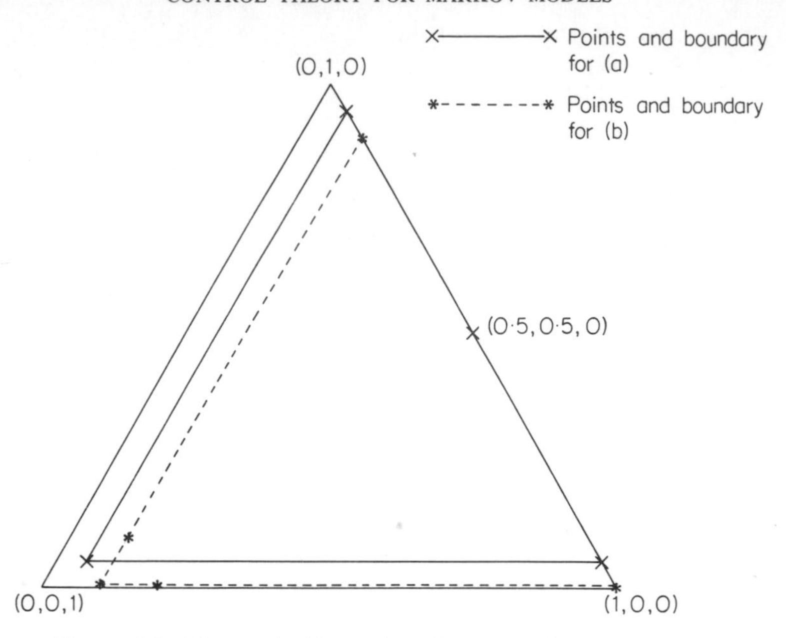

Figure 4.5 The attainable regions for promotion control with $k = 3$, $\mathbf{w} = (0{\cdot}1, 0{\cdot}1, 0{\cdot}2)$ and (a) $\mathbf{r} = (0{\cdot}5, 0{\cdot}5, 0)$, (b) $\mathbf{r} = (1, 0, 0)$

Corresponding results for an expanding or contracting system can be found by straightforward extension of the methods used elsewhere in this section.

Bounds on the time to reach an attainable structure

The attainability and maintainability conditions already given serve to identify certain structures as attainable and others as non-attainable. However, even if $\mathbf{q}^*$ is attainable it may take so long to reach it as to make it unattainable for practical purposes. We shall now obtain some simple bounds on the number of steps required to attain a given goal, assuming that the size of the organization is constant. In general, they seem likely to be rather poor bounds but their simplicity makes them very easy to calculate.

Let T^* be the smallest number of steps in which the goal can be reached. Then, if control is to be by recruitment, the sequence of recruitment vectors $\{r(1), r(2), \ldots, r(T^*)\}$ must satisfy

$$\mathbf{q}^* = \mathbf{q}(0) \prod_{j=1}^{T^*} \{\mathbf{P} + \mathbf{w}'\mathbf{r}(j)\}. \tag{4.40}$$

The leading term in the expansion of the product is $\mathbf{q}(0)\mathbf{P}^{T^*}$ and the remaining terms are all positive. Hence T^* must be such that

$$\mathbf{q} \geq \mathbf{q}(0)\mathbf{P}^{T^*} \tag{4.41}$$

so that if T' is the smallest value of T for which $\mathbf{q}^* \geq \mathbf{q}(0)\mathbf{P}^T$ then it is clear that $T^* \geq T'$. The vector $\mathbf{q}(0)\mathbf{P}^T$ gives the expected numbers, expressed as proportions of the original strength, who would be in the grades in the absence of all recruitment. The inequality says that the target cannot be reached until all of these numbers have fallen below the corresponding target values.

No other simple bounds have been found without making further restrictions on the structure of $\mathbf{P}$ or the values of the wastage rates. Suppose first that the wastage rates are all equal to w. This is not a very reasonable assumption in most applications but it leads to some useful insights. Under these conditions the expected proportion of losses at each time point is $\sum_{i=1}^{k} q_i(T)w = w$, which is independent of $\mathbf{q}(T)$, and the expected recruitment to grade i is $r_i w$. Assume further that $\mathbf{P}$ is such that no transitions are possible into grade 1 from elsewhere in the system. We can now find the minimum number of steps to reach the goal in grade 1; this will obviously be a lower bound for T^*. The expected proportion in grade 1 at time T is composed of the survivors of the original stock in that grade plus the survivors of all subsequent intakes. That is

$$q_1(T) = p_{11}^T q_1(0) + w\{p_{11}^{T-1} r_1(1) + p_{11}^{T-2} r_1(2) + \dots + r_1(T)\}. \tag{4.42}$$

All of the terms here are non-negative so that q_1^* can be achieved as soon as it lies between the maximum and minimum values of $q_1(T)$ given by (4.42). The minimum occurs when $r_1(1) = r_1(2) = \dots = r_1(T) = 0$ and the maximum when they are all equal to one. Thus, if T_1^* denotes the minimum number of steps before the goal is attained in grade 1, it is given by the smallest value of T satisfying

$$p_{11}^T q_1(0) \leq q_1^* \leq p_{11}^T q_1(0) + w\{p_{11}^{T-1} + p_{11}^{T-2} + \dots + 1\}$$

or

$$0 \leq \frac{q_1^* - p_{11}^T q_1(0)}{w} \leq \frac{1 - p_{11}^T}{1 - p_{11}}. \tag{4.43}$$

In general this is not likely to be a close bound, but when $k = 2$ it determines T^* precisely because the size of the upper grade is the complement of the lower. As an example, consider the system for which

$$\mathbf{P} = \begin{pmatrix} 0{\cdot}6 & 0{\cdot}2 \\ 0 & 0{\cdot}8 \end{pmatrix}, \quad \mathbf{w} = (0{\cdot}2, 0{\cdot}2).$$

Suppose

$$q_1(0) = 0{\cdot}4, \qquad q_2(0) = 0{\cdot}6$$
$$q_1^* \;= 0{\cdot}2, \qquad q_2^* \;= 0{\cdot}8.$$

The expression

$$\frac{q_1^* - p_{11}^T q_1(0)}{w} = \frac{0{\cdot}2 - (0{\cdot}6)^T(0{\cdot}6)}{0{\cdot}2}.$$

The value of this is below the lower bound if $T = 1$ or 2, but it becomes positive at $T = 3$. Hence the goal can be achieved in threee steps. Any set of $r_1(j)$'s satisfying (4.42) will suffice to achieve the goal. Equation (4.43) also enables us to determine an attainability condition for grade 1. Both bounds are monotonic functions of T and so if the inequality is not satisfied at both $T = 0$ and $T = \infty$ then q_1^* is unattainable from $q_1(0)$. This condition may be expressed as

$$q_1^* \leq \max \{w + p_{11}q_1(0), w/(1 - p_{11})\}. \tag{4.44}$$

In the case of the example this becomes

$$q_1^* \leq \max \{0{\cdot}2 + 0{\cdot}6q_1(0), 0{\cdot}5\}.$$

Thus if $q_1(0) > 0{\cdot}5$ any q_1^* is attainable if it satisfies

$$q_1^* \leq 0{\cdot}2 + 0{\cdot}6q_1(0),$$

whereas if $q_1(0) \leq 0{\cdot}5$ the condition is $q_1^* \leq 0{\cdot}5$. There is thus an upper limit to the size of the lower grade which can be attained.

In a special but interesting case an upper bound for T^* can be found. The case arises when the grades are single years of age or seniority and when the time interval between moves is also a year (or other common time unit). At each step an individual then moves up a grade or leaves the system and so $\mathbf{P}$ has entries on the super-diagonal only and zeros elsewhere. On reaching grade k leaving is certain. In this case it is possible to determine a recruitment strategy which will attain any goal in k steps; hence T^* cannot exceed k. Choose $\mathbf{r}(T)$ to be any admissible vector with zero in the first T positions for $T = 1, 2, \ldots, k - 1$). This ensures that all original members of the system and all those recruited at times $1, 2, \ldots, k - 1$ will have left the system at time k. Any desired structure can then be attained by choosing $\mathbf{r}(k) = \mathbf{q}^*$. Of course, this is hardly likely to be a very sensible strategy in practice, but the result serves the purpose of showing that no more than k steps are needed.

Similar bounds may be obtained when control is by promotion. Promotion control cannot be exercised in the age distribution example, but otherwise we can obtain two lower bounds similar to (4.41) and (4.43). Exploiting the dualism between $\mathbf{P}$ and $\mathbf{w'r}$, there must now be a sequence of matrices $\{\mathbf{P}(j)\}$ such that

$$\mathbf{q}^* = \mathbf{q}(0) \prod_{j=1}^{T^*} \{\mathbf{P}(j) + \mathbf{w'r}\}.$$

For this to hold T^* must be at least as large as T' where T' is the smallest T such that

$$\mathbf{q}^* \geq \mathbf{q}(0)[\mathbf{w'r}]^T. \tag{4.45}$$

The equivalent to (4.42) is now

$$q_1(T) = p_{11}^T(0)q_1(0) + w\{p_{11}^{T-1}(1)r_1 + p_{11}^{T-2}(2)r_1 + \ldots + p_{11}(T-1)r_1 + r_1\}. \tag{4.46}$$

The minimum value of this expression which occurs when all of the $p_{11}(j)$'s are zero is rw_1. The maximum occurs when all of the $p_{11}(j)$'s are equal to $1 - w$, in which case

$$\begin{aligned} q_1(T) &= q_1(0)(1-w)^T + wr_1\{(1-w)^{T-1} + (1-w)^{T-2} + \ldots + 1\} \\ &= r_1 + (1-w)^T\{q_1(0) - r_1\}. \end{aligned} \tag{4.47}$$

Thus the goal cannot be reached in grade 1 until

$$wr_1 \leq q_1^* \leq r_1 + (1-w)^T\{q_1(0) - r_1\}. \tag{4.48}$$

Only the upper bound involves T. The lower bound arises from the fact that q_1^* can never be attained if the input to that grade exceeds the target. The behaviour of the upper bound depends on the relative sizes of $q_1(0)$ and r_1. If $q_1(0) \geq r_1$ the bound is non-increasing in T and so if a large value of q_1^* can be included within the limits at all it will be included for $T = 1$. If $q_1(0) < r_1$ the bound increases with T to its limit of r_1. Thus, provided $q_1^* < r_1$ there will be a T such that the goal can be attained. The smallest value of T for which (4.48) is satisfied will then be a lower bound for T^*.

4.4 CONTROL STRATEGIES

Direct methods

Here we shall describe a number of *ad hoc* methods for finding sequences of recruitment vectors or transition matrices which take a system from a starting structure $\mathbf{q}(0)$ to a goal structure $\mathbf{q}^*$. The discussion is deterministic in the sense that it is based on the expected path which the system would follow if the prescribed policy were adopted. In practice all manner of chance variation will combine to move the system away from its charted course. A partial way of coping with stochastic variation is to recompute the strategy at each step. In this way we only take the first step of any projected path before having the opportunity to modify the strategy in the light of the current situation.

We shall not normally consider goals outside the maintainable region because there would be no means of remaining there when we arrived. As an interim measure we might wish to aim for a non-maintainable target with the intention of making more fundamental changes which would enable us to maintain the structure once it was reached. Nevertheless, we shall proceed on the understanding that all goals are maintainable.

The result expressed in (4.25) provides one simple control strategy since it tells us how to find a constant policy (that is not changing with time) which will cause the structure to converge on the required goal. Since such a strategy does not depend on the current position but only on the goal no adjustment is required to cope with stochastic variation. The main question to be answered about such a strategy is how efficient it is in reaching the goal. Later in this section we shall be in a position to make comparisons with other strategies and provide a partial answer to this question.

Davies (1973) has computed control strategies when $k = 3$ when control is by recruitment, promotion or both. His method consists of two stages. First the minimum number of steps in which the goal can be attained is determined. In the course of Section 4.3 we found sets of points which could be reached in one step from a given $\mathbf{q}(0)$ or from which a given $\mathbf{q}^*$ can be reached. Davies extends these methods to give the regions which can be reached in n steps. The method for finding T^* is then to compute these regions successively until $\mathbf{q}^*$ first lies within the n-step region (a modification involves working back from the goal simultaneously but the principle is the same). The second part of the method is then to find a sequence of control parameters which leads to the goal in the minimum number of steps. In general, there are infinitely many such strategies and Davies chooses a path between vertices of the n-step attainable regions. In the case of recruitment control, for example, this means recruiting into a single grade as far as possible.

The essentials of the second stage can be illustrated by considering control by recruitment. Suppose that we have found T^* and wish to determine the recruitment vectors $\{\mathbf{r}(1), \mathbf{r}(2), \ldots, \mathbf{r}(T^*)\}$. First we construct, by trial and error, a sequence of structures $\mathbf{q}(1), \mathbf{q}(2), \ldots, \mathbf{q}(T^* - 1)$ such that

$$\mathbf{q}(j) \geq \mathbf{q}(j-1)\mathbf{P} \qquad (j = 1, 2, \ldots, T^* - 1).$$

Having found such a sequence the required recruitment vectors are computed from

$$\mathbf{r}(j) = \{\mathbf{q}(j) - \mathbf{q}(j-1)\mathbf{P}\}/\mathbf{q}(j-1)\mathbf{w}' \qquad (j = 1, 2, \ldots, T^* - 1). \tag{4.49}$$

This method is only feasible if k is small, preferably 2 or 3, and T^* is not too large. It is useful for gaining insight into the form of optimal strategies, but it does not provide an algorithm which can be programmed to cope with large problems.

Mathematical programming methods: free time

Davies' method is essentially one of finding a feasible solution to a set of restraints by trial and error. By formulating the problem in programming terms it becomes possible to draw on the extensive theory of mathematical programming to do this automatically. Like all numerical approaches the solutions for special cases which such a method produces give little insight into the structure of the solution. In this book the emphasis is on what can be learnt from model-building and analysis, and we shall not therefore go into the numerical side of computations in any detail. Instead, we shall concentrate on the formulation of the control problem in programming terms and on the general deductions which can be made about the properties of the strategies to which it leads. The term 'free time' in the title of this section is taken from control theory and refers to the fact that the number of steps to reach the solution is not known in advance. In the next section we shall discuss the corresponding fixed time problem in which the aim is to get as near the goal as possible in a fixed number of steps.

Control by Recruitment

In formulating our problem in programming terms it is desirable to work in terms of numbers rather than proportions. Accordingly we start with the equation

$$\mathbf{n}(T+1) = \mathbf{n}(T)\mathbf{P} + \mathbf{f}(T+1) \tag{4.50}$$

where $\mathbf{f}(T+1) = R(T+1)\mathbf{r}$ is the vector of the numbers recruited at time $T+1$. In this chapter we shall not distinguish notationally between stock numbers and their expectations. We shall suppose that the total size of the organization, $N(T)$, is given, but this will be taken care of through the restraints to be imposed below. The problem is to find a T^* and a sequence of vectors $\mathbf{f}(1), \mathbf{f}(2), \ldots, \mathbf{f}(T^*)$ such that T^* is the smallest T for which $\mathbf{n}(T) = \mathbf{n}^*$, the target structure. The restraints which the model imposes on the unknowns are

$$\mathbf{n}(T+1) = \mathbf{n}(T)\mathbf{P} + \mathbf{f}(T+1) \qquad (T = 0, 1, 2, \ldots, T^* - 2) \tag{4.51a}$$

$$\mathbf{n}^* = \mathbf{n}(T^* - 1)\mathbf{P} + \mathbf{f}(T^*) \tag{4.51b}$$

$$\sum_{i=1}^{k} n_i(T) = N(T) \qquad (T = 1, 2, \ldots, T^* - 1) \tag{4.51c}$$

$$\begin{aligned} &\mathbf{n}(T) \geq 0 \qquad (T = 1, 2, \ldots, T^* - 1); \\ &\mathbf{f}(T) \geq \mathbf{0} \qquad (T = 1, 2, \ldots, T^*). \end{aligned} \tag{4.51d}$$

If $T^* = 1$, (4.51a) and (4.51c) disappear. These equations are linear in the $n_j(T)$'s and the $f_j(T)$'s, and we may therefore use the standard methods of mathematical programming to find a feasible solution. Since T^* is unknown we must put $T^* = 1, 2, \ldots$ in turn until the first value is found for which a feasible solution exists. Notice that a solution to the restraints (4.51) will give not only the vectors $\mathbf{f}(T)$ which we set out to find but also the intermediate structures $\mathbf{n}(T)$ through which the system passes. When T has been found there will normally be many solutions and it is therefore open to us to select one set by minimizing some function of economic or social interest. One possible choice in the manpower context is to minimize the total salary bill. If the average salary cost in grade i is c_i the total salary cost associated with a strategy is

$$\sum_{T=1}^{T^*-1} \sum_{i=1}^{k} c_i n_i(T) \tag{4.52}$$

and then we have a standard linear programming problem. (The argument which follows is not affected if future costs are discounted.) A minimization algorithm would be used to minimize (4.52) as soon as a T^* had been found for which a feasible solution exists. A practical difficulty is that the number of variables is typically very large so that the size of problem which can be solved is limited by the size of computer available.

The test for the existence of a feasible solution is trivial when $T^* = 1$ but of considerable practical interest. In this case only (4.51b) is operative and we have to find whether there is an $\mathbf{f}(1) \geq 0$ such that

$$\mathbf{n}^* = \mathbf{n}(0)\mathbf{P} + \mathbf{f}(1). \tag{4.53}$$

If there is it will be given by

$$\mathbf{f}(1) = \mathbf{n}^* - \mathbf{n}(0)\mathbf{P}. \tag{4.54}$$

The condition for $\mathbf{f}(1)$ to be non-negative is, of course, the condition for the one-step attainability given earlier by (4.27). The fixed time methods in the following section take this as their starting point.

We can make some interesting deductions about the kind of recruitment strategies which minimize the cost function (4.52) by enumerating the number of unknowns and equations in (4.51). Altogether the vectors $\mathbf{n}(T)(T = 1, 2, \ldots, T^* - 1)$ and $\mathbf{f}(T)(T = 1, 2, \ldots, T^*)$ give $2kT^* - k$ variables in all. The number of equations may be enumerated as follows:

(4.51a)	$k(T^* - 1)$
(4.51b)	k
(4.51c)	$T^* - 1$
	$kT^* + T^* - 1$

A well-known result in the theory of linear programming states that the number of non-zero variables in a basic feasible solution cannot exceed $kT^* + T^* - 1$. Unless degeneracy occurs it will be equal to that number. If all the elements of $\mathbf{n}(0)$ are positive it is clear from (4.51a) that all subsequent vectors $\mathbf{n}(T)$ will have positive elements and that this accounts for $k(T^* - 1)$ of those available. The remaining $(kT^* + T^* - 1) - (kT^* - k) = T^* + k - 1$ non-zero values (at most) must therefore be distributed among the kT^* positions in the $\mathbf{f}(T)$ vectors. It therefore follows that at least $(k - 1)(T^* - 1)$ of the recruitment elements will be zero, which means that recruitment will tend to be concentrated at a few levels.

The argument may be carried further to obtain an idea of how the non-zero elements will be distributed. Suppose that we have reached time $T^* - 1$. We then have a problem which we know can be solved in one step. Hence the accounting procedure used above can be applied with $T^* = 1$ and we deduce that all of the elements in the last vector in the recruitment vector will be positive. This leaves $T^* + k - 1 - k$ non-zero elements to be distributed among the preceding $T^* - 1$ vectors. Provided that $N(T)$ does not decrease sufficiently fast for redundancy to occur *some* recruits will be needed at each T, so there will have to be at least one non-zero element in each vector. However, with only $T^* - 1$ non-zero elements available there can be no more than one per vector. Hence a typical recruitment strategy minimizing (4.52) will consist of recruiting into one grade only at each time point until the last step when recruits will be taken in at every level. This argument needs slight modification if degeneracy occurs or if the initial vector $\mathbf{n}(0)$ contains zeros (as it does in a numerical example considered later). The kind of strategy which the programming formulation dictates might be acceptable in practice if recruitment was always at the same level but if it called for repeated changes in the recruitment level it would be very difficult to implement. This suggests that we ought to impose further restrictions designed to limit the range of feasible solutions to those which could be implemented.

If we decided to drop the $(T^* - 1)$ restrictions of (4.51c), implying that there is no need to control the intermediate size of the organization, there could be no more than kT^* non-zero elements in the optimal solution. Of these only $kT^* - k(T^* - 1) = k$ could be elements of the recruitment vectors. In general, therefore, there would be no recruitment at all until $T^* - 1$. The absurdity of this strategy in most practical situations highlights the need for some restrictions of the kind of (4.51c).

Control by Promotion

The problem in this case is to find a sequence of matrices $\mathbf{P}(T)$ which, when coupled with a given recruitment vector $\mathbf{r}$, will take the system from $\mathbf{n}(0)$ to $\mathbf{n}^*$ in the shortest possible time. The basic equation may now be written

$$\mathbf{n}(T + 1) = \mathbf{n}(T)\mathbf{P}(T) + \mathbf{n}(T)\mathbf{w}'\mathbf{r} + M(T + 1)\mathbf{r} \tag{4.55}$$

where $M(T + 1) = N(T + 1) - N(T)$. The unknowns are the vector $\mathbf{n}(T)$ and the matrices $\mathbf{P}(T)$, but the basic equation is not linear in these variables. We must therefore re-cast the problem in a form which makes the restrictions linear. One method which we shall illustrate, when $M(T) = 0$ for all T, takes advantage of the duality relationship between $\mathbf{P}$ and $\mathbf{w}'\mathbf{r}$ but is only satisfactory if $\mathbf{P}(T)$ is unrestricted.

We rewrite (4.55) in the form

$$\mathbf{n}(T + 1) = \mathbf{n}(T)[\mathbf{w}'\mathbf{r}] + \mathbf{v}(T) \tag{4.56}$$

where $\mathbf{v}(T)$ contains the total flows into each grade excluding recruitment. The problem can therefore be formulated and solved exactly as for control by recruitment by replacing $\mathbf{P}$ by $\mathbf{w}'\mathbf{r}$ and $\mathbf{f}(T + 1)$ by $\mathbf{v}(T)$. By this means a value of T^* would be found and a sequence of vectors $\{\mathbf{v}(T)\}$ related to the required transition probabilities by

$$\mathbf{v}(T) = \mathbf{n}(T)\mathbf{P}(T). \tag{4.57}$$

The corresponding relationship in the recruitment case was

$$\mathbf{r}(T + 1) = \mathbf{f}(T + 1)/\mathbf{n}(T)\mathbf{w}'$$

and the method of calculation automatically ensured that the elements of $\mathbf{r}(T + 1)$ were non-negative and added up to one. In the present case the fact that the elements of the vectors $\mathbf{v}(T)$ are non-negative and have the right sum is not sufficient, in general, to ensure that an admissible $\mathbf{P}(T)$ can be found. An essentially similar situation arose in Section 4.3 in the argument centring on (4.34). If $\mathbf{P}(T)$ is unrestricted there will, in general, be an infinity of solutions satisfying (4.57) and we are at liberty to choose among them. If $\mathbf{P}(T)$ is triangular or super-diagonal in form it may not be possible to find a solution. Nevertheless, the T^* found by this method will certainly be a lower bound to the minimum number of steps to attain the goal and the programme may be useful for this purpose alone.

A more direct approach which requires no special assumptions about the form of $\mathbf{P}(T)$ is to express all the internal flows as numbers rather than proportions. It is easier to count the equations and to recognize linear dependence in the equations if we temporarily abandon matrix notation. Let $n_{ij}(T) = n_i(T)p_{ij}$ be the number flowing from grade i to grade j in $(T, T+1)$. Then we have to find a T^* and values of the variables $n_{ij}(T)(i, j = 1, 2, \ldots, k)$ and $n_i(T)(i = 1, 2, \ldots, k)$ such that

$$n_j(T+1) = \sum_{i=1}^{k} n_{ij}(T) + r_j\left\{\sum_{i=1}^{k} n_i(T)w_i + M(T+1)\right\}$$
$$(j = 1, 2, \ldots, k;\ T = 0, 1, \ldots, T^* - 2) \qquad (4.58a)$$

$$n_j^* = \sum_{i=1}^{k} n_{ij}(T^*-1) + r_j\left\{\sum_{i=1}^{k} n_i(T^*-1)w_i + M(T^*)\right\}$$
$$(j = 1, 2, \ldots, k) \qquad (4.58b)$$

$$\sum_{i=1}^{k} n_{ij}(T) = n_i(T)(1 - w_i)$$
$$(i = 1, 2, \ldots, k-1; \qquad T = 0, 1, \ldots, T^* - 1) \qquad (4.58c)$$

$$\sum_{j=1}^{k} n_j(T) = N(T) \qquad (T = 1, 2, \ldots, T^* - 1) \qquad (4.58d)$$

$$\begin{aligned} \mathbf{n}(T) &\geq \mathbf{0} \qquad (T = 1, 2, \ldots, T^* - 1); \\ n_{ij}(T) &\geq 0 \qquad (i, j = 1, 2, \ldots k;\ T = 0, 1, \ldots T^* - 1). \end{aligned} \qquad (4.58e)$$

The $N(T)$'s are assumed to be given as before and (4.58d) drops out if $T^* = 1$. The case $i = k$ has been excluded from (4.58c) because it is implied by (4.58a) together with (4.58c). By summing both sides of (4.58a) over j and subtracting (4.58c) summed over i we obtain the missing equation.

As in the case of recruitment control a good deal can be deduced from an enumeration of the number of variables and equations. The number of variables will depend on whether $\mathbf{P}$ has a special structure; the enumeration is made for three structures in the table below:

Variables	*Full matrix*	*Upper-triangular form*	*Super-diagonal form*
$n_j(T)$'s	$k(T^* - 1)$	$k(T^* - 1)$	$k(T^* - 1)$
$n_{ij}(T)$'s	k^2T^*	$\frac{1}{2}k(k + 1)T^*$	$(2k - 1)T^*$
Total number	$k(k + 1)T^* - k$	$\frac{1}{2}k(k + 3)T^* - k$	$3kT^* - T^* - k$

The number of equations is as follows:

Set	*Number of equations*
4·58a	$k(T^* - 1)$
4·58b	k
4·58c	$T^*(k - 1)$
4·58d	$T^* - 1$
Total	$2kT^* - 1$

In general, therefore, there will be $2kT^* - 1$ non-zero elements in a basic feasible solution. Again, this is small compared with the total number of elements so an optimal control strategy (in the sense of minimizing a cost function such as 4.52) will consist of a sequence of matrices with a large number of zeros. If recruitment takes place into every grade the $n_j(T)$'s must have non-zero elements in every position. This leaves $(2kT^* - 1) - k(T^* - 1) = kT^* + k - 1$ to be distributed among the T^* **P** matrices. At the penultimate stage, when $T^* = 1$, there are $2k - 1$ non-zero elements for the k^2 places in the last transition matrix. This leaves $k(T^* - 1)$ non-zero elements for the preceding $(T^* - 1)$ matrices. Every matrix must have at least one element in each row so we conclude there will have to be exactly one element in each row. This implies that at every step all of the people in a grade who do not leave will either stay where they are or be moved *en bloc* to another grade. If the matrix is to be of super-diagonal form this means that either everyone is promoted or no-one is promoted. Extreme policies like this kind are clearly not practicable. However, there will be other feasible (not basic feasible) solutions which reach the goal in the same number of steps. It may be possible to find one among these which is more acceptable but it will not have minimum cost. If recruitment is not allowed into every grade then some $n_j(T)$'s may be zero and so more non-zero variables will be available for the transition matrices. This hardly improves the situation because it will be equally impracticable to empty whole grades from time to time.

These results emphasize the limitations of the linear programming approach to control in the simple form we have adopted. However, they clearly demonstrate that acceptable ways of changing the structure by promotion control are unlikely to be either the quickest or the cheapest. Perhaps the main use of the linear programming algorithm is for determining whether a goal is attainable in a reasonably small number of steps. Even this result must be interpreted with caution because of the stochastic nature of the flows. The programming algorithm will give the number of steps if the system keeps to its expected path. In reality the number of steps will be a random variable.

Mathematical programming methods: fixed time

In practice, the target structure may not be something which has to be attained precisely but rather an indication of the direction in which we should aim. What is then required is to get somewhere near the target in a reasonably short time. Indeed there may be a fixed time available in which to do the best we can.

Let the time available be denoted by T^* so that the problem is now to get as close to the goal as possible in T^* steps. To formulate this requirement mathematically we must define closeness in terms of a measure of distance between two structures. Suppose such a measure is denoted by $D(\mathbf{n}^*, \mathbf{n}(T^*))$ for the situation at time T^*. The problem is then to minimize this distance with respect to the sequence of control parameters and subject to the appropriate restraints. Formally, then, we have the following two problems:

1. Control by recruitment

$$\text{Minimize:} \quad D(\mathbf{n}^*, \mathbf{n}(T^*))$$

Subject to:

$$\mathbf{n}(T+1) = \mathbf{n}(T)\mathbf{P} + \mathbf{f}(T+1) \qquad (T = 0, 1, \ldots, T^* - 1) \tag{4.59a}$$

$$\sum_{i=1}^{k} n_i(T) = N(T) \qquad (T = 1, 2, \ldots, T^*) \tag{4.59b}$$

$$\mathbf{n}(T) \geq \mathbf{0} (T = 1, 2, \ldots, T^*); \mathbf{f}(T+1) \geq 0 \qquad (T = 0, 1, \ldots, T^* - 1). \tag{4.59c}$$

2. Control by promotion

$$\text{Minimize:} \quad D(n^*, n(T^*))$$

Subject to:

$$n_j(T+1) = \sum_{i=1}^{k} n_{ij}(T) + r_j \left\{ \sum_{i=1}^{k} n_i(T) w_i + M(T+1) \right\}$$

$$(j = 1, 2, \ldots, k; T = 0, 1, \ldots, T^* - 1) \tag{4.60a}$$

$$\sum_{i=1}^{k} n_{ij}(T) = n_i(T)(1 - w_i)$$

$$(i = 1, 2, \ldots, k - 1; T = 0, 1, \ldots, T^* - 1) \qquad (4.60b)$$

$$\sum_{i=1}^{k} n_i(T) = N(T) \qquad (T = 1, 2, \ldots, T^*). \qquad (4.60c)$$

The ease with which these problems can be solved depends on the form of the distance function. This function ought to reflect the penalties attached to having the various grades over or under strength. A fairly general function which will be the basis of the analysis in this section is

$$D_a = \sum_{i=1}^{k} W_i|n_i^* - n_i(T^*)|^a \qquad (a > 0) \qquad (4.61)$$

where $W_1, W_2, \ldots, W_k$ are a set of non-negative weights chosen to reflect the importance attached to the correct manning of each grade. A large value of a gives greater relative weight to large discrepancies. When $a = 1$ the problems posed in (4.59) and (4.60) can be converted into linear programming form; when $a = 2$ they are quadratic programmes. In the remainder of this section we shall explore a special case of the general problem which combines practical interest with sufficient mathematical simplicity to provide explicit solutions.

We first assume that the organization is to be of constant size. This enables us to express the problem in terms of the proportions in the grades and so dispense with (4.59b) and (4.60c). Secondly, we shall assume that $T^* = 1$. At first sight this appears to be an unduly severe restriction until we recall that the stochastic nature of the flows will require us to recalculate the strategy at each step. Putting $T^* = 1$ means that, at each step, we shall move as far as possible towards the goal. In view of the uncertainties of the manpower planning environment it may be highly desirable as well as mathematically convenient to make as much progress towards the goal at each step. Our third restriction is that we shall only consider the case of control by recruitment. This has wider applications than promotion control because of its relevance to the control of age distributions. Recruitment strategies determined by programming methods are also less apt to be impractical than those involving promotion.

The particular case we have decided to examine in detail may therefore be expressed as follows:

Minimize:

$$D_a = \sum_{i=1}^{k} W_i|q_i^* - q_i(1)|^a \text{ with respect to } \mathbf{r}(1)$$

Subject to:

$$\mathbf{q}(1) = \mathbf{q}(0)\mathbf{P} + \mathbf{q}(0)\mathbf{w}'\mathbf{r}(1) \qquad (\mathbf{r}(1) \geq \mathbf{0}). \tag{4.62}$$

Now

$$\begin{aligned}\mathbf{q}^* - \mathbf{q}(1) &= \mathbf{q}^* - \mathbf{q}(0)\{\mathbf{P} + \mathbf{w}'\mathbf{r}(1)\} \\ &= \mathbf{q}(0)\mathbf{w}'\left[\frac{\mathbf{q}^* - \mathbf{q}(0)\mathbf{P}}{\mathbf{q}(0)\mathbf{w}'} - \mathbf{r}(1)\right] \\ &\propto \mathbf{y} - \mathbf{r}(1)\end{aligned}$$

where $\mathbf{y} = \{\mathbf{q}^* - \mathbf{q}(0)\mathbf{P}\}/\mathbf{q}(0)\mathbf{w}'$. Note that $\sum_{i=1}^{k} \mathbf{y}_i = 1$. Thus, instead of D_a, we can minimize the same function of the differences $\mathbf{y}_i - r_i$. Making the appropriate substitutions our problem finally becomes:

Minimize:

$$D_a = \sum_{i=1}^{k} W_i|y_i - r_i|^a \qquad (a > 0)$$

Subject to:

$$\sum_{i=1}^{k} r_i = 1 \qquad (\mathbf{r} \geq \mathbf{0}) \tag{4.63}$$

where, for brevity, we have dropped the argument of $\mathbf{r}$. If the goal is attainable in one step the minimum of D_a will be zero and the solution is $r_i = y_i$. If at least one y_i is negative the solution is the 'nearest' $\mathbf{r}$ to $\mathbf{y}$.

We first show that $r_i = 0$ for each i such that $y_i \leq 0$. (This is also true for a much wider class of distance functions.) Let $\sum^+$ denote summation over those i for which $y_i > 0$ and $\sum^-$ summation over the remaining i. Then

$$D_a = \sum{}^+ W_i(y_i - r_i)^a + \sum{}^- W_i(|y_i| + r_i)^a.$$

D_a is decreased if any r_i in the first summation is increased up to the point when $r_i = y_i$ and it is increased if any r_i in the second summation is increased. Hence we should make the r_i's in the second sum as small as possible. There is nothing to prevent them all being made zero because $\sum^+ y_i > 1$ and so we can increase the r_i's in the first sum without the need for $r_i > y_i$ for any i. Hence we can replace D_a in (4.63) by $\sum^+ W_i(y_i - r_i)^a$ and add the restrictions $r_i \leq \min(y_i, 1)$.

The case a = 1

$$D_1 = \sum{}^+ W_i y_i - \sum{}^+ W_i r_i.$$

The minimum will occur when $\sum^+ W_i r_i$ is a maximum subject to

$0 \leq r_i \leq \min(y_i, 1)$, $\sum^+ r_i = 1$. The solution is to take the grade with the largest W_i and make that r_i as large as possible. If this r_i is less than one we go to the grade with the next largest W_i and make its r_i as large as the restraints allow, and so on. In the special case when $W_i = W$ for all i, D_1 is independent of the r_i's (because $\sum^+ r_i = 1$) and so any solution satisfying $r_i \leq y_i$ if $y_i > 0$ and $r_i = 0$ if $y_i \leq 0$ yields the same minimum value. A particular member of this class of solutions which will be investigated numerically is one which takes $r_i \propto y_i$ if $y_i > 0$. In practical terms this means that each grade which is undermanned, after leavers have been taken out, receives the same proportion of its needs. This strategy embodies a principle of 'fairness' which might commend itself on other than mathematical grounds.

The case a = 2

$$D_2 = \sum^+ W_i(y_i - r_i)^2.$$

This has to be minimized with respect to the r_i's subject to $\sum^+ r_i = 1$ and $r_i \geq 0$ for all i. Let

$$\phi = \sum^+ W_i(y_i - r_i)^2 + 2\alpha \sum^+ r_i$$

where α is an undetermined multiplier. Then, if $\mathbf{r}^*$ is the minimizing value of $\mathbf{r}$, at this point

$$\left.\begin{aligned} \frac{\partial \phi}{\partial r_i} &= 0 \quad \text{if } r_i^* > 0 \\ \frac{\partial \phi}{\partial r_i} &> 0 \quad \text{if } r_i^* = 0 \end{aligned}\right\}. \tag{4.64}$$

For the function ϕ defined above this implies

$$\left.\begin{aligned} r_i^* &= y_i - \frac{\alpha}{W_i} \quad \text{if } y_i > \alpha/W_i \\ &= 0 \qquad\qquad \text{if } y_i \leq \alpha/W_i \end{aligned}\right\} \tag{4.65}$$

where α must be chosen such that $\sum^+ r_i^* = 1$. If, for example, the W_i's are equal to one and

$$y_1 = 0{\cdot}7, \qquad y_2 = -0{\cdot}5, \qquad y_3 = 0{\cdot}1, \qquad y_4 = 0{\cdot}4, \qquad y_5 = 0{\cdot}3$$

it may be verified that $\alpha = 0{\cdot}133$ and hence that

$$r_1^* = 0{\cdot}567, \qquad r_2^* = r_3^* = 0, \qquad r_4^* = 0{\cdot}267, \qquad r_5^* = 0{\cdot}167.$$

The same result may be reached by the following algorithm.

Compute:

$$\left.\begin{aligned} y_i' &= 0 \text{ if } y_i \leq 0 \\ y_i' &= y_i - \left(\sum{}^{+} y_i - 1\right)/W_i \sum_j{}^{+} W_j^{-1} \end{aligned}\right\}. \tag{4.66}$$

If $\mathbf{y}' \geq 0$ then $\mathbf{r} = y'$; otherwise treat the y_i''s as original y_i's and repeat the process until a non-negative vector is obtained. Similar strategies can be devised for other values of a, and it may easily be verified that if the W_i's are equal they are all equivalent to that described for $a = 2$. Note that the solution given by (4.65) also minimizes D_1 but it will tend to have fewer non-zero elements in the control vectors than the strategy which takes $r_i \propto y_i$.

Numerical comparison of strategies

We take first an example with $k = 3$ using the transition matrix and wastage rates which have been the basis of earlier examples, viz.

$$\mathbf{P} = \begin{pmatrix} 0{\cdot}5 & 0{\cdot}4 & 0 \\ 0 & 0{\cdot}6 & 0{\cdot}3 \\ 0 & 0 & 0{\cdot}8 \end{pmatrix}, \mathbf{w} = (0{\cdot}1, 0{\cdot}1, 0{\cdot}2).$$

As the goal we take (0·286, 0·286, 0·428) and as the starting point (0, 0, 1). These structures may seem rather unlikely ones but the goal is the least top-heavy structure which can be maintained with these parameters. The starting point represents the most top-heavy structure possible. All points on the path from $\mathbf{q}(0)$ to $\mathbf{q}^*$ can, of course, be regarded as new starts, so the calculation is more comprehensive than appears at first sight. In Table 4.1a we have compared the sequence of structures given by the following strategies.

- (a) Fixed — This is the constant vector which will lead the structure to converge on the goal—in this case $\mathbf{r} = (1, 0, 0)$.
- (b) $S_1^{(1)}$ — This is the strategy minimizing D_1 with $r_i \propto y_i$ if $y_i > 0$.
- (c) $S_1^{(2)}$ — This is another strategy minimizing D_1. In this case the grade with the largest $y_i(y_{(1)}$, say) has $r = \min(1, y_{(1)})$, the grade with the next largest $y_i(y_{(2)}$, say) has its $r = \min(1 - r_{(1)}, y_{(2)})$, and so on.
- (d) S_2 — This is the strategy minimizing D_2 at each step. It also minimizes D_1.
- (e) D — This is the strategy following Davies (1973) which attains the goal in the minimum number of steps using $\mathbf{r} = (1, 0, 0)$ as far as possible.

(f) *L.P.* This is the linear programming solution in which the salary costs for the grades are in the ratios 1:2:3 (in this example it is identical to *D*).

Table 4.1b gives the sequences of recruitment vectors for each strategy.

Only the *D* and *L.P.* strategies actually reach the goal in the minimum number of steps (seven), but all of the others are close. Indeed, the most striking thing about Table 4.1a is the similarity of the trajectories for the last few steps. In the early stages some of the strategies (fixed, $S_1^{(2)}$, *D* and *L.P.*) concentrate on building up the bottom grade at the expense of the middle one, whereas the others spread the recruits more evenly. The *L.P.* strategy concentrates on the bottom grade because salaries are lower. If the salary ratios were changed the trajectory would change; if the salary costs were the same in each grade any basic feasible solution would yield the same salary bill. Interesting differences between the strategies are revealed by Table 4.1b.

Table 4.1a** **A numerical comparison of control strategies with** $k = 3$**: the grade structures

T	*Strategy*	$q_1(T)$	$q_2(T)$	$q_3(T)$	T	*Strategy*	$q_1(T)$	$q_2(T)$	$q_3(T)$
0		0	0	1					
1	Fixed	0·200	0	0·800	5	Fixed	0·300	0·253	0·447
	$S_1^{(1)}$	0·100	0·100	0·800		$S_1^{(1)}$	0·223	0·268	0·508
	$S_1^{(2)}$	0·200	0	0·800		$S_1^{(2)}$	0·286	0·228	0·486
	S_2	0·100	0·100	0·800		S_2	0·253	0·253	0·495
	L.P. and *D*	0·200	0	0·800		*L.P.* and *D*	0·300	0·253	0·447
2	Fixed	0·280	0·080	0·640	6	Fixed	0·295	0·272	0·433
	$S_1^{(1)}$	0·151	0·179	0·670		$S_1^{(1)}$	0·237	0·276	0·487
	$S_1^{(2)}$	0·100	0·260	0·640		$S_1^{(2)}$	0·286	0·257	0·457
	S_2	0·165	0·165	0·670		S_2	0·264	0·264	0·472
	L.P. and *D*	0·280	0·080	0·640		*L.P.* and *D*	0·295	0·272	0·433
3	Fixed	0·304	0·160	0·536	7	Fixed	0·291	0·281	0·428
	$S_1^{(1)}$	0·182	0·228	0·590		$S_1^{(1)}$	0·247	0·280	0·473
	$S_1^{(2)}$	0·214	0·196	0·590		$S_1^{(2)}$	0·286	0·271	0·443
	S_2	0·207	0·207	0·586		S_2	0·272	0·272	0·457
	L.P. and *D*	0·304	0·160	0·536		*L.P.* and *D*	0·286	0·286	0·429
4	Fixed	0·306	0·218	0·477		Goal	0·286	0·286	0·429
	$S_1^{(1)}$	0·205	0·254	0·540					
	$S_1^{(2)}$	0·266	0·203	0·531					
	S_2	0·235	0·235	0·531					
	L.P. and *D*	0·306	0·218	0·477					

Table 4.1b A numerical comparison of control strategies with** $k = 3$**: the recruitment vectors $\mathbf{r}'(T)$

	T							
Strategy	1	2	3	4	5	6	7	*Relative cost*
	0·500	0·559	0·641	0·719	0·782	0·831	0·868	
$S_1^{(1)}$	0·500	0·441	0·359	0·281	0·218	0·169	0·132	1·06
	0	0	0	0	0	0	0	
	1·000	0	1·000	1·000	1·000	0·963	0·981	
$S_1^{(2)}$	0	1·000	0	0	0	0·037	0·019	1·04
	0	0	0	0	0	0	0	
	0·500	0·639	0·747	0·827	0·883	0·922	0·949	
S_2	0·500	0·361	0·253	0·173	0·117	0·078	0·051	1·05
	0	0	0	0	0	0	0	
	1	1	1	1	1	1	0·965	
L.P. and *D*	0	0	0	0	0	0	0·032	1·00
	0	0	0	0	0	0	0·003	

Notes (a) In the case of $S_1^{(2)}$, y_1 and y_2 were both equal to 0·286 in the first period. There was no single largest value so all recruits were placed in the lower grade. If the recruits had been equally divided the early stages of $S_1^{(2)}$ would have been different.

(b) When interpreting the relative costs in the last column it must be remembered that only the *L.P.* and *D* strategies (which, here, are identical) actually achieve the target.

At any stage the three strategies $S_1^{(1)}$, $S_2^{(2)}$ and S_2 give structures which are all the same distance from the goal in the sense of D_1, but $S_1^{(1)}$ and, to some extent, S_2 allow more recruitment at the middle level than $S_1^{(2)}$. This suggests that the class of strategies minimizing D_1 is sufficiently wide to permit considerable flexibility in the choice of policy. The sequential strategies have more non-zero elements than the *L.P.*, which shows that the former are not minimum cost solutions. The last column in the table gives the cost of each strategy as a ratio to the *L.P.* cost, assuming the same salary ratios for the grades.

For this example, Davies has shown that the goal can be reached in four steps if both the $\mathbf{P}$ matrix and the $\mathbf{r}$ vector are subject to control. His solution proceeds as above as far as $T = 3$. At that point the promotion matrix

$$\mathbf{P}(3) = \begin{pmatrix} 0{\cdot}5 & 0{\cdot}4 & 0 \\ 0 & 0{\cdot}9 & 0 \\ 0 & 0 & 0{\cdot}8 \end{pmatrix}$$

and the recruitment vector $\mathbf{r}(4) = (0{\cdot}871, 0{\cdot}129, 0)$ take the structure to the goal in one further step.

When k is larger it tends to take longer to reach the goal. In Tables 4.2a and b we illustrate the case $k = 5$ with parameters

$$\mathbf{P} = \begin{pmatrix} 0{\cdot}65 & 0{\cdot}20 & 0 & 0. & 0 \\ 0 & 0{\cdot}75 & 0{\cdot}15 & 0 & 0 \\ 0 & 0 & 0{\cdot}75 & 0{\cdot}15 & 0 \\ 0 & 0 & 0 & 0{\cdot}85 & 0{\cdot}10 \\ 0 & 0 & 0 & 0 & 0{\cdot}95 \end{pmatrix}, \mathbf{w} = (0{\cdot}15, 0{\cdot}10, 0{\cdot}10, 0{\cdot}05, 0{\cdot}05).$$

The goal is again the vertex of the maintainable region giving the least top-heavy structure reached by the fixed strategy $\mathbf{r} = (1, 0, 0, 0, 0)$. As a starting point we take the more sharply tapering structure $\mathbf{q}(0) = (0{\cdot}40, 0{\cdot}30, 0{\cdot}15, 0{\cdot}10, 0{\cdot}05)$. The strategies compared in these tables are the same as those in Tables 4.1a and b, with the exception of D and the *L.P.* strategies. The strategies $S_1^{(1)}$, $S_1^{(2)}$ and S_2 initially concentrate their recruitment at the top and then switch with varying degrees of abruptness to recruitment at the

***Table 4.2a** A numerical comparison of control strategies with $k = 5$: the grade structures*

T	*Strategy*	$q_1(T)$	$q_2(T)$	$q_3(T)$	$q_4(T)$	$q_5(T)$
0		0·40	0·30	0·15	0·10	0·05
1	$S_1^{(1)}$	0·284	0·290	0·158	0·115	0·154
	$S_1^{(2)}$	0·260	0·290	0·158	0·108	0·185
	S_2	0·260	0·290	0·158	0·108	0·185
2	$S_1^{(1)}$	0·251	0·260	0·162	0·122	0·206
	$S_1^{(2)}$	0·282	0·255	0·162	0·115	0·187
	S_2	0·265	0·255	0·162	0·115	0·204
5	$S_1^{(1)}$	0·268	0·203	0·148	0·136	0·244
	$S_1^{(2)}$	0·306	0·217	0·150	0·132	0·196
	S_2	0·287	0·208	0·148	0·132	0·256
10	$S_1^{(1)}$	0·282	0·203	0·128	0·136	0·251
	$S_1^{(2)}$	0·311	0·209	0·132	0·136	0·212
	S_2	0·300	0·199	0·127	0·134	0·239
	Goal	0·306	0·204	0·122	0·122	0·245

***Table 4.2b** A numerical comparison of control strategies with* $k = 3$*: the recruitment vectors* $\mathbf{r}'(T)$

Strategy	T: 1	2	3	4	5	6	7	8	9	10
	0·186	0·580	0·792	0·914	0·954	0·942	0·941	0·945	0·949	0·953
	0	0	0	0	0·025	0·058	0·059	0·055	0·051	0·047
$S_1^{(1)}$	0	0	0	0	0	0	0	0	0	0
	0·060	0·044	0	0	0	0	0	0	0	0
	0·754	0·416	0·208	0·086	0·022	0	0	0	0	0
	0	1	1	1	1	1	1	1	1	1
	0	0	0	0	0	0	0	0	0	0
$S_1^{(2)}$	0	0	0	0	0	0	0	0	0	0
	0	0	0	0	0	0	0	0	0	0
	1	0	0	0	0	0	0	0	0	0
	0	0·849	0·927	0·949	0·966	0·978	0·987	0·993	0·997	1
	0	0	0	0	0	0	0	0	0	0
S_2	0	0	0	0	0	0	0	0	0	0
	0	0	0	0	0	0	0	0	0	0
	1	0·159	0·073	0·051	0·034	0·022	0·013	0·007	0·003	0

bottom. In the early stages, therefore, they offer a considerable advantage over the fixed strategy if an early approach to the desired structure is required.

The final tables, Tables 4.3a and b, show a case in which $S_1^{(1)}$ and S_2, but not $S_1^{(2)}$, reach the goal in the minimum number of steps for this problem. The parameters of the system are the same as in the last example but the goal is now (0·05, 0·10, 0·15, 0·30, 0·40). This would be a rather unlikely goal in practice, but it illustrates the fact that the sequential strategies are capable of reaching a goal in the minimum number of steps rather than converging on the goal as they appeared to be doing in the other examples. In this case the goal is outside the maintainable region

***Table 4.3a** A numerical comparison of control strategies with $k = 5$: the grade structure*

T	*Strategy*	$q_1(T)$	$q_2(T)$	$q_3(T)$	$q_4(T)$	$q_5(T)$
0		0·20	0·20	0·20	0·20	0·20
1	$S_1^{(1)}$	0·130	0·190	0·180	0·231	0·269
	$S_1^{(2)}$	0·130	0·190	0·180	0·200	0·300
	S_2	0·130	0·190	0·180	0·205	0·296
2	$S_1^{(1)}$	0·085	0·152	0·162	0·261	0·340
	$S_1^{(2)}$	0·085	0·152	0·162	0·287	0·315
	S_2	0·085	0·152	0·162	0·251	0·351
5	$S_1^{(1)}$	0·048	0·098	0·148	0·297	0·410
	$S_1^{(2)}$	0·029	0·101	0·151	0·301	0·417
	S_2	0·048	0·098	0·148	0·298	0·410
10	$S_1^{(1)}$	0·041	0·088	0·136	0·286	0·449
	$S_1^{(2)}$	0·021	0·100	0·148	0·276	0·455
	S_2	0·038	0·088	0·138	0·288	0·450
	Goal	0·05	0·10	0·15	0·30	0·40

Note: $S_1^{(1)}$ and S_2 achieve the goal at $T = 4$.

***Table 4.3b** A numerical comparison of grade structures with $k = 5$: the recruitment vectors $\mathbf{r}'(T)$*

	T			
Strategy	1	2	3	4
$S_1^{(1)}$	0	0	0	0·192
	0	0	0	0·036
	0	0	0·052	0·267
	0·345	0·391	0·486	0·457
	0·655	0·609	0·462	0·050
$S_1^{(2)}$	0	0	0	0·123
	0	0	0	0
	0	0	0	0·310
	0	1	0·113	0·566
	1	0	0·887	0
S_2	0	0	0	0·191
	0	0	0	0·036
	0	0	0	0·310
	0·050	0·550	0·628	0·439
	0·950	0·450	0·372	0·024

CHAPTER 5

Continuous Time Models for Stratified Social Systems

5.1 INTRODUCTION

We now turn to the development of continuous time versions of the Markov chain models introduced in Chapters 2 and 3. This has been anticipated to some extent by the discussion of occupational mobility in Chapter 2. However, we supposed there that the system could only be observed at fixed intervals of time so that only the discrete aspects of the process were of direct interest. Here we shall begin a fuller study of continuous time models, beginning with those based on the theory of Markov processes.

The choice, in practice, between the discrete and continuous time versions of a model is partly a matter of realism and partly one of convenience. On grounds of realism models for the movement of employees in a firm should usually be in continuous time, but a discrete approximation often offers considerable computational advantages. On the other hand, continuous models tend to be more amenable to mathematical analysis and this may count in their favour even when a discrete model is more realistic. In this chapter we shall not simply give a parallel treatment in continuous time of the various problems discussed in the last three chapters. Instead, we shall concentrate on those aspects of the modelling process where a continuous time approach offers advantages.

The development of a Markov chain is determined by a set of transition probabilities. In continuous time, when changes can occur at any time, it is necessary to base the development of the theory on infinitesimal transition probabilities. Let the states of the system be labelled $S_1, S_2, \ldots, S_k$; then the probability of a transtion from S_i to $S_j (i \neq j)$ in $(T, T + \delta T)$ is termed an *infinitesimal transition probability*. The set of such probabilities for all pairs of states determines the future of the process. We shall assume that these probabilities can be written in the form

$$Pr\{S_i \rightarrow S_j \text{ in } (T, T + \delta T)\} = r_{ij}(T)\delta T + o(\delta T) \qquad (i \neq j). \qquad (5.1)$$

For our purpose the term $o(\delta T)$ can be ignored and we shall omit it from future specifications of infinitesimal probabilities. The function $r_{ij}(T)$ will be referred to as the *rate*, or *intensity*, of transition between S_i and S_j at time T or, more briefly, as the transition rate. The possible dependence of the rates on T has been made explicit in the notation. In particular applications they may be constant or depend on other variables such as the length of time spent in the present state. When complications of this latter kind occur the Markov property is lost and the analysis becomes more difficult. Some examples of this kind will occur at the end of the chapter.

Transition rates, or intensities, are frequently used in actuarial and reliability studies though under various names; perhaps the most familiar and expressive is the term 'force of mortality'. In our present terminology this is the transition intensity at age T from the state 'life' to the state 'death'. These intensities can often be estimated in practice and have a direct intuitive meaning. They therefore provide a natural starting point for the construction of stochastic models of social phenomena which can, in principle at least, be observed continuously in time.

It is instructive to consider the link between the present formulation and the model used for occupational mobility in Chapter 2. The latter was basically a continuous time process although it had to be treated as discrete because of the nature of the data available. The connexion is made apparent if $r_{ij}(T)$ is expressed in the form

$$\left.\begin{aligned} r_{ij}(T) &= \lambda_i(T)p_{ij.T}, \\ \text{with} \qquad \sum_{j=1}^{k} p_{ij.T} &= 1 \qquad (i = 1, 2, \ldots, k). \end{aligned}\right\} \tag{5.2}$$

When considered in this form $\lambda_i(T)\delta T$ is the probability that a member of S_i changes state at time T and $p_{ij.T}$ is the probability that the transition is to S_j. The need for separating the two processes in Chapter 2 arose because the transition matrix $\{p_{ij.T}\}$ was the only part which could be estimated from the available data. In the applications now to be discussed it is the rates $\{r_{ij}(T)\}$ which are of fundamental interest.

A number of social applications of continuous time Markov processes have been made. Coleman (1964b) used such a model to study social mobility using the data analysed in Chapter 2. The continuous model which he used requires a somewhat different viewpoint from that which we adopted earlier. Instead of using a transition matrix to describe changes in class from one generation to the next, Coleman argued as follows. Until the son becomes independent of his father his class may be taken to be the same as his father's. Subsequently, as the son becomes older, he is subject to social pressures which tend to change his class. If it is assumed that these forces can be represented by constant transition rates then mobility will be represented by a continuous time Markov process. Coleman tested the model using both Glass and Hall's data and that given by Svalagosta (1959). A general Markov model in which transitions between all classes are possible could not be used because it would have more parameters than could be estimated from the data. It was therefore assumed that transitions were possible into adjacent classes only. This model did not provide a good fit to either set of data but, in any case, the data were not in a suitable form. A satisfactory test of the model could only have been made if each son had been exposed to the risk of transition for the same length of time. For the data collected by Glass and Svalagosta this could, at best, have only been approximately true.

Another model in which the transition rates themselves are of interest is one for survival after treatment for cancer, proposed by Fix and Neyman (1951). We shall use their model as a basis for the discussion of the next section. Zahl (1955) has also given an introductory account of the application of the theory to follow-up studies in cancer research and other related fields. A rather similar model was proposed by Herbst (1963) to explain the form of the distribution of completed length of service for employees in a firm. In this case the states of the system were psychological states through which individuals were supposed to pass before departure. This application will be discussed, along with others for durations, in Chapter 6.

Continuous time Markov models have also been used for manpower systems. Seal (1945) and Vajda (1947) introduced models of considerable generality, special cases of which are Markovian. Models of this kind will occupy the major part of the present chapter. An application to occupational mobility has been made by Kuhn and coworkers (1973).

Sverdrup (1965) applied a Markov model to the flow of people between different states of health such a 'able', 'disabled' and 'dead'. A full discussion of such applications will be found in Chiang (1968). The utilization of health services has been the subject of a paper by Schach and Schach (1972). The estimation of transition rates is not of particular relevance to us here because our purpose is to demonstrate the possibilities of mathematical analysis. However, the problem is an important one and some results are given in Sverdrup (1965), Hoem (1971), Zahl (1955) and Kuhn and coworkers (1973). A brief discussion will be given in this chapter as the need arises.

The theory of Markov processes in continuous time is given in most of the basic texts on stochastic theory. Of these books those by Bharucha–Reid (1960), Cox and Miller (1965) and Moran (1968) are particularly recommended. Authors who have applied the theory to social phenomena have often derived such results as they needed by heuristic arguments without reference to formal theory. However, in order to give a unified account of their work and to lay a foundation for new developments we shall set our exposition within the framework of the general theory of Markov processes.

5.2 SOME BASIC THEORY OF MARKOV PROCESSES

Transient theory

The quantities of greatest interest to us are the transition probabilities which we denote by $p_{ij}(T)(i, j = 1, 2, \ldots, k)$. For given i and j this is the probability that an individual who is in S_i at time zero is in S_j at time $T(T > 0)$. We define

$$\begin{aligned} p_{ij}(0) &= 0 \quad \text{if } i \neq j \\ &= 1 \quad \text{if } i = j. \end{aligned}$$

These transition probabilities should not be confused with the infinitesimal transition probabilities, $r_{ij}(T)\delta T$, which are used to define the process.

The first step is to set up a system of differential equations relating the $p_{ij}(T)$'s to the $r_{ij}(T)$'s. Readers who are familiar with the derivation of the equations for the birth and death process will recognize the following argument as a straightforward generalization. Consider the probability $p_{ij}(T + \delta T)$. Let us suppose that we have found $p_{ij}(T)$ for all i and j. In addition we know the probabilities of the various transitions in $(T, T + \delta T)$ because the process is defined in terms of them. We then have

$$p_{ij}(T + \delta T) = p_{ij}(T)\left\{1 - \delta T \sum_{\substack{h=1 \\ h \neq j}}^{k} r_{jh}(T)\right\} + \sum_{\substack{h=1 \\ h \neq j}}^{k} p_{ih}(T) r_{hj}(T)\, \delta T$$

$$(i, j = 1, 2, \ldots, k). \qquad (5.3)$$

Subtracting $p_{ij}(T)$ from both sides of (5.3), dividing by δT and allowing δT to tend to zero gives

$$\frac{\mathrm{d}p_{ij}(T)}{\mathrm{d}T} = -p_{ij}(T) \sum_{\substack{h=1 \\ h \neq j}}^{k} r_{jh}(T) + \sum_{\substack{h=1 \\ h \neq j}}^{k} p_{ih}(T) r_{hj}(T) \qquad (i, j = 1, 2, \ldots, k). \qquad (5.4)$$

The transition rate $r_{ij}(T)$ has not been defined for $i = j$. However, the set of equations (5.4) can be written more compactly if we define

$$r_{jj}(T) = -\sum_{\substack{h=1 \\ j \neq h}}^{k} r_{jh}(T). \qquad (5.5)$$

We then have

$$\frac{\mathrm{d}p_{ij}(T)}{\mathrm{d}T} = \sum_{h=1}^{k} p_{ih}(T)r_{hj}(T) \qquad (i, j = 1, 2, \ldots, k). \tag{5.6}$$

A further notational simplification, which also facilitates the solution of the system, is obtained by adopting a matrix representation as follows. Let

$$\mathbf{P}(T) = \{p_{ij}(T)\}, \qquad \mathbf{R}(T) = \{r_{ij}(T)\}.$$

Then (5.6) becomes

$$\frac{\mathrm{d}\mathbf{P}(T)}{\mathrm{d}T} = \mathbf{P}(T)\mathbf{R}(T). \tag{5.7}$$

The differential operator on the left-hand side of (5.7) is to be understood as applying to each element in the matrix $\mathbf{P}(T)$. Equation (5.7) can be used to yield a system of differential equations for $\bar{\mathbf{n}}(T)$, the vector of expected grade sizes at time T. This follows when it is recognized that

$$\bar{\mathbf{n}}(T) = \mathbf{n}(0)\mathbf{P}(T). \tag{5.8}$$

Post-multiplying both sides of (5.7) by $\mathbf{n}(0)$ gives

$$\frac{\mathrm{d}\bar{\mathbf{n}}(T)}{\mathrm{d}T} = \bar{\mathbf{n}}(T)\mathbf{R}(T). \tag{5.9}$$

Systems of first-order linear differential equations have been extensively studied and methods are available for their solution (see, for example, Frazer, Duncan and Collar, 1946). The simplest case occurs if the transition rates are constant. Since this is the case with the main models described in this chapter we shall, in the remainder of this section, assume constant rates. Later we shall solve the equations in one case when the rates are not all constants. When the transition rates are constant we shall omit the T from the notation, writing r_{ij} for the rate and $\mathbf{R}$ for the matrix.

A solution to (5.7) when $\mathbf{R}$ is constant is suggested by consideration of the special case when R and $P(T)$ are scalars. In that case the solution is

$$P(T) = \mathrm{e}^{RT} = \sum_{i=0}^{\infty} (RT)^i/i!. \tag{5.10}$$

In the matrix case we may easily verify that the matrix series on the right-hand side of (5.10) satisfies (5.7) if we define $\mathbf{R}^0 = \mathbf{I}$. It will be the solution if the series converges and this is always so if k is finite. In a similar manner it follows that the solution of (5.8) is

$$\bar{\mathbf{n}}(T) = \mathbf{n}(0)\exp(\mathbf{R}T), \tag{5.11}$$

where we use the exponential notation for the infinite matrix series. A direct solution of either set of equations using this representation would involve the summation of k^2 infinite series, one for each element in the matrix. This is hardly feasible as a practical method unless T is very small. An alternative method is available using Sylvester's theorem (see Chapter 3, Section 3.2), which enables us to express the solution as a finite series of k terms. In general the matrix $\mathbf{R}$ admits the spectral representation

$$\mathbf{R} = \sum_{i=1}^{k} \lambda_i \mathbf{A}_i$$

where $\{\lambda_i\}$ are the eigenvalues of $\mathbf{R}$ and $\{\mathbf{A}_i\}$ is the associated spectral set. The representation is only valid if all the λ's are distinct. It then follows from Sylvester's theorem that

$$\begin{aligned} \exp(\mathbf{R}T) &= \sum_{i=1}^{k} e^{\lambda_i T} \mathbf{A}_i \\ &= \mathbf{P}(T), \end{aligned} \tag{5.12}$$

using the matrix form of (5.10). Likewise we have

$$\mathbf{n}(T) = \sum_{i=1}^{k} e^{\lambda_i T} \mathbf{n}(0) \mathbf{A}_i. \tag{5.13}$$

The behaviour of the solution, especially for large T, will clearly depend critically on the eigenvalues. Before proceeding to a discussion of the practical steps needed to obtain a complete solution we shall therefore obtain some general results about their values.

The eigenvalues are obtained by solving the equation

$$\begin{vmatrix} r_{11} - \lambda & r_{21} & \cdots & r_{k1} \\ r_{12} & r_{22} - \lambda & \cdots & r_{k2} \\ \vdots & \vdots & \ddots & \vdots \\ r_{1k} & r_{2k} & \cdots & r_{kk} - \lambda \end{vmatrix} = 0. \tag{5.14}$$

The value of the determinant is unchanged if we replace the first row by a row whose elements are the column sums. Each element in the first row of the new determinant so formed is $-\lambda$ because of (5.5). The equation (5.14) is therefore always satisfied when $\lambda = 0$. It is clear from (5.12) that all roots must be zero or negative otherwise there would be some T for which the transition probabilities did not lie between zero and one.

Since we are mainly interested in the vector $\mathbf{n}(T)$ we shall consider the solution given by (5.13). It follows from the foregoing remarks that $\bar{n}_i(T)$'s can be expressed in the form

$$\bar{n}_i(T) = c_{i1} + \sum_{j=2}^{k} c_{ij}\, e^{\lambda_j T} \qquad (i = 1, 2, \ldots, k) \tag{5.15}$$

where $\lambda_j < 0$ ($j = 2, 3, \ldots, k$). This form is valid provided that all the λ's are distinct. It holds also, with a slight modification discussed below, if $\lambda = 0$ is a multiple root. The coefficients $\{c_{ij}\}$ in (5.15) may be found by first determining the spectral set $\{\mathbf{A}_i\}$. An equivalent, but more direct approach, is the following. Substituting $\bar{\mathbf{n}}(T)$ from (5.15) into (5.9), we have

$$\sum_{j=1}^{k} c_{ij}\lambda_j\, e^{\lambda_j T} = \sum_{j=1}^{k} r_{ji} \sum_{h=1}^{k} c_{jh}\, e^{\lambda_h T} \qquad (i = 1, 2, \ldots, k) \tag{5.16}$$

where $\lambda_1 = 0$. Equating coefficients of $e^{\lambda_h T}$ we find that the c_{ij}'s must satisfy the equations

$$\sum_{j=1}^{k} r_{ji} c_{jh} = \lambda_h c_{ih} \qquad (i = 1, 2, \ldots, k; h = 1, 2, \ldots, k). \tag{5.17}$$

Although there are k^2 equations here for the same number of unknowns they are not independent. In fact, if we sum each side of (5.17) over i we obtain zero in each case. In order to determine the c's we therefore require k further equations. These arise from the necessity of ensuring that the initial conditions are satisfied. Thus, setting $T = 0$ in (5.15), we have

$$\sum_{j=1}^{k} c_{ij} = n_i(0) \qquad (i = 1, 2, \cdots, k). \tag{5.18}$$

We shall solve the equations in the case of particular applications later.

The theory given above covers the case when all the λ's are distinct. If multiplicities occur among the roots the form of the solution can be determined by an appropriate limiting operation on (5.15). For example, suppose that $\lambda_2 = \lambda_3 = \lambda$. The coefficients $\{c_{ij}\}$ are functions of the λ's and we must determine their limits as $\lambda_2 \to \lambda_3$. The terms involving λ_2 and λ_3 in the exponent require special attention. In a typical case we have

$$\left.\begin{aligned} &\lim_{\lambda_2 \to \lambda_3 = \lambda} \{c_{i2}\, e^{\lambda_2 T} + c_{i3}\, e^{\lambda_3 T}\} \\ &\quad = e^{\lambda T} \lim_{\lambda_2 \to \lambda_3} \{c_{i2} + c_{i3}\, e^{(\lambda_3 - \lambda_2)T}\} \\ &\quad = e^{\lambda T} \lim_{\lambda_2 \to \lambda_3} \{c_{i2} + c_{i3} + c_{i3}((\lambda_3 - \lambda_2)T + O(\lambda_3 - \lambda_2)^2)\}. \end{aligned}\right\} \tag{5.19}$$

Since the limit in curly brackets cannot be infinite in any meaningful problem, the pair of terms corresponding to $j = 2$ and $j = 3$ in (5.15) must be replaced by a single term of the form

$$(d_{i2} + d_{i3}T)\,e^{\lambda T}. \tag{5.20}$$

In general, if there is a root of multiplicity m with common value λ, the terms corresponding to that root in (5.15) must be replaced by the following expression

$$\sum_{j=2}^{m+1} d_{ij}T^{j-2}\,e^{\lambda T}. \tag{5.21}$$

A particularly simple and important case occurs when the multiple root has the value zero. The general solution then has the form

$$\bar{n}_i(T) = \sum_{j=1}^{m} d_{ij}T^{j-1} + \sum_{j=m+1}^{k} c_{ij}\,e^{\lambda_j T} \qquad (i = 1, 2, \ldots, k). \tag{5.22}$$

It may be shown that, in this case, $d_{ij} = 0$ for $j > 1$ and all i. The necessity for this may be seen by considering the limit of $\bar{n}_i(T)$ as T tends to infinity. Under these conditions the second sum in (5.22) vanishes and the first sum will tend to $\pm\infty$ according to the sign of d_{im}. Since $\bar{n}_i(T)$ possesses a finite limit we conclude that $\bar{n}_i(T)$ must have the form

$$\bar{n}_i(T) = d_{i1} + \sum_{j=m+1}^{k} c_{ij}\,e^{\lambda_j T} \qquad (i = 1, 2, \ldots, k). \tag{5.23}$$

If we substitute this in (5.9) the equations for $\{d_{i1}\}$ and $\{c_{ij}\}$ are identical with those given in (5.17), if we delete those equations with $h \leq m - 1$ and replace c_{im} by d_{i1}.

Limiting behaviour

The limiting behaviour of $\bar{n}_i(T)$ as $T \to \infty$ can be studied directly by reference to the results given earlier in this chapter. Referring to (5.15) and noting that $\lambda_j < 0\,(j = 2, 3, \ldots, k)$ it is clear that

$$\lim_{T\to\infty} \bar{n}_i(T) = c_{i1} \qquad (i = 1, 2, \ldots, k). \tag{5.24}$$

In the case when $\lambda = 0$ is a multiple root we merely have to replace c_{i1} by d_{i1} on the right-hand side of (5.24).

If we are interested only in the limiting values we can avoid the necessity of calculating the c_{ij}'s for $j > 1$. Two methods are available according to whether or not there are terminal states in the system. Suppose first that there are no terminal states; then transitions are possible both into and out of every state. Since we know that the $\bar{n}_i(T)$'s approach limits we may deduce that their derivatives vanish at infinity. It then follows from (5.9) that if the transition rates are constant the limiting vector of expected state sizes must satisfy

$$\bar{\mathbf{n}}(\infty)\mathbf{R} = \mathbf{0}, \tag{5.25}$$

where $\mathbf{0}$ is a row vector of zeros. We have the additional equation

$$\sum_{j=1}^{k} \bar{n}_j(\infty) = N$$

which, with (5.25), determines the limits uniquely. It may also be shown that

$$\lim_{T\to\infty} p_{ij}(T) = \bar{n}_j(\infty)/N \qquad (i, j = 1, 2, \ldots, k), \tag{5.26}$$

indicating that the probability of ultimate transition to S_j is independent of the original state.

The equations (5.25) are satisfied by the limiting state sizes even if there are terminal states. In this case, however, they no longer determine the expectations uniquely, since each terminal state gives rise to a row of zeros in the $\mathbf{R}$ matrix. The limiting expectations for the non-terminal states do not appear in the equations (5.25) because their coefficients all vanish. We shall not enter into a full discussion of the treatment in this case. For fairly small systems a simple direct method is available which will be sufficiently illustrated by examples.

5.3 APPLICATION TO SURVIVAL AFTER TREATMENT FOR CANCER

Description of the problem

We shall first apply the theory to a closed system with two terminal (or absorbing) states arising in follow-up studies of cancer treatment. A patient who has been treated for a disease may, at any subsequent time, be in one of a number of states. These states might be, for example, 'health', 'relapsed' and 'death'; the precise classification used will obviously depend on the objects of the enquiry and the kind of records available. A stochastic model for the post-treatment history of patients treated for cancer was developed by Fix and Neyman (1951) and discussed in more general terms by Zahl (1955). Fix and Neyman used the model to provide measures of the effectiveness of a treatment and we shall describe how this can be done below. The basic situation is of wide occurrence and there are many other possible applications, some of which have already been mentioned.

In Fix and Neyman's model there were four states. The description of the states and the allowable transitions are indicated in Figure 5.1. The authors emphasized the difficulty of defining 'recovery' and also pointed

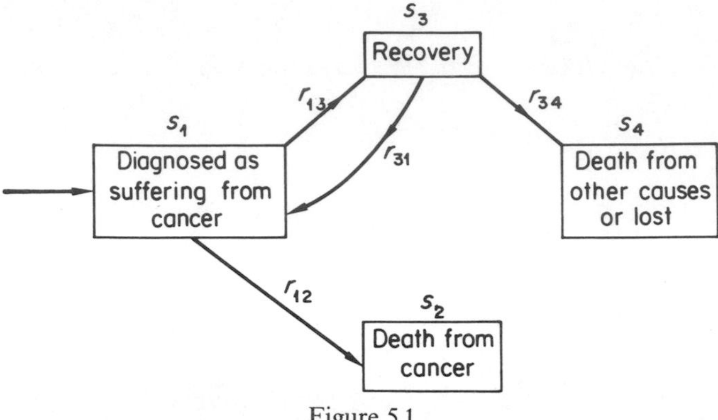

Figure 5.1

out that it might be desirable to subdivide some of the states. For example, S_4 might be divided into those who died from natural causes and those who were lost from observation. It might also be thought desirable to allow transitions between S_1 and S_4. We shall not digress to discuss these technical points since we are using the example primarily to illustrate the application of the theory of Markov processes in a social context.

The first objective in the cancer application was to estimate the transition rates. These were then used to provide measures of survival which did not suffer from the disadvantages associated with the commonly used measures. One such measure is the 'T-year survival rate'. This is the proportion of those receiving treatment who survive for at least T years. This measure would be satisfactory if cancer was the only cause of death and if all persons were under observation for the full T years. In practice this is not the case, and the T-year survival rate may be misleading. To see this we merely note that the rate would be higher if loss or death from other causes were eliminated since then more people would survive to ultimately die of cancer. The observed value of the rate thus depends not only upon the risk of cancer death but also on other risks which have nothing to do with cancer. If a 'treatment' and 'control' group were compared by means of the crude rate the comparison would be valueless if the two groups were subject to different risks from other causes. To overcome this difficulty it is customary to calculate net rates which make allowances for such differences. Our purpose in introducing this example is to show that the stochastic model provides a more satisfactory basis for estimating net rates than the 'actuarial' method.

In their model, Fix and Neyman treated the transition rates between the states as constants. However, it is well known that the force of natural mortality for human populations is not constant but, after infancy, increases with age. It does not increase very rapidly over the middle years of life and if T is short relative to the normal life span the assumption of constancy may be quite adequate. In any event we shall show that data can be collected in such a way that the validity of the assumption can be tested. The force of mortality from various kinds of cancer has been extensively studied. The survival time following treatment appears to be highly skew; Boag (1949), for example, has suggested that it can often be adequately fitted by a skew lognormal distribution. In this case the lognormal distribution is not easy to distinguish from the exponential which would arise if the death rate were constant. The assumption of a constant rate for death from cancer is thus probably not unrealistic. Direct evidence on the true nature of the transition rates between S_1 and S_3 (recovery) and between S_3 and S_1 is lacking, but it does seem plausible to assume a constant loss rate, at least in the case of those who are lost to observation.

In the model we assume that there are N people in state S_1 at time zero and none elsewhere. The numbers in the four groups at any subsequent time T will be random variables which we denote by $n_j(T)(j = 1, 2, 3, 4)$; $\bar{n}_j(T)$ is the expectation of $n_j(T)$. By observing these random variables at one or more points in time it will be possible to estimate the transition rates. Using these estimates it will then be possible to predict the numbers in the various states at future times. More important, it will be possible to estimate what these numbers would have been if death from cancer had been the only risk.

Application of the theory

The matrix $\mathbf{R}$ in the cancer case has the form

$$\mathbf{R} = \begin{pmatrix} r_{11} & r_{12} & r_{13} & 0 \\ 0 & 0 & 0 & 0 \\ r_{31} & 0 & r_{33} & r_{34} \\ 0 & 0 & 0 & 0 \end{pmatrix} \tag{5.27}$$

where $r_{11} = -(r_{12} + r_{13})$ and $r_{33} = -(r_{31} + r_{34})$. The equation for the eigenvalues is $|\mathbf{I}\lambda - \mathbf{R}| = 0$ or

$$\lambda^2(r_{11} - \lambda)(r_{33} - \lambda) - \lambda^2 r_{13} r_{31} = 0. \tag{5.28}$$

This equation clearly has a double root equal to zero; the two remaining roots, which we label λ_3 and λ_4, are

$$\lambda_3, \lambda_4 = \tfrac{1}{2}\{r_{11} + r_{33} \pm \sqrt{(r_{11} - r_{33})^2 + 4r_{13}r_{31}}\}, \tag{5.29}$$

taking the positive sign for λ_3 and the negative for λ_4. It then follows from (5.23) that

$$\bar{n}_i(T) = d_{i1} + c_{i3}\,e^{\lambda_3 T} + c_{i4}\,e^{\lambda_4 T} \qquad (i = 1, 2, 3, 4). \tag{5.30}$$

The next step is to set down and solve the simultaneous equations for the coefficients. We first put $i = 1$ and let h take the values 2, 3 and 4, thus obtaining

$$\left.\begin{aligned} r_{11}d_{11} + r_{31}d_{31} &= 0 \\ r_{11}c_{13} + r_{31}c_{33} &= \lambda_3 c_{13} \\ r_{11}c_{14} + r_{31}c_{34} &= \lambda_4 c_{14} \end{aligned}\right\}. \tag{5.31}$$

Three further sets of equations are obtained for $i = 2$, 3 and 4 as follows:

$i = 2$

$$\left.\begin{aligned} r_{12}d_{11} &= 0 \\ r_{12}c_{13} &= \lambda_3 c_{23} \\ r_{12}c_{14} &= \lambda_4 c_{24} \end{aligned}\right\} \tag{5.32}$$

$i = 3$

$$\left.\begin{aligned} r_{13}d_{11} + r_{33}d_{31} &= 0 \\ r_{13}c_{13} + r_{33}c_{33} &= \lambda_3 c_{33} \\ r_{13}c_{14} + r_{33}c_{34} &= \lambda_4 c_{34} \end{aligned}\right\} \tag{5.33}$$

$i = 4$

$$\left.\begin{aligned} r_{34}d_{31} &= 0 \\ r_{34}c_{33} &= \lambda_3 c_{43} \\ r_{34}c_{34} &= \lambda_4 c_{44} \end{aligned}\right\}. \tag{5.34}$$

It follows at once that $d_{11} = d_{31} = 0$ and hence the first equation in each group can be ignored in what follows. The initial conditions require that all members of the system are in S_1 at time zero. Let us suppose, therefore, that $n_1(0) = 1$ and $n_i(0) = 0$, $i > 1$. If $n_1(0) = N$ the appropriate values of $\bar{n}_i(T)$ can be obtained simply by multiplying by N those obtained assuming $n_1(0) = 1$. In addition to the equations listed above we now have

$$\left.\begin{aligned} d_{11} + c_{13} + c_{14} &= 1 \\ d_{21} + c_{23} + c_{24} &= 0 \\ d_{31} + c_{33} + c_{34} &= 0 \\ d_{41} + c_{43} + c_{44} &= 0 \end{aligned}\right\}. \tag{5.35}$$

To solve the equations we proceed as follows. Adding both sides of (5.31) and using the initial conditions we obtain

$$\lambda_3 c_{13} + \lambda_4 c_{14} = r_{11}. \tag{5.36}$$

A similar operation on (5.32) yields

$$\lambda_3 c_{23} + \lambda_4 c_{24} = r_{12}. \tag{5.37}$$

But this equation can be expressed in terms of c_{13} and c_{14} from (5.32) to give

$$c_{13} + c_{14} = 1. \tag{5.38}$$

The pair of simultaneous equations (5.36) and (5.38) may then be solved, giving

$$c_{13} = \frac{\lambda_4 - r_{11}}{(\lambda_4 - \lambda_3)}, \qquad c_{14} = \frac{r_{11} - \lambda_3}{(\lambda_4 - \lambda_3)}, \tag{5.39}$$

and hence

$$c_{23} = \frac{r_{12}(\lambda_4 - r_{11})}{\lambda_3(\lambda_4 - \lambda_3)}, \qquad c_{24} = \frac{r_{12}(r_{11} - \lambda_3)}{\lambda_4(\lambda_4 - \lambda_3)}. \tag{5.40}$$

If this whole procedure is repeated on (5.33) and (5.34) we obtain

$$\left.\begin{aligned} c_{33} &= \frac{-r_{13}}{(\lambda_4 - \lambda_3)}, & c_{34} &= \frac{r_{13}}{(\lambda_4 - \lambda_3)} \\ c_{43} &= \frac{-r_{13}r_{34}}{\lambda_3(\lambda_4 - \lambda_3)}, & c_{44} &= \frac{r_{13}r_{34}}{\lambda_4(\lambda_4 - \lambda_3)} \end{aligned}\right\}. \tag{5.41}$$

Only two constraints remain to be determined; these are d_{21} and d_{41}. Using the initial conditions we find

$$d_{21} = -c_{23} - c_{24} = -r_{12}r_{33}/\lambda_3\lambda_4, \tag{5.42}$$

$$d_{41} = -c_{42} - c_{43} = r_{13}r_{34}/\lambda_3\lambda_4. \tag{5.43}$$

We now turn to consider how these results may be used to make valid comparisons of survival rates. When $N = 1$, $\bar{n}_i(T)$ may be regarded as the probability of being in S_i at time T. Thus $\bar{n}_2(T)$ and $\bar{n}_4(T)$ may be interpreted as the crude risks of death from cancer and natural causes respectively. However, $\bar{n}_4(T)$ also depends on the force of natural mortality, and, as we pointed out above, this reduces its value as a measure of risk. We really require a net measure of risk from which the effect of natural mortality is eliminated. The actuarial approach to the problem defines a net rate of mortality due to cancer by the formula

$${}_A\bar{n}_2(T) = \bar{n}_2(T)/\{1 - \tfrac{1}{2}\bar{n}_4(T)\}. \tag{5.44}$$

This measure purports to give the expected number of cancer deaths that would occur in $(0, T)$ if natural mortality did not exist. The derivation of (5.44) will be clearer if we write it in the form

$$\bar{n}_2(T) = {}_A\bar{n}_2(T) - \tfrac{1}{2}\bar{n}_4(T)_A\bar{n}_2(T). \tag{5.45}$$

The second term on the right-hand side of (5.45) is an estimate of the number of people who would have died from cancer in the period had they not in fact died from natural causes. It is obtained by assuming that the probability of death from cancer being preceded by death from natural causes is one half. Our model provides an alternative method of estimating net rates. We can eliminate the effect of natural mortality by putting $r_{34} = 0$ in the model. The net risk may then be written

$$\bar{n}_2^0(T) = \frac{r_{12}r_{31}}{\lambda_3^0\lambda_4^0} + \frac{r_{12}(\lambda_4^0 - r_{11})}{\lambda_3^0(\lambda_4^0 - \lambda_3^0)}\,e^{\lambda_3^0 T} + \frac{r_{12}(r_{11} - \lambda_3^0)}{\lambda_4^0(\lambda_4^0 - \lambda_3^0)}\,e^{\lambda_4^0 T} \tag{5.46}$$

where the superfix on $\bar{n}_2(T)$, λ_3 and λ_4 denotes that r_{34} has been set equal to zero.

The use of these results may be illustrated by two numerical examples. We assume the following transition intensities

	r_{12}	r_{13}	r_{31}	r_{34}
Example 1	1·0	2·0	0·5	0·2
Example 2	0·5	0·5	0·5	0·5

Substituting these values in (5.29) we find, for Example 1,

$$\left.\begin{aligned}\bar{n}_2(T) &= 0{\cdot}6364 - 0{\cdot}3764\,e^{-0\cdot3260T} - 0{\cdot}2600\,e^{-3\cdot3740T}\\ \bar{n}_2^0(T) &= 1 - 0{\cdot}7344\,e^{-0\cdot1492T} - 0{\cdot}2657\,e^{-3\cdot3508T}\end{aligned}\right\} \tag{5.47}$$

and, for Example 2,

$$\left.\begin{aligned}\bar{n}_2(T) &= \tfrac{2}{3} - \tfrac{1}{2}e^{-\frac{1}{2}T} - \tfrac{1}{6}e^{-(3/2)T}\\ \bar{n}_2^0(T) &= 1 - 0{\cdot}7236\,e^{-0\cdot1910T} - 0{\cdot}2764\,e^{-1\cdot3090T}\end{aligned}\right\}. \tag{5.48}$$

One unsatisfactory feature of the actuarial risk is seen from its limiting behaviour as $T \to \infty$. Instead of approaching one, as we would expect a reasonable measure to do, it tends to a limit less than one in both cases. Inspection of (5.44) shows that this result always holds. It also appears to be generally true that ${}_A\bar{n}_2(T) < \bar{n}_2^0(T)$ if (T) is sufficiently large. Some numerical values are given in Table 5.1.

This example provides a good illustration of the use of a stochastic model for measuring a social phenomenon. It also shows that the use of 'common-

Table 5.1 A comparison of net risks of cancer death calculated by (a) the actuarial method and (b) using the stochastic model

	T	0·5	1	2	5	∞
Example 1	$\bar{n}_2(T)$	0·269	0·356	0·440	0·563	0·636
	$\bar{n}_2^0(T)$	0·269	0·358	0·455	0·652	1·000
	${}_A\bar{n}_2(T)$	0·272	0·370	0·477	0·656	0·778
Example 2	$\bar{n}_2(T)$	0·199	0·326	0·474	0·626	0·667
	$\bar{n}_2^0(T)$	0·199	0·328	0·486	0·721	1·000
	${}_A\bar{n}_2(T)$	0·201	0·338	0·515	0·733	0·800

sense' corrections to crude measures may seriously underestimate the quantity being measured. These arguments presuppose that the model used provides an adequate description of the phenomenon. If, in fact, the transition intensities are not constant the simpler actuarial estimate may be preferable because it is 'distribution-free'. As we show below, rough methods are available for testing the adequacy of the model.

We have conducted the foregoing discussion as if the transition rates were known. In practice they will not be known and must therefore be estimated from the data. A general method for estimating the intensities was proposed by Zahl (1955). For our present purposes the simpler method of Fix and Neyman will suffice. At time T we can observe the numbers of original patients in each of the four states. These numbers may be treated as estimates of the $\bar{n}_i(T)$'s, which in turn are functions of the unknown parameters. In the present case this method would yield four equations for estimating the four unknown parameters. Unfortunately the equations are not linearly independent because

$$\sum_{i=1}^{4} \bar{n}_i(T) = N,$$

the total number observed. The situation would be even worse if there were other non-zero intensities in $\mathbf{R}$. The difficulty can be overcome if the state of the system can be observed at several points in time. An alternative method is to observe some additional feature of the system by, for example, adopting the proposal of Fix and Neyman to count the number of returns to S_1 in $(0, T)$. If sufficient observational material is available it will not only be possible to estimate all the parameters but also to test the fit of the model. A preliminary discussion of the estimation problem was given by Fix and Neyman, but this is an area which requires further investigation.

The limiting structure, $\bar{\mathbf{n}}(\infty)$, can be derived directly without recourse to the full treatment just described, but (5.30) yields the result immediately.

From equations (5.42) and (5.43) we have

$$\left.\begin{aligned} \bar{n}_2(\infty) &= d_{21} = r_{12}(r_{31} + r_{34})/\lambda_3\lambda_4 \\ \bar{n}_4(\infty) &= d_{41} = r_{13}r_{34}/\lambda_3\lambda_4 \end{aligned}\right\}. \tag{5.49}$$

The remaining limiting expectations are zero. The relative values of $\bar{n}_2(\infty)$ and $\bar{n}_4(\infty)$ thus depend in a simple way on the transition rates. The form of this dependence can be most clearly seen by writing the ratio as follows:

$$\frac{n_2(\infty)}{n_4(\infty)} = \frac{r_{12}}{r_{13}}\left(1 + \frac{r_{31}}{r_{34}}\right) \tag{5.50}$$

in which r_{12}/r_{13} is the ratio of the intensities out of the state 'diagnosed as suffering from cancer' and r_{31}/r_{34} is the ratio of intensities out of the state 'recovery'. A high recovery rate r_{13} tends to increase the proportion of patients who die of 'other causes', but this effect will be counteracted to some extent if there is also a high rate of relapse, r_{31}.

We have already pointed out that the model was originally developed to provide a basis for measuring the effect of a treatment. One such measure is provided by $\bar{n}_2^0(T)$, the net proportion who would die of cancer if death from other causes was eliminated. Fix and Neyman argue that $\bar{n}_2^0(T)$ is not the only nor necessarily the most appropriate measure of survival. A discussion of this point would be outside the scope of the book but we mention it in order to observe that the quantities $\bar{n}_2(T)$ or $\bar{n}_2^0(T)$ are likely to be useful in constructing alternative measures. For example, Fix and Neyman suggest the use of the expected normal life in a period $(0, T)$ if cancer were the only risk of death. Since $\bar{n}_2^0(T)$ is the distribution function of 'normal' life in the absence of other risks the expectation may be written

$$e_2 = T\{1 - \bar{n}_2^0(T)\} + \int_0^T x\,\frac{\mathrm{d}\bar{n}_2^0(x)}{\mathrm{d}x}\,\mathrm{d}x$$

or

$$e_2 = \int_0^T \{1 - \bar{n}_2^0(x)\}\,\mathrm{d}x. \tag{5.51}$$

5.4 A MANPOWER SYSTEM WITH GIVEN INPUT

The model applied to a hierarchical system

Let the expected number of entrants to the system in $(x, x + \delta x)$ be $R(x)\delta x$. As in Chapter 3, suppose that the proportion of recruits allocated to grade i is p_{0i} $(i = 1, 2, \ldots, k)$. The expected number of those entering the system at time x who will be in grade j at time $T\,(T > x)$ is then

$$\sum_{i=1}^{k} p_{0i}p_{ij}(T - x)R(x)\,\delta x.$$

The total expected number in the jth grade at time T is then obtained by integrating this expression with respect to x and adding the contribution from those who were in the system at $T = 0$. Thus we have

$$\bar{n}_j(T) = \sum_{i=1}^{k} \left\{ p_{0i} \int_0^T p_{ij}(T - x)R(x)\,\mathrm{d}x + n_i(0)p_{ij}(T) \right\} \qquad (j = 1, 2, \ldots, k + 1). \quad (5.52)$$

The expected number who will have left is denoted by $\bar{n}_{k+1}(T)$ and (5.52) includes this case also. If input occurs at discrete points in time the equation is easily modified by replacing the integral by a sum. For convenience we shall treat the case where input occurs continuously. Equation (5.52) is the continuous analogue of the equations numbered (3.1a) and (3.1b) in Chapter 3.

Continuous time models for hierarchical systems were first proposed by Seal (1945) and Vajda (1948). Their models were non-Markovian, but both authors discussed some special cases which coincide with those derived from our general theory. We consider a system which can be represented diagrammatically as shown in Figure 5.2. This system has one terminal state

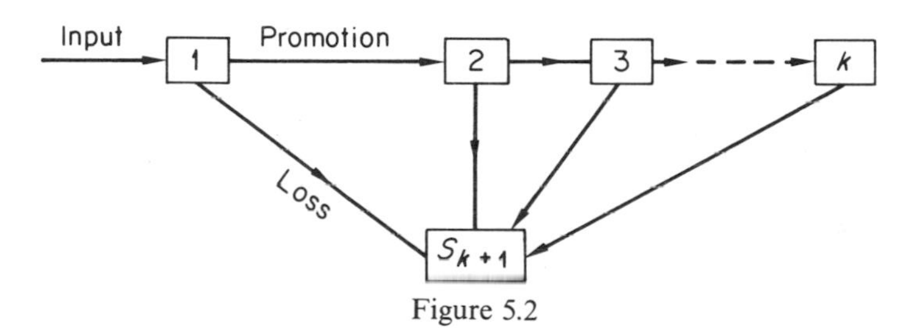

Figure 5.2

which we have labelled S_{k+1}. Promotion takes place only from one grade to the one above it and new entrants all go into grade 1. The matrix of transition intensities for the system we have described has the form

$$\mathbf{R} = \begin{pmatrix} r_{11} & r_{12} & 0 & \cdots & 0 & r_{1,k+1} \\ 0 & r_{22} & r_{23} & \cdots & 0 & r_{2,k+1} \\ 0 & 0 & r_{33} & \cdots & 0 & r_{3,k+1} \\ \vdots & \vdots & \vdots & \ddots & & \vdots \\ 0 & 0 & 0 & \cdots & \cdots & r_{k+1,k+1} \end{pmatrix} \qquad (5.53)$$

where

$$r_{ii} = -(r_{i,i+1} + r_{i,k+1}), \qquad i < k$$

$$r_{kk} = -r_{k,k+1}, \qquad r_{k+1,k+1} = 0.$$

The simple triangular structure of **R** enables us to obtain explicit formulae for the eigenvalues and the coefficients $\{c_{ij}\}$ which appear in the expressions for the transition probabilities $\{p_{ij}(T)\}$. We find at once that $\lambda_i = r_{ii}$ $(i = 1, 2, \ldots, k + 1)$. The equations which determine the c's obtained from (5.18) are

$$\left.\begin{aligned} r_{i-1,i}c_{i-1,h} + r_{ii}c_{ih} &= r_{hh}c_{ih} \qquad (i, h = 1, 2, \ldots, k + 1) \\ \text{and} \qquad \sum_{h=1}^{k+1} c_{ih} &= 1 \quad \text{if } i = 1 \\ &= 0 \quad \text{if } i > 1. \end{aligned}\right\} \tag{5.54}$$

The initial conditions represented by this last pair of equations follow from the fact that all new entrants begin their careers in grade 1. The set of equations (5.54) may be solved to give

$$\left.\begin{aligned} c_{ih} &= \prod_{j=1}^{i-1} r_{j,j+1} \Bigg/ \prod_{\substack{j=1 \\ j \neq h}}^{i} (r_{hh} - r_{jj}) \qquad (i = 2, 3, \ldots, k+1; h = 1, 2, \ldots, i) \\ c_{11} &= 1 \\ c_{ih} &= 0 \text{ otherwise.} \end{aligned}\right\} \tag{5.55}$$

We shall be interested only in $\bar{n}_j(T)$ for $j \leq k$, in which case

$$\bar{n}_j(T) = \int_0^T R(x)p_{1j}(T - x)\,\mathrm{d}x + \sum_{i=0}^{j} n_i(0)p_{ij}(T) \qquad (j = 1, 2, \ldots, k). \tag{5.56}$$

The coefficients obtained from (5.54) give

$$p_{1j}(T) = \sum_{h=1}^{k} c_{jh}\,\mathrm{e}^{r_{hh}T} \tag{5.57}$$

and this may be substituted in (5.56). A similar expression can be found for $p_{ij}(T)$, $i > 1$ and all j, using the appropriate initial conditions but they can easily be deduced from those for $p_{1j}(T)$ when we have a simple hierarchy. An entrant who starts his career in grade i of a k-state system is in the same position as one who starts in grade 1 in a $(k - (i - 1))$-stage system. By replacing k by $k - i + 1$ and relabelling the transition intensities the required expressions are obtained. We give an example below. It is obvious that $p_{ij}(T) = 0$ if $i > j$, which is why the upper limit of summation in the last term of (5.56) is j.

The model which we have described is slightly more general than the Markov version of Vajda's (1948) model. He assumed a constant rate of input to the system and a constant loss rate; his results may thus be obtained from ours by putting $R(x) = R$ and $r_{i,k+1} = r_{k+1}$, say, for $i \leq k$. We have also given the expected grade sizes for all T, whereas Vajda discussed only the limiting case.

As we have pointed out on several occasions, the theory which we have used above requires that the quantities r_{ii} $(i = 1, 2, \ldots, k + 1)$ are all distinct. In the case we are discussing $r_{ii} = -(r_{i,i+1} + r_{i,k+1})$ for $i \leq k$, so that equalities between the r_{ii} would occur if the total loss rates for certain grades were equal. A case of particular interest in which this happens occurs if $r_{i,i+1} = r$ and if $r_{i,k+1} = r_{k+1}$ for $i < k$. This corresponds to a situation in which promotion rates and loss rates are the same for all grades except the last. The appropriate modifications to the theory may be obtained by allowing the eigenvalues r_{ii} $(i < k)$ to approach one another in (5.57). The resulting expression for $p_{1j}(T)$ is then

$$p_{1j}(T) = \frac{r^{j-1}}{(j-1)!} T^{j-1} \mathrm{e}^{-(r+r_{k+1})T} \qquad (j = 1, 2, \ldots, k-1). \tag{5.58}$$

If $j = k$ the expression is slightly more complicated but we shall see below that it is not needed. For $p_{ij}(T)$ we use the device mentioned above and find that

$$p_{ij}(T) = \frac{r^{j-i}}{(j-i)!} T^{j-i} \mathrm{e}^{-(r+r_{k+1})T} \qquad (i \leq j < k). \tag{5.59}$$

Using (5.58) and (5.59) we may determine $\bar{n}_j(T)$ for $j < k$ by substitution in (5.56); $\bar{n}_k(T)$ is then obtained from

$$\bar{n}_k(T) = \sum_{j=1}^{k} \bar{n}_j(T) - \sum_{j=1}^{k-1} \bar{n}_j(T). \tag{5.60}$$

The total size at time T is obtained by summing both sides of (5.56) over j and using the fact that

$$\sum_{j=1}^{k} p_{ij}(x) = 1 - p_{i,k+1}(x) = 1 - \mathrm{e}^{-r_{k+1}x}.$$

This then gives

$$\sum_{j=1}^{k} \bar{n}_j(T) = \int_0^T R(x)\, \mathrm{e}^{-r_{k+1}(T-x)}\, \mathrm{d}x + N(0)\, \mathrm{e}^{-r_{k+1}T} \tag{5.61}$$

where

$$N(0) = \sum_{j=1}^{k} n_i(0).$$

We illustrate the theory by taking $k = 3$,

$$r_{i,i+1} = 1 \qquad (i = 1, 2), \qquad r_{i,4} = 2 \quad (i = 1, 2, 3)$$

for which

$$p_{ij}(T) = \frac{1}{(j-i)!} T^{j-i} \mathrm{e}^{-3T} \qquad (i = 1, 2; i \leq j < 3). \tag{5.62}$$

Let the input be an exponential function of time with $R(x) = 100\,\mathrm{e}^{\alpha x}$ and let $n_1(0) = 100$, $n_2(0) = 50$, $n_3(0) = 20$. We leave the parameter α unspecified in order to examine the effect which it has on the solution.

Equation (5.56) yields

$$\left.\begin{aligned} \bar{n}_1(T) &= \frac{100}{3+\alpha}\mathrm{e}^{\alpha T} + 100\,\mathrm{e}^{-3T}\left\{\frac{2+\alpha}{3+\alpha}\right\} \\ \bar{n}_2(T) &= \frac{100}{(3+\alpha)^2}\mathrm{e}^{\alpha T} + 100\,\mathrm{e}^{-3T}\left\{T + \frac{1}{2} - \frac{T}{3+\alpha} - \frac{1}{(3+\alpha)^2}\right\} \end{aligned}\right\}. \tag{5.63}$$

The expectation $\bar{n}_3(T)$ is now obtained by subtracting $\bar{n}_1(T) + \bar{n}_2(T)$ from

$$\sum_{j=1}^{3} \bar{n}_j(T) = \frac{100\,\mathrm{e}^{\alpha T}}{(2+\alpha)} + 100\,\mathrm{e}^{-2T}\left\{1{\cdot}7 - \frac{1}{2+\alpha}\right\}. \tag{5.64}$$

The crucial dependence of $\bar{n}_j(T)$ on α is evident from these formulae. If $\alpha > 0$ the input increases exponentially with time and so the expected sizes of the grades. If $\alpha < 0$ the expected grade sizes tend to zero with increasing time. The reader should compare these results with those obtained in Chapter 3 for the discrete time model with geometric input. In the case of constant input, obtained by putting $\alpha = 0$, the $\bar{n}_j(T)$'s approach finite limits as $T \to \infty$. These limits are

$$\bar{n}_1(\infty) = 33{\cdot}3, \qquad \bar{n}_2(\infty) = 11{\cdot}1, \qquad \bar{n}_3(\infty) = 5{\cdot}6,$$

the overall size being 50. This last result could have been found directly from the fact that the average time spent in the system is $r_{i4}^{-1} = \frac{1}{2} (i = 1, 2, 3)$ and that the input is 100 per unit time. The limiting size is, of course, independent of the initial size and the parameter values we have chosen bring about a reduction in overall size from 170 to 50. Inspection of (5.63) shows that the lowest grade reaches its limit more rapidly than the middle and highest grades. We observed the same phenomenon with the discrete time model.

It is interesting to observe that the *relative* sizes of the $\bar{n}_j(T)$'s for large T depend upon α in a simple fashion. If $\alpha > -2$ then

$$\left.\begin{aligned}\bar{n}_1(T) &\sim \frac{100}{3+\alpha}\,\mathrm{e}^{\alpha T} \qquad \bar{n}_2(T) \sim \frac{100}{(3+\alpha)^2}\,\mathrm{e}^{\alpha T} \\ \bar{n}_3(T) &\sim \frac{100}{(2+\alpha)(3+\alpha)^2}\,\mathrm{e}^{\alpha T}\end{aligned}\right\}. \tag{5.65}$$

The effect of increasing the rate of input is to increase the sizes of the lower two grades relative to the highest. Conversely, a decreasing rate of input leads to a concentration at the upper end of the hierarchy. Had we repeated the calculation with large values of k we should have found that, for large T, the grade sizes were in geometric progression except for the kth grade. In particular, this result holds for the case of constant input obtained by putting $\alpha = 0$. In this case we find

$$\left.\begin{aligned}\bar{n}_j(\infty) &= \frac{R}{r}\left(\frac{r}{r+r_{k+1}}\right)^j \qquad (j = 1, 2, \ldots, k-1) \\ \bar{n}_k(\infty) &= \frac{R}{r_{k+1}} - \sum_{j=1}^{k-1} \bar{n}_j(\infty) = \frac{R}{r_{k+1}}\left(\frac{r}{r+r_{k+1}}\right)^{k-1}\end{aligned}\right\} \tag{5.66}$$

where R is the rate of input to the system. It thus follows that *a hierarchical system with constant promotion rates throughout, a constant loss rate and a constant input rate will tend to a geometric structure*. The exception to this general rule is that the kth grade will be larger than the term in the geometric series corresponding to $j = k$. A system of the kind that we have discussed will thus have a tendency to become 'top-heavy'. If $R(x)$ has the exponential form considered earlier the same result holds for the relative expected grade sizes, as we have already noted in our example. Later in this chapter we shall consider how far this general conclusion remains true when the rather special assumptions of the present model are relaxed.

Stochastic variation of the grade sizes

Our analysis so far has been wholly concerned with the expected values of the grade sizes. It is also desirable to have some knowledge about the distribution of the grade sizes and, in particular, their variances and covariances. There is as yet no continuous version of Pollard's method which we used in the discrete time case but Staff and Vagholkar's method extends quite easily. It only yields a tractable result if the input is a Poisson process, but this enables us to establish a continuous analogue of the earlier result about the conditions under which the grade sizes have Poisson distributions.

We require the joint distribution of the grade sizes at time T, given that the input to the system is a Poisson process with rate R. As before, we shall find the joint distribution of these numbers excluding the survivors from the initial stock. This will then enable us to deduce the asymptotic distribution by observing that all of the initial stock will ultimately be lost to the system.

Suppose that up to time T individuals enter the system at times T_1, $T_2, \ldots, T_m$. Then we shall first find the generating function of the numbers in the grades conditional upon $T_1, T_2, \ldots, T_m$ and m. The unconditional generating function will then be found by averaging the conditional generating function with respect to the joint distribution of the T's and m. We shall do this in two stages using the fact that, given m and T, $T_1, T_2, \ldots, T_m$ may be regarded as a random sample from the rectangular distribution with density

$$f(x) = \frac{1}{T} \qquad (0 \leq x \leq T)$$

and that m has a Poisson distribution with mean RT.

Let $X_{hj} = 1$ if the individual recruited at T_h is in grade j at time T and be zero otherwise. We have, in terms of the transition probabilities, that

$$Pr\{X_{hj} = 1\} = \sum_{i=1}^{k} p_{0i} P_{ij}(T - T_h). \tag{5.67}$$

Hence the probability generating function of $X_{h1}, X_{h2}, \ldots, X_{h,k+1}$, given T_h and m, is

$$g(z_{h1}, z_{h2}, \ldots, z_{h,k+1} | T_h, m) = \sum_{j=1}^{k+1} z_{hj} \sum_{i=1}^{k} p_{0i} P_{ij}(T - T_h) \tag{5.68}$$

and the joint probability generating function of $\{X_{hj}: h = 1, 2, \ldots, m;$ $j = 1, 2, \ldots, k + 1\}$ is

$$g(z_{11}, z_{12}, \ldots, z_{r,k+1} | \mathbf{T}, m) = \prod_{m=1}^{n} \sum_{j=1}^{k+1} z_{rj} \sum_{i=1}^{k} p_{0i} P_{ij}(T - T_h). \tag{5.69}$$

We require the (conditional) distribution of the sums $\sum_{h=1}^{m} X_{hj}$, and their generating functions are obtained by putting $z_{hj} = z_j$, say, for $h = 1, 2, \ldots, m$, giving

$$g(z_1, z_2, \ldots, z_{k+1} | \mathbf{T}, m) = \prod_{h=1}^{m} \sum_{j=1}^{k+1} z_j \sum_{i=1}^{k} p_{0i} P_{ij}(T - T_h). \tag{5.70}$$

Averaging with respect to $\mathbf{T}$,

$$\begin{aligned} g(z_1, \ldots, z_{k+1}|m) &= \frac{1}{T^n}\int_0^T \cdots \int_0^T g(z_1, \ldots, z_{k+1}|\mathbf{T}, m)\,\mathrm{d}T_1 \ldots \mathrm{d}T_m \\ &= \prod_{h=1}^{m} \sum_{j=1}^{k+1} z_j \sum_{i=1}^{k} p_{0i}\frac{1}{T}\int_0^T p_{ij}(T - T_h)\,\mathrm{d}T_h \\ &= \left[\frac{1}{T}\sum_{j=1}^{k+1} z_i \sum_{i=1}^{k} p_{0i}\int_0^T p_{ij}(T - x)\,\mathrm{d}x\right]^m. \end{aligned} \tag{5.71}$$

Finally,

$$\begin{aligned} g(z_1, z_2, \ldots, z_{k+1}) &= \sum_{m=0}^{\infty} \frac{(RT)^m}{m!}\mathrm{e}^{-RT} g(z_1, \ldots, z_{k+1}|m) \\ &= \exp\left. -RT + R\left[\sum_{j=1}^{k+1} z_i \sum_{i=1}^{k} p_{0i}\int_0^T p_{ij}(T - x)\,\mathrm{d}x\right]\right\} \\ &= \exp\left\{R\sum_{j=1}^{k+1}\sum_{i=1}^{k} p_{0i}\int_0^T p_{ij}(T - x)\,\mathrm{d}x(z_j - 1)\right\}. \end{aligned} \tag{5.72}$$

This is the probability generating function of a set of $(k + 1)$ independent Poisson variates. Hence, excluding the survivors from the original stock, the numbers in the grades at time T will be independently distributed in the Poisson form with means

$$R\sum_{i=1}^{k} p_{0i}\int_0^T p_{ij}(T - x)\,\mathrm{d}x \qquad (j = 1, 2, \ldots, k + 1), \tag{5.73}$$

which checks with (5.52).

The result may be extended to the case when the input is a time-dependent Poisson process. In that case the input times $T_1, T_2, \ldots, T_r$ will be treated like a random sample from the distribution with density

$$f(x) = \frac{R(x)}{\int_0^T R(x)\,\mathrm{d}x} \qquad (0 \leq x \leq T)$$

and r will have a Poisson distribution with mean $\int_0^T R(x)\,\mathrm{d}x$.

We have therefore shown that, after a sufficiently long time, the grade sizes will have Poisson distributions. The Poisson approximation should be good as soon as the original stock is fairly small. In a hierarchical system this will usually occur more quickly in the lower grades. If the input is not a Poisson process the argument breaks down at the point where we treat the T's as independent observations from a known distribution and so turn the product $\prod_{h=1}^{r}$ into a power. Pollard's calculations for the discrete case suggest that this step may not be critical and that the Poisson distribution may remain a good approximation, but further research on this point is required.

5.5 A MANPOWER SYSTEM WITH GIVEN GROWTH RATE

General theory

The theory in this case may be derived from the basic equation (5.52). Instead of being given the input function $R(x)$ we now start from a function which specifies the rate of expansion. Let us suppose that the organization is to be increased in size by an amount $M(T)\,\delta T$ between time T and $T + \delta T$. (The increase in size cannot, of course, be a continuous variable but it is convenient to treat it as such. If the increases are realizations of a stochastic process $M(T)\,\delta T$ would be interpreted as the expected increase.) The input function is thus determined by the need to replace losses and to provide for the increase in overall size. Thus we have

$$R(T)\,\delta T = M(T)\,\delta T + \bar{n}_{k+1}(T + \delta T) - \bar{n}_{k+1}(T)$$

or, as $\delta T \to 0$,

$$R(T) = M(T) + \frac{d\bar{n}_{k+1}(T)}{dT}. \tag{5.74}$$

Using this relationship we may eliminate $R(T)$ from (5.52) and obtain a set of equations for the expected grade sizes which depend only on known functions as follows:

$$\bar{n}_j(T) = \sum_{i=1}^{k} p_{0i}\left\{\int_0^T p_{ij}(T-x)M(x)\,dx + \int_0^T p_{ij}(T-x)\frac{d\bar{n}_{k+1}(x)}{dx}\,dx\right\}$$
$$+ \sum_{i=1}^{k} n_i(0)p_{ij}(T) \qquad (j = 1, 2, \ldots, k+1). \tag{5.75}$$

By putting $j = k + 1$ in this equation we have an integral equation for $\bar{n}_{k+1}(T)$. Substitution of $\bar{n}_{k+1}(T)$ in (5.75) for $j = 1, 2, \ldots, k$ then gives expressions for the remaining grade sizes. This procedure is facilitated by the introduction of Laplace transforms. The Laplace transform of a continuous function $f(x)$, defined for $x \geq 0$, is as follows:

$$f^*(s) = \int_0^{\infty} f(x)\,e^{-sx}\,dx. \tag{5.76}$$

We shall also require the following well-known results about transforms.

(a) If $h(T) = \int_0^T f(t - x)g(x)\,dx$

then

$$h^*(s) = f^*(s)g^*(s).$$

(b) The transform of the derivative of $f(x)$ is $sf^*(s) - f(0)$.

Using these definitions and results we take the Laplace transform of each side of (5.75) and obtain

$$\begin{aligned}\bar{n}_j^*(s) &= \sum_{i=1}^{k} p_{0i}\{p_{ij}^*(s)M^*(s) + s\bar{n}_{k+1}^*(s)p_{ij}^*(s)\} + \sum_{i=1}^{k} n_i(0)p_{ij}^*(s)\\ &= \{M^*(s) + s\bar{n}_{k+1}^*(s)\}\sum_{i=1}^{k} p_{0i}p_{ij}^*(s) + \sum_{i=1}^{k} n_i(0)p_{ij}^*(s)\end{aligned}$$

$$(j = 1, 2, \ldots, k + 1). \qquad (5.77)$$

Setting $j = k + 1$ gives

$$\bar{n}_{k+1}^*(s) = \frac{\left\{M^*(s)\sum_{i=1}^{k} p_{0i}p_{i,k+1}^*(s) + \sum_{i=1}^{k} n_i(0)p_{i,k+1}^*(s)\right\}}{1 - s\sum_{i=1}^{k} p_{0i}p_{i,k+1}^*(s)}. \qquad (5.78)$$

This expression may now be substituted in (5.77) to give

$$\bar{n}_j^*(s) = \frac{\left\{s\sum_{i=1}^{k} n_i(0)p_{i,k+1}^*(s) + M^*(s)\right\}\sum_{i=1}^{k} p_{0i}p_{ij}^*(s)}{1 - s\sum_{i=1}^{k} p_{0i}p_{i,k+1}^*(s)} + \sum_{i=1}^{k} n_i(0)p_{ij}^*(s)$$

$$(j = 1, 2, \ldots, k). \qquad (5.79)$$

This formula appears to be rather formidable, especially since it has to be inverted before yielding what we require. However, some simplifications may be possible in particular applications. For example, if demotions are ruled out then $p_{ij}(T)$, and hence $p_{ij}^*(s)$, are zero for $j < i$. If the loss rate is independent of the grade, $p_{i,k+1}^*(s) = p_{k+1}^*(s)$, say, and the denominator simplifies to $1 - sp_{k+1}^*(s)$. Finally, if $p_{01} = 1$, $p_{0i} = 0$ $(i > 1)$, each sum in (5.79) is replaced by its first term.

Even without these simplifications the problem of inversion is not intractable. The form of $p_{ij}(T)$ makes its transform $p_{ij}^*(s)$ a rational algebraic fraction.

If $M^*(s)$ has this form also, $\bar{n}_j^*(s)$ can be inverted by general methods described, for example, in Feller (1968) (Vol. I, Chapter 11). As we shall see below it is possible to determine the limiting behaviour of the system from (5.79) without inverting the transform.

An illustration

Consider a system with $k = 3$, $p_{01} = 1$, $p_{02} = p_{03} = 0$, $r_{12} = r_{23} = r$ and $r_{14} = r_{24} = r_{34} = r_4$, say. Assume also that $n_1(0) = N$, $n_2(0) = n_3(0) = 0$. The Laplace transform of $\bar{n}_j(T)$ is then, from (5.79),

$$\begin{aligned}\bar{n}_j^*(s) &= \frac{\{Nsp_4^*(s) + M^*(s)\}p_{1j}^*(s)}{1 - sp_4^*(s)} + Np_{1j}^*(s) \\ &= \{N + M^*(s)\}p_{1j}^*(s)/(1 - sp_4^*(s)) \qquad (j = 1, 2, 3).\end{aligned}$$

The probabilities $\{p_{1j}(T)\}$ are given, under the above assumptions, by (5.58) for $j = 1$ and 2. Hence

$$p_{1j}^*(s) = \frac{r^{j-1}}{(r + r_4 + s)^j} \qquad (j = 1, 2). \tag{5.80}$$

Under the assumption of a constant loss rate

$$p_{1,4}(T) = 1/s - 1/(s + r_4). \tag{5.81}$$

At this point we must specify the function $M(T)$; we shall again assume exponential growth, setting $M(T) = M\,e^{\alpha T}$ $(M > 0)$. On this assumption $M^*(s) = M/(s - \alpha)$ if $s > \alpha$, otherwise the transform does not exist.

Taking first $j = 1$ we find

$$\bar{n}_1^*(s) = (s + r_4)(Ns - N\alpha + M)/s(s - \alpha)(r + r_4 + s). \tag{5.82}$$

Resolution into partial fractions gives

$$\bar{n}_1^*(s) = \frac{A_1}{s} + \frac{B_1}{s - \alpha} + \frac{C_1}{s + r + r_4}, \tag{5.83}$$

provided that $\alpha \neq 0$. The coefficients in the expansion are

$$A_1 = \left(\frac{N\alpha - M}{\alpha}\right)\left(\frac{r_4}{r + r_4}\right)$$

$$B_1 = \frac{M}{\alpha}\frac{(r_4 + \alpha)}{(r + r_4 + \alpha)}$$

$$C_1 = \frac{r\{N(r + r_4) + N\alpha - M\}}{(r + r_4)(r + r_4 + \alpha)}.$$

We may note that $A_1 + B_1 + C_1 = N$. If $\alpha = 0$ the partial fraction representation is

$$\bar{n}_1^*(s) = \frac{A_1'}{s} + \frac{B_1'}{s^2} + \frac{C_1'}{(s + r + r_4)} \tag{5.84}$$

where

$$A_1' = \frac{Mr}{(r + r_4)^2} + \frac{Nr_4}{(r + r_4)}$$

$$B_1' = \frac{Mr_4}{(r + r_4)}$$

$$C_1' = \frac{r\{N(r + r_4) - M\}}{(r + r_4)^2}.$$

Inverting the leading expressions in (5.83) and (5.84) we find

$$\left.\begin{aligned} \bar{n}_1(T) &= A_1 + B_1\,\mathrm{e}^{\alpha T} + C_1\,\mathrm{e}^{-(r+r_4)T} \quad (\alpha \neq 0) \\ &= A_1' + B_1'T + C_1\,\mathrm{e}^{-(r+r_4)T} \quad (\alpha = 0) \end{aligned}\right\}. \tag{5.85}$$

The last term in each part of the right-hand side of (5.85) tends to zero as T increases. For large values of T the behaviour of $\bar{n}_1(T)$ depends critically on the value of α. Three cases must be distinguished as follows:

$$\begin{aligned} \bar{n}_1(T) &\sim B_1\,\mathrm{e}^{\alpha T} \quad \text{if } \alpha > 0 \\ &\sim B_1'T \quad \text{if } \alpha = 0 \\ &\sim A_1 \quad \text{if } \alpha < 0. \end{aligned}$$

The determination of $\bar{n}_2(T)$ follows similar lines to that of $\bar{n}_1(T)$ and yields a solution of the form

$$\bar{n}_2(T) - A_2 + B_2\,\mathrm{e}^{\alpha T} + C_2\,\mathrm{c}^{-(r+r_4)T} + D_2T\,\mathrm{c}^{-(r+r_4)T} \quad (\alpha \neq 0). \tag{5.86}$$

If $\alpha = 0$ the second term in this equation is replaced by one which is linear in T and the coefficients have different values. Their determination is left to the reader. The approach to the limit is somewhat slower for $\bar{n}_2(T)$ than for $\bar{n}_1(T)$ because of the factor T which is present in the last term. The general form of $\bar{n}_3(T)$ is identical with (5.86) and a similar remark about the rate of approach to the limit applies. We take up the general question of limiting behaviour in this model in the following section.

Limiting behaviour of $\bar{\mathbf{n}}(T)$

The limit of $\bar{\mathbf{n}}(T)$ can be found from (5.79) without inverting the transform. This follows from the fact that

$$\lim_{T\to\infty} \bar{n}_j(T) = \lim_{s\to 0} s\bar{n}_j^*(s) \qquad (j = 1, 2, \ldots, k) \tag{5.87}$$

if $\bar{n}_j^*(s)$ exists for all $s > 0$. In order to evaluate the limit we need the following results about the behaviour of the Laplace transforms of the transition probabilities near the origin $s = 0$:

$$\left.\begin{aligned} &\text{(a)} \qquad p_{ij}^*(s) = \int_0^\infty p_{ij}(T)\,\mathrm{d}T + O(s) \qquad (j \leq k) \\ &\text{(b)} \qquad p_{i,k+1}^*(s) = s^{-1} + O(1) \end{aligned}\right\}. \tag{5.88}$$

These results are a consequence of the fact that the transition probabilities are finite series of exponential terms. It now follows from (5.87) that

$$\lim_{T\to\infty} \bar{n}_j(T) = C\{N + \lim_{s\to 0} M^*(s)\} \sum_{i=1}^{k} p_{0i} \int_0^\infty p_{ij}(T)\,\mathrm{d}T. \tag{5.89}$$

The value of C is most easily found using the result

$$\sum_{j=1}^{k} \bar{n}_j^*(s) = \frac{N}{s} + \frac{M^*(s)}{s}. \tag{5.90}$$

Thus summing both sides of (5.89) over j we find

$$C = \left\{ \sum_{i=1}^{k} p_{0i} \int_0^\infty \sum_{j=1}^{k} p_{ij}(T)\,\mathrm{d}T \right\}^{-1}.$$

The limit given by (5.89) will be finite if

$$\lim_{s\to 0} M^*(s) = \int_0^\infty M(T)\,\mathrm{d}T$$

is finite; that is if the total increase in size is finite. In practice this must always be so, but it is instructive, nevertheless, to consider infinite increases as we did in earlier in this section. However, whether or not the total increase is finite the relative expected grade sizes are given by

$$q_j(\infty) = \sum_{i=1}^{k} p_{0i} \int_0^\infty p_{ij}(T)\,\mathrm{d}T \Big/ \sum_{i=1}^{k} p_{0i} \int_0^\infty \sum_{j=1}^{k} p_{ij}(T)\,\mathrm{d}T$$

$$(j = 1, 2, \ldots, k) \tag{5.91}$$

where the sum

$$\sum_{j=1}^{k} p_{ij}(T) = 1 - p_{i,k+1}(T).$$

If the loss rate is constant and independent of the grade, with intensity r_{k+1}, we have

$$1 - p_{i,k+1}(T) = e^{-r_{k+1}T}$$

and hence

$$\int_0^\infty \{1 - p_{i,k+1}(T)\}\, dT = \frac{1}{r_{k+1}}.$$

The denominator of (5.91) may thus be replaced by r_{k+1}^{-1}.

The foregoing argument rests on the assumption that $M^*(s)$ exists for all $s > 0$. It therefore fails, for example, if $M(T) = M\, e^{\alpha T}$ with $\alpha > 0$ since the transform exists only for $s > \alpha$. We shall pursue this point further in the next section. For the present we state the important conclusion that *if $M^*(s)$ exists for all $s > 0$, then the relative expected grade sizes approach limits, independent of $M(T)$, given by* (5.91). The condition that $M^*(s)$ exist for $s > 0$ imposes a restriction on the rate of growth comparable to condition A of Chapter 3. This is not immediately obvious since the condition is stated in terms of the Laplace transform. We therefore proceed to express it in an alternative form which makes its nature clearer.

Reference to the standard works on the Laplace transform, for example Widder (1946) (Theorem 2.2a), shows that the existence of $M^*(s)$ for all $s > 0$ implies that

$$\lim_{T\to\infty} e^{-\beta T} \int_0^T M(x)\, dx = 0 \tag{5.92}$$

for all $\beta > 0$. The condition will certainly be satisfied if $M(T)$ is proportional to any power of T but not if it increases exponentially with time. The similarity between (5.92) and condition A of Chapter 3 becomes evident if we note that it is equivalent to

$$\lim_{T\to\infty} \left\{ -\beta T + \log \int_0^T M(x)\, dx \right\} = -\infty \tag{5.93}$$

for all $\beta > 0$, which holds if, and only if,

$$\lim_{T\to\infty} \frac{d}{dT}\left\{\log \int_0^T M(x)\, dx\right\} < \beta \quad \text{for all } \beta > 0. \tag{5.94}$$

The left-hand side of this inequality may be expressed as

$$\lim_{T\to\infty} M(T) \Big/ \int_0^T M(x)\,\mathrm{d}x$$

and hence for the condition to hold we must have

$$\lim_{T\to\infty} \frac{M(T)}{\displaystyle\int_0^T M(x)\,\mathrm{d}x} = 0; \tag{5.95}$$

this should be compared with condition A. We shall call (5.95) condition A'.

Illustration of limiting behaviour

We illustrate the foregoing theory by applying it to the example discussed above. The expression for $\bar{n}_1(T)$ is given in equation (5.85). If $\alpha < 0$ condition A' is satisfied and we find from (5.89), or directly from (5.85), that

$$\bar{n}_1(\infty) = A_1 = \left(\frac{M - N\alpha}{-\alpha}\right)\frac{r_4}{(r + r_4)}.$$

In a similar manner we obtain

$$\bar{n}_2(\infty) = \left(\frac{M - N\alpha}{-\alpha}\right)\frac{rr_4}{(r + r_4)^2} \tag{5.96}$$

and

$$\bar{n}_3(\infty) = \left(\frac{M - N\alpha}{-\alpha}\right)\frac{r^2}{(r + r_4)^2}.$$

The relative grade sizes can be obtained from these equations by omitting the factor $(M - N\alpha)/(-\alpha)$.

If $\alpha = 0$ the grade sizes increase without limit and we find, for large values of T, that

$$\left.\begin{aligned} \bar{n}_1(T) &\sim TMr_4/(r + r_4) \\ n_2(T) &\sim TMrr_4/(r + r_4)^2 \\ \bar{n}_3(T) &\sim TMr^2/(r + r_4)^2 \end{aligned}\right\}. \tag{5.97}$$

It should be noted that the relative expectations are the same for $\alpha = 0$ as when $\alpha < 0$; this is as it should be since condition A' is satisfied in both cases.

The case $\alpha > 0$ is particularly interesting because the rate of growth is then too rapid for condition A' to hold. Referring again to (5.85) and (5.86) we see that the term in $e^{\alpha T}$ will become dominant and hence that

$$\bar{n}_j(T) \sim B_j e^{\alpha T} \qquad (j = 1, 2, 3) \tag{5.98}$$

as $T \to \infty$. The relative expectations $\{q_i(\infty)\}$ are thus given by

$$\left.\begin{aligned} q_1(\infty) &= (r_4 + \alpha)/(r + r_4 + \alpha) \\ q_2(\infty) &= r(r_4 + \alpha)/(r + r_4 + \alpha)^2 \\ q_3(\infty) &= r^2/(r + r_4 + \alpha)^2 \end{aligned}\right\}. \tag{5.99}$$

These expressions should be compared with those derived from equations (5.96), (5.97) and (5.98); the latter are obtained from (5.99) by deleting α. *The effect of the rapid rate of growth is to increase the relative size of the lowest grade and decrease that of the highest.* In fact if α is very large the lowest grade becomes dominant, as we observed with the discrete time model. In the limit as $\alpha \to \infty$, $\mathbf{q}(\infty) \to \mathbf{p}_0$ which, in the present example, is $(1, 0, 0)$.

5.6 SYSTEMS WITH GIVEN INPUT AND LOSS RATE DEPENDING ON LENGTH OF SERVICE

Basic theory

Our initial formulation of the continuous time Markov model allowed the transition intensities to be functions of time. The basic equations for the transition probabilities and the expected grade sizes were given in (5.7) and (5.8). Since that point we have restricted attention to the special case where the transition intensities were constant. This assumption appears to be sufficient for the application to cancer survival. In the context of the hierarchical systems considered here the time parameter refers to the length of service or seniority of an individual member. In assuming that the transition intensities are constant we are supposing that promotion policies and decisions to leave an organization are independent of length of service. Both assumptions seem implausible but there is little firm evidence on the way in which promotion rates depend on seniority. On the other hand, there is a large body of data to show (see Chapter 6) that the leaving intensity is not constant but decreases as length of service increases. We therefore proceed to investigate the effect of relaxing the assumption of a constant loss rate.

General methods are available for the solution of the basic equations of the Markov process. It is easy to obtain an explicit solution in one case which is of particular interest to us. To do this we restrict attention to systems in which there are no demotions. The matrix $\mathbf{R}(T)$ is then triangular. The next step is to consider an organization of k grades; those who leave the system are considered, as before, to be in state S_{k+1}. Finally, we suppose that all transition intensities are constants except for those into S_{k+1}. The latter intensities are supposed to depend only on length of service and not on the grade of the individual. The matrix of transition intensities thus has the following form:

$$\mathbf{R}(T) = \begin{pmatrix} r_{11}(T) & r_{12} & \cdots & r_{1k} & r_{k+1}(T) \\ 0 & r_{22}(T) & \cdots & r_{2k} & r_{k+1}(T) \\ \vdots & \vdots & & \vdots & \vdots \\ 0 & 0 & \cdots & r_{kk}(T) & r_{k+1}(T) \\ 0 & 0 & \cdots & 0 & 0 \end{pmatrix}. \quad (5.100)$$

The differential equations for the transition probabilities $\{p_{ij}(T)\}$ are thus

$$\frac{\mathrm{d}p_{ij}(T)}{\mathrm{d}T} = \sum_{h=1}^{j-1} r_{hj}p_{ih}(T) + r_{jj}(T)p_{ij}(T) \qquad (i = 1, 2, \ldots, k; j = 1, 2, \ldots, k) \quad (5.101)$$

$$\begin{aligned} \frac{\mathrm{d}p_{i,k+1}(T)}{\mathrm{d}T} &= \sum_{h=1}^{k} r_{k+1}(T)p_{ih}(T) \\ &= r_{k+1}(T)\{1 - p_{i,k+1}(T)\} \qquad (i = 1, 2, \ldots, k). \end{aligned} \quad (5.102)$$

The set of equations given by (5.102) can be solved immediately, using the initial condition $p_{i,k+1}(0) = 0$, to give

$$p_{i,k+1}(T) = 1 - \exp\left\{-\int_0^T r_{k+1}(x)\,\mathrm{d}x\right\} \qquad (i = 1, 2, \ldots, k). \quad (5.103)$$

This probability is, of course, the distribution function of completed length of service and it could have been found directly. In fact, the appropriate choice of $r_{k+1}(T)$ for a given problem would usually be made on the basis of the observed distribution $p_{i,k+1}(T)$.

The system of equations (5.101) cannot be solved in the manner of Section 5.2 because some of its coefficients are not constants. It can, however, be transformed into a set which has constant coefficients as follows. Let

$$\left.\begin{aligned} r'_{ij} &= r_{ij} \qquad (i \neq j) \\ r'_{jj} &= r_{jj}(T) + r_{k+1}(T) \qquad (j = 1, 2, \ldots, k) \end{aligned}\right\}. \quad (5.104)$$

Note that r'_{jj} is a constant for all j in consequence of the definition of $r_{jj}(T)$. We introduce a new set of probabilities, denoted by $\{p'_{ij}(T)\}$, as follows:

$$p_{ij}(T) = p'_{ij}(T)\{1 - p_{i,k+1}(T)\} \qquad (i, j = 1, 2, \ldots, k). \tag{5.105}$$

The probability $p'_{ij}(T)$ defined in this equation is the conditional probability of the transition from grade i to grade j in $(0, T)$, given that no loss occurs in the same interval. Substituting from (5.105) into (5.101) we obtain

$$\frac{\mathrm{d}p'_{ij}(T)}{\mathrm{d}T} = \sum_{h=1}^{j} r'_{hj} p'_{ih}(T) \qquad (i, j = 1, 2, \ldots, k). \tag{5.106}$$

This system of equations can be solved by the methods of Section 5.2. By first finding $p'_{ij}(T)$ from (5.106) we are able to determine $p_{ij}(T)$ from (5.105).

If the probabilities $\{p_{ij}(T)\}$ have been computed for the system on the assumption of a constant loss rate it is a simple matter to obtain the probabilities for a time-dependent loss rate. To illustrate the procedure let $i = 1$. Then we may write

$$p'_{1j}(T) = \sum_{h=1}^{k} c'_{jh}\, \mathrm{e}^{r'_{hh}T} \qquad (j = 1, 2, \ldots, k). \tag{5.107}$$

Reference to (5.55) shows that the coefficients $\{c'_{jh}\}$ are the same as those for the system with constant loss rate because $r_{jj} - r_{hh} = r'_{jj} - r'_{hh}$ and $r_{j,j+1} = r'_{j,j+1}$. Using (5.105) we then have

$$p_{1j}(T) = \sum_{h=1}^{k} c_{jh} \exp\left\{r'_{hh}T - \int_0^T r_{k+1}(x)\,\mathrm{d}x\right\} \tag{5.108}$$

where the coefficients $\{c_{jh}\}$ are those previously computed on the basis of a constant loss rate. The only alteration required by our generalization is in the form of the exponent where $r_{hh}T$ is replaced by

$$r'_{hh}T - \int_0^T r_{k+1}(x)\,\mathrm{d}x.$$

Illustration of the theory

We are now in a position to assess the likely consequences of erroneously assuming the loss rate to be constant. We shall illustrate the procedure using the example first introduced in Section 5.4. In that case we took $k = 3$, $r_{12} = r_{23} = r = 1$ and $r_{14} = r_{24} = r_{34} = r_4 = 2$. The discussion will be confined to the case of a constant rate of input with $R = 100$ and will be concerned solely with the limiting grade structure $\bar{\mathbf{n}}(\infty)$. We first note that the expected total size of the system depends only on the mean of the distribution of completed length of service. We shall therefore choose $r_4(T)$ to be such that the mean length of stay in the system is $\frac{1}{2}$ as before. In the case of a constant loss rate the length of stay has an exponential distribution. In our examples below we shall consider two extreme departures from this form.

Consider first a mixed exponential distribution with density function

$$f(T) = \tfrac{1}{2}\{\lambda_1 \mathrm{e}^{-\lambda_1 T} + \lambda_2 \mathrm{e}^{-\lambda_2 T}\}. \tag{5.109}$$

In order that this shall give the same mean length of service as the earlier example we must choose λ_1 and λ_2 to satisfy

$$\lambda_1^{-1} + \lambda_2^{-1} = 1. \tag{5.110}$$

To obtain the transition probabilities from those already found for a constant loss rate we require the result that, for the mixed exponential distribution,

$$\exp\left\{-\int_0^T r_{k+1}(x)\,\mathrm{d}x\right\} = \tfrac{1}{2}\{\mathrm{e}^{-\lambda_1 T} + \mathrm{e}^{-\lambda_2 T}\}.$$

The case $\lambda_1 = \lambda_2 = 2$ yields the formulae for the exponential distribution of length of stay. Making the necessary substitutions in (5.108) we find

$$\left.\begin{aligned} \bar{n}_1(\infty) &= 50\left\{\frac{1}{\lambda_1 + 1} + \frac{\lambda_1 - 1}{2\lambda_1 - 1}\right\} \\ \bar{n}_2(\infty) &= 50\left\{\frac{1}{(\lambda_1 + 1)^2} + \frac{(\lambda_1 - 1)^2}{(2\lambda_1 - 1)^2}\right\} \\ \bar{n}_3(\infty) &= 50 - \bar{n}_1(\infty) - \bar{n}_2(\infty) \end{aligned}\right\} \tag{5.111}$$

where λ_2 has been eliminated using (5.110) and $\lambda_1 \geq 1$. As λ_1 varies between one and infinity, $\bar{\mathbf{n}}(\infty)$ takes on all its possible values. It may readily be shown that the expected grade sizes given by (5.111) have extreme values at $\lambda_1 = 2$, which is the exponential case, at $\lambda_1 = 1$ and at $\lambda_1 = \infty$. The last two cases are equivalent because $f(T)$ is symmetrical in λ_1 and λ_2. The extreme structures attainable are given in the first two columns of Table 5.2. We shall discuss the results below.

At the opposite extreme to that considered in the last paragraph we may suppose that the length of stay is a constant. That is we take

$$\left.\begin{aligned} r_4(T) &= 0 \qquad (T \leq r_4^{-1}) \\ &= \infty \qquad (T > r_4^{-1}) \end{aligned}\right\}. \tag{5.112}$$

***Table 5.2** Grade structures for the example under various extreme assumptions about the loss intensity*

	$\lambda_1 \to 1$ or ∞	$\lambda_1 = 2$	*Fixed length of stay*
$\bar{n}_1(\infty)$	25·0	33·3	39·4
$\bar{n}_2(\infty)$	12·5	11·1	9·0
$\bar{n}_3(\infty)$	12·5	5·6	1·6
$N(\infty)$	50·0	50·0	50·0

In this case, with $r_4 = 2$, the limiting grade sizes will be given by

$$\bar{n}_j(\infty) = 100 \int_0^{\frac{1}{2}} p_{1j}^0(x)\,\mathrm{d}x \quad (j = 1, 2, 3) \tag{5.113}$$

where $p_{1j}^0(x)$ is obtained from (5.58) by setting $r_{k+1} = 0$. The range of integration is restricted because

$$\exp\left\{-\int_{\frac{1}{2}}^{T} r_4(x)\,\mathrm{d}x\right\} = 0 \quad \text{for } T > \tfrac{1}{2}.$$

Substituting the numerical values for r we obtain

$$\left.\begin{aligned} \bar{n}_1(\infty) &= 100(1 - \mathrm{e}^{-\frac{1}{2}}) = 39{\cdot}4 \\ \bar{n}_2(\infty) &= 100(1 - \tfrac{3}{2}\mathrm{e}^{-\frac{1}{2}}) = 9{\cdot}0 \\ \bar{n}_3(\infty) &= 50 - \bar{n}_1(\infty) - \bar{n}_2(\infty) = 1{\cdot}6 \end{aligned}\right\}. \tag{5.114}$$

The foregoing results are brought together in Table 5.2.

The figures in Table 5.2 represent, in a certain sense, the greatest variation in structure that can occur. They relate only to one particular example but calculations for other cases suggest that the general pattern revealed here is typical. It can be seen from the table that the greater the variability of the length of service distribution the greater the size of the highest grade. In absolute terms, the change in the highest grade is roughly balanced by that in the lowest, but, in relative terms, the highest grade depends most critically on the assumption of constant loss rate. In practice, length of service distributions have been found to be highly skew and they have been successfully graduated by mixed exponential distributions. With constant promotion rates we would therefore expect the limiting structure to be more like that in the first column of the table with a relatively large number of members in the highest grade. *Any factor which tends to increase the variability of length of service with the mean held constant is therefore likely to increase the size of the higher grades at the expense of the lower.*

In the preceding discussion we have considered a constant rate of input. We would have reached similar conclusions had we dealt with expanding organizations with a rate of growth satisfying condition A', because the limiting structures are the same in both cases. A fuller generalization allowing the promotion rates to depend on length of stay would be of great interest.

5.7 HIERARCHICAL SYSTEMS WITH GIVEN INPUT AND PROMOTION RATES DEPENDING ON SENIORITY

A direct approach

Although empirical evidence about how promotion rates depend on seniority is less abundant than that relating to leaving, it is common for firms to take seniority into account when making promotions. In this section, therefore, we shall try to throw some light on the effect of seniority-dependent promotion rates on our earlier conclusions.

In the discussion of discrete time models in Chapter 2 we encountered a similar problem in trying to accommodate the principle of cumulative inertia. In that case we expanded the state space so that seniority within a class became part of the specification of the class. This was possible because seniority was measured in the same units as the basic time scale. Ginsberg (1971) extended this idea by making use of the theory of semi-Markov processes to construct continuous time models. In this approach the transitions are governed by a probability transition matrix and the times at which transitions take place by probability distributions—one for each possible transition. Such a process can be specified in terms of transition rates expressed as functions of the time spent in a given state; this is the approach we shall adopt here to emphasize the continuity with what has gone before.

We consider initially a closed system of hierarchical form in which promotion is into the next highest grade and recruitment is into the lowest. The only transition probabilities that we require are then $p_{1j}(T)$ $(j = 1, 2, \ldots, k)$, where k is the number of grades. When these have been found we can go back to (5.52) or (5.75), as the case may be, and obtain the expected grade sizes for the open system in which we are interested. Let $r_{j,j+1}(\tau)$ $(j = 1, 2, \ldots, k - 1)$ denote the promotion intensity from grade j to grade $j + 1$ for a person who entered grade j a time τ ago. As before, $r_{j,k+1}(T)$ is the intensity of loss for an individual with present length of service T. It is now possible to set up an integral equation for $p_{1j}(T)$ as follows. Consider first the case $j = 1$; $p_{11}(T)$ is the probability that an entrant to grade 1 has neither been promoted or lost in time T. The total intensity acting on a member of grade 1 at time T is $r_{12}(T) + r_{1,k+1}(T)$ and hence

$$p_{11}(T) = \exp\left\{-\int_0^T (r_{12}(x) + r_{1,k+1}(x))\,\mathrm{d}x\right\}. \tag{5.115}$$

The case $j = 2$ is now approached by observing that $p_{12}(T)$ may be written

$$p_{12}(T) = \int_0^T p_{11}(x)r_{12}(x)p_{22}(T - x)\,\mathrm{d}x. \tag{5.116}$$

We obtain this expression by first computing the probability conditional upon x, the time of transfer to grade 2, and then integrating over x from 0 to T. The probability $p_{22}(T - x)$ is obtained in a manner similar to that used for $p_{11}(T)$. The total intensity at time T acting on a person who entered grade 2 at time $T - \tau$ is $r_{2,k+1}(T) + r_{23}(\tau)$ and hence

$$p_{22}(T - x) = \exp\left\{-\int_{T-\tau}^T \{r_{2,k+1}(y) + r_{23}(y - T + \tau)\}\,\mathrm{d}y\right\}. \tag{5.117}$$

In general we have

$$p_{1j}(T) = \int_0^T p_{11}(x)r_{12}(x)p_{2j}(T - x)\,\mathrm{d}x \qquad (j = 2, 3, \ldots, k). \tag{5.118}$$

The probabilities $\{p_{1j}(T)\}$ may be obtained recursively since $p_{2j}(T - x)$ can obviously be found from $p_{1,j-1}(T)$, as in the case $j = 2$ above.

Equations (5.115) and (5.118) provide, in principle, the means of obtaining a complete solution to the problem. In practice, it is not easy to obtain explicit solutions, especially for large values of j. Since we shall present an alternative method which is sufficiently general for most purposes a detailed discussion of these integral equations will not be necessary. The chief value of the direct approach is that it easily yields a general solution for $j = 1$ which, in turn, leads to $n_1(T)$. We shall now use it for that purpose.

Suppose that there is a maximum time that may be spent in the lowest grade. If promotion or loss has not occurred by that time then promotion follows automatically. Such an assumption requires that the promotion intensity $r_{12}(\tau)$ becomes infinite at $\tau = b$, where b is the maximum length of service in the grade. A simple, increasing function of length of service having this property was suggested by Vajda (1947) who took

$$r_{12}(\tau) = c/(b - \tau) \qquad (0 \leq \tau < b; c > 0). \tag{5.119}$$

We shall assume that the loss intensity has a similar form with

$$r_{1,k+1}(\tau) = u/(v - \tau) \qquad (0 \leq \tau < v; u > 0) \tag{5.120}$$

where v is the maximum time that can be spent in the organization. (In grade 1 total length of service and seniority within the grade are synonymous. Hence we may use either T or τ to denote it.) If (5.120) obtains, the length of completed service distribution associated with this loss rate has the density function

$$f(\tau) = \frac{u}{v}\left(1 - \frac{\tau}{v}\right)^{u-1} \qquad (0 \leq \tau < v), \tag{5.121}$$

and the mean length of service is

$$\mu = v/(u + 1). \tag{5.122}$$

Under the foregoing assumptions we find from (5.115) that

$$p_{11}(T) = \left(\frac{b - T}{b}\right)^c \left(\frac{v - T}{v}\right)^u \qquad (0 \leq T < \min(b, v)). \tag{5.123}$$

The expected size of the lowest grade may now be found by substituting from the last equation into (5.52). Assuming a constant rate of input we have, for the kind of hierarchy being considered in this section,

$$\left.\begin{aligned} \bar{n}_1(T) &= R\int_0^T \left(1 - \frac{x}{b}\right)^c \left(1 - \frac{x}{v}\right)^u \mathrm{d}x + n_1(0)\left(1 - \frac{T}{b}\right)^c\left(1 - \frac{T}{v}\right)^u \\ &\qquad\qquad (0 \leq T < \min(b, v)) \\ &= R\int_0^{\min(b,v)} \left(1 - \frac{x}{b}\right)^c \left(1 - \frac{x}{v}\right)^u \mathrm{d}x \qquad (\min(b, v) \leq T < \infty). \end{aligned}\right\} \tag{5.124}$$

It is obvious from this equation that the limiting value of $\bar{n}_1(T)$ is attained at $T = \min(b, v)$ and is given by the second of the two integrals.

We may now compare these results with those obtained when we assumed constant promotion and loss rates. To do this meaningfully we must arrange that the Markov system has the same average length of stay as the one considered above. This is achieved by taking

$$r_{1,k+1} = (u + 1)/v, \qquad r_{12} = (c + 1)/b. \tag{5.125}$$

If we restrict the comparison to the limiting case we have shown that, for the Markov system, $\bar{n}_1(\infty) = R/(r_{12} + r_{1,k+1})$. This has to be compared with the second expression in (5.124), which may now be written

$$\bar{n}_1(\infty) = \int_0^{\min(b,v)} \left(1 - \frac{x}{b}\right)^{r_{12}b-1} \left(1 - \frac{x}{v}\right)^{r_{1,k+1}v-1} \mathrm{d}x. \tag{5.126}$$

If both v and b tend to infinity

$$\bar{n}_1(\infty) \to R \int_0^{\infty} \mathrm{e}^{-x(r_{12}+r_{1,k+1})}\,\mathrm{d}x = R/(r_{12} + r_{1,k+1})$$

which agrees with the known result for constant transition rates. The greatest divergence between the two assumptions will occur when b and v are both small. The extreme case occurs when

$$b = r_{12}^{-1} \quad \text{and} \quad v = r_{1,k+1}^{-1}$$

(smaller values of b or v would make u or c negative). In this case

$$\bar{n}_1(\infty) = R \min(r_{12}^{-1}, r_{1,k+1}^{-1}). \tag{5.127}$$

It thus follows that, for any system with transition rates given by (5.119) and (5.120),

$$R/(r_{12} + r_{1,k+1}) \leq \bar{n}_1(\infty) \leq R \min(r_{12}^{-1}, r_{1,k+1}^{-1}). \tag{5.128}$$

The effect of making promotion chances increase with length of service is thus to increase the expected size of the lowest grade. Since the total expected size does not depend on the form of the promotion intensity we further conclude that the *relative* expected size of the lowest grade will also be increased.

As b and v increase, the approach to the lower bound of the inequalities (5.128) is quite rapid, as may be seen from the fact that when $b = v$

$$\bar{n}_1(\infty) = R/(r_{12} + r_{1,k+1} - v^{-1}). \tag{5.129}$$

We may therefore conclude that the assumption of a constant promotion intensity may not be crucial, at least for the lowest grade. This investigation could be pursued for the second lowest grade but we shall use an alternative method.

A method involving hypothetical grades

This method has been widely adopted in the theory of queues for dealing with non-Markovian systems. In essence it involves the replacement of the actual system by a Markov system of greater complexity. It depends on the following argument. Suppose that we have two grades in series with constant promotion rates as shown in Figure 5.3 and that there are no losses. The

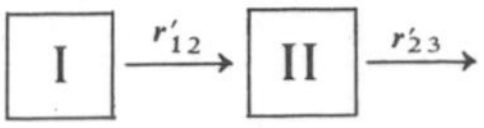

I $\xrightarrow{r'_{12}}$ II $\xrightarrow{r'_{23}}$

Figure 5.3

lengths of stay in stages I and II will now be exponential with means $1/r'_{12}$ and $1/r'_{23}$ respectively. Suppose now that the transitions between I and II are not observable and that the members of the two grades cannot be distinguished. The distribution of length of stay for the combined grade is thus the sum of two independent exponential variates. There is little loss in generality if we assume that $r'_{12} = r'_{23} = r$, say, in which case the length of service in the combined grade has density function

$$f(T) = r^2 T \mathrm{e}^{-rT} \qquad (T \geq 0). \tag{5.130}$$

Promotions from the combined grade will now appear *as if* the promotion rate was

$$\begin{aligned} r_{12}(T) &= f(T) \Big/ \int_T^\infty f(x)\,\mathrm{d}x = r\left(\frac{rT}{rT+1}\right) \\ &= 2r_{12}\left(\frac{r_{12}T}{r_{12}T + \frac{1}{2}}\right) \end{aligned} \tag{5.131}$$

where r_{12} is the constant rate which would lead to the same mean length of stay. We have thus produced a Markov system which behaves, when viewed in a particular way, as a non-Markov system with $r_{12}(T)$ given by this equation.

The argument just given may be generalized. Instead of two stages we may consider g stages for which we find

$$f(T) = \frac{r^g}{(g-1)!} T^{g-1} \mathrm{e}^{-rT} \qquad (T \geq 0). \tag{5.132}$$

Under these circumstances $r_{12}(T)$ is always an increasing function of T with rate of increase depending on g, and it always tends to a limit as $T \to \infty$.

The method we adopt is to replace each grade of the actual system by an appropriate number, g, of what will be termed subgrades. The loss rates must be the same for each subgrade within a given grade. The expected sizes of the subgrades are determined by the standard theory for Markov processes. Those for the actual grades are then simply obtained by summation over the relevant subgrades. We illustrate the method on the kind of simple hierarchy discussed in Section 5.4 for the case of constant transition rates. Again we restrict attention to the limiting behaviour. Let us assume that the loss rate is the same for all grades and denoted by r_{k+1} and that there is a constant rate of input R. Although not necessary, it is convenient to assume that the same value of g is appropriate for each grade. Our system is to be replaced by one of gk grades with constant and equal promotion intensities. To facilitate comparison with the results given in (5.66) we denote these intensities by gr. Let $\bar{z}_j(\infty)$ denote the limiting expected size of the jth subgrade; it may be obtained from (5.66) by replacing r by rg and k by gk. Thus

$$\left.\begin{aligned} \bar{z}_j(\infty) &= \frac{R}{gr}\left(\frac{gr}{gr + r_{k+1}}\right)^j \qquad (j = 1, 2, \ldots, gk - 1) \\ \text{and} \\ \bar{z}_{gk}(\infty) &= \frac{R}{r_{k+1}}\left(\frac{gr}{gr + rk}\right)^{gk-1}. \end{aligned}\right\} \qquad (5.133)$$

The expected number in the jth grade of the original organization is then

$$\left.\begin{aligned} \bar{n}_j(\infty) &= \frac{R}{gr}\sum_{i=g(j-1)+1}^{gj}\left(\frac{gr}{gr + r_{k+1}}\right)^i \\ &= \frac{R}{r_{k+1}}\left(\frac{gr}{gr + r_{k+1}}\right)^{g(j-1)}\left\{1 - \left(\frac{gr}{gr + r_{k+1}}\right)^g\right\} \\ &\qquad (j = 1, 2, \ldots, k - 1). \end{aligned}\right\} \qquad (5.134)$$

If $g = 1$ this reduces to the expression given in (5.66) for constant promotion rates. By increasing g we increase the dependence of promotion on length of service in the grade. In the extreme case as $g \to \infty$ all promotions take place after a fixed length of service r^{-1}, when we find

$$\left.\begin{aligned} \bar{n}_j(\infty) &= \frac{R}{r_{k+1}}\exp\left\{-\frac{r_{k+1}}{r}(j-1)\right\}\left[1 - \exp\left\{-\frac{r_{k+1}}{r}\right\}\right] \\ &\qquad (j = 1, 2, \ldots, k - 1) \\ \text{and} \\ \bar{n}_k(\infty) &= R/r_{k+1} - \sum_{j=1}^{k-1}\bar{n}_j(\infty). \end{aligned}\right\} \qquad (5.135)$$

An interesting feature revealed by these results is that the relative expected grade sizes, except the last, form a geometric progression, whatever g. The constant factor of the progression varies between $(1 + r_{k+1}/r)$ for the constant promotion intensity and $\exp(-r_{k+1}/r)$ for promotion after a fixed length of service. Some numerical values are given in Table 5.3.

Table 5.3 Comparison of the constant factors appropriate for the two extreme promotion rules

r_{k+1}/r	2	1	$\frac{1}{2}$	$\frac{1}{5}$	0
$\left(1+\frac{r_{k+1}}{r}\right)^{-1}$	0·333	0·500	0·667	0·833	1·000
$\exp\left\{-\frac{r_{k+1}}{r}\right\}$	0·135	0·368	0·607	0·819	1·000

The importance of the assumption about the promotion rates thus depends upon the ratio r_{k+1}/r. If the ratio is small, meaning that promotion is much more likely than loss, the assumption is not critical. On the other hand, if r_{k+1}/r is large the kind of assumption we make will be much more important. This point is illustrated by Table 5.4 where we have compared grade structures for high and low values of the ratio.

Table 5.4 The relative grade structures $\mathbf{q}(\infty)$ *for promotion after a fixed or random length of stay when* $k = 4$

	Grade	1	2	3	4
$\frac{r_{k+1}}{r} = 2$	Random	0·667	0·222	0·074	0·037
	Fixed	0·865	0·117	0·016	0·002
$\frac{r_{k+1}}{r} = \frac{1}{2}$	Random	0·333	0·222	0·148	0·296
	Fixed	0·393	0·239	0·145	0·223

The effect of making the promotion intensities increasing functions of seniority is to increase the relative sizes of the lower groups at the expense of the higher. A similar result was obtained when we allowed the loss rate to depend on total length of service. Another similarity between the two cases is that the size of the higher grades depends more critically on the assumptions. In general, therefore, we may expect the Markov model to overestimate the sizes of the higher grades and underestimate the lower.

Although it seems more realistic to suppose that promotion rates are increasing functions of seniority the method of substages can be adapted for use when they are decreasing functions. One way of doing this is to set up Markov systems with grades in parallel. The idea can be demonstrated on the system shown in Figure 5.4 with two stages in parallel and constant

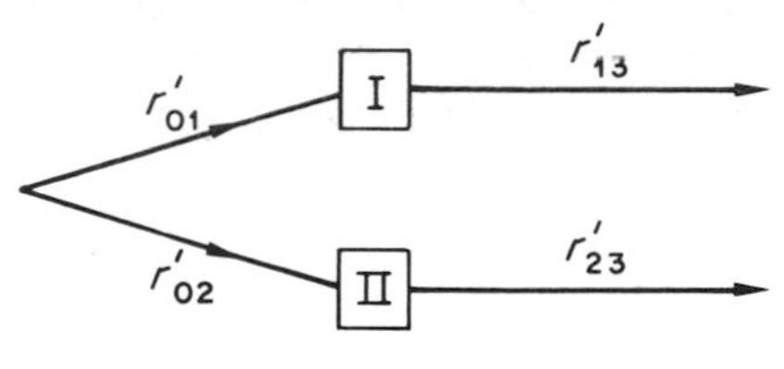

Figure 5.4

transition intensities. The length of stay of those members passing through I will be exponential with parameter r'_{13}. Those who pass through II will have an exponential length of stay distribution with parameter r'_{23}. If we are unable to distinguish the two stages the apparent density function of length of stay will be a mixture of the two exponential densities. More precisely, the density function will be

$$f(T) = \left(\frac{r'_{01}}{r'_{01} + r'_{02}}\right) r'_{13}\, \mathrm{e}^{-r'_{13}T} + \left(\frac{r'_{02}}{r'_{01} + r'_{02}}\right) r'_{23}\, \mathrm{e}^{-r'_{23}T} \qquad (T \geq 0). \quad (5.136)$$

It can easily be shown that the apparent intensity of loss for the combined grades,

$$f(T) \Big/ \int_T^\infty f(x)\,\mathrm{d}x,$$

is a strictly decreasing function of T for this distribution.

We now show how this result may be used to study a three-grade hierarchy with promotion intensities of the kind leading to the density function of (5.136). Consider the Markov system shown in Figure 5.5 with the possible routes of transfer indicated by arrows. Each pair of grades within the dotted rectangles is assumed to have the same loss rate. If the rectangles represent the grades of the actual system, this Markov system will have decreasing promotion intensities when its subgrades are combined. The transition

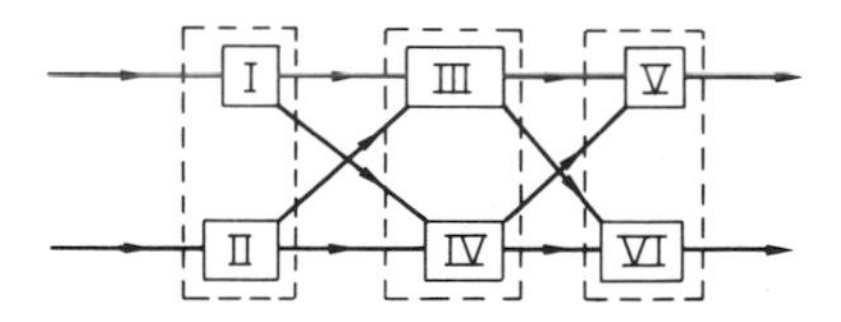

Figure 5.5

rates can be chosen to give the required average throughputs both for the individual grades and for the system as a whole. It should be noticed that the matrix of transition intensities arising from the system described above is triangular so that much of the simplicity associated with simple hierarchical systems is retained. The general effect of introducing this kind of dependence is to make the higher grades relatively larger than they would have been with constant promotion intensities.

CHAPTER 6

Models for Duration

6.1. INTRODUCTION

In many social situations the duration of some particular activity is a useful measure of behaviour. Thus, for example, the length of human life, the age at marriage or the spacing between successive births in a family are all of fundamental interest to the demographer. The length of time that a patient survives after treatment for illness, such as cancer, is often used as a measure of the efficacy of the treatment. The duration of a strike, the length of time a firm remains in business and the time that an individual spends in his job, at his address, in absence from work or in unemployment are all further examples of durations which are of social and economic interest. These diverse examples, and many others like them, have two things in common which make them appropriate subjects for stochastic analysis. First, they are all subject to a high degree of variability and, secondly, the pattern of that variability as revealed by the frequency distribution is often simple in form and relatively stable over time. As in all other applications considered so far, our object in constructing stochastic models for durations is to learn something about the underlying social process. The characteristic feature of the models discussed in this chapter is that their point of contact with the real world is in the frequency distribution of the duration. We shall aim to learn about the process by comparing frequency distributions predicted by our models with those observed in practice. Modelling must be carried out in relation to a particular application. We shall therefore concentrate on the duration of a person's job, since this has been the subject of much interesting work; it is also required as a basic ingredient for the renewal models of manpower systems introduced in Chapter 7.

The loss of individuals from organizations has been a feature of some of the models of earlier chapters, where it was described by a leaving rate or intensity. However, we were then interested in the properties of the system, such as its size and shape, and wastage was just one of several factors affecting the situation. Now we are interested in the loss, or wastage, process itself.

Some of the models to be described were not originally devised for the manpower application. Equally, others now used to describe length of service may well find application in other fields. What is important is the philosophy and methodology of the approach to model-building which we shall aim to exemplify in this particular application. Thus the reader who has no particular interest in the process of labour wastage should nevertheless find ideas for work in other fields.

Durations have been studied in other areas outside the social sciences. The best known and most developed is in reliability theory and industrial life-testing. There is a close parallel between much of the work that has been done in this field and the subject matter of this chapter. Where appropriate we shall draw freely on the terminology and methodology used in life-testing, but there are important differences between the social and industrial applications. In life-testing the exponential distribution has a central role because the failure risk of many types of equipment is approximately independent of age. Where the exponential distribution fails the Weibull or the gamma distributions are often used. None of these distributions plays a significant part in the applications we shall consider, although the exponential distribution is our point of departure in the next section. In other words, the kinds of model which have been found to satisfactorily describe the life of industrial components are rarely suitable for the durations of social processes. The second point of difference is fundamental to the statistical analysis of the data but rather marginal to our present interest. In social applications data must be collected by observing the process in its 'natural' state. To experiment is usually out of the question. In industrial applications the emphasis is on designing life-test experiments with all the advantages which this offers for efficient statistical analysis. The distinction is not a sharp one, but in social situations a great deal of statistical ingenuity must often be exercised to extract enough information from the fragmentary data to reconstruct the form of the life distribution. Models for the leaving process are very useful in devising appropriate statistical methods for this purpose. In spite of these important differences the basic objects of the exercise are much the same in all fields of application and great benefit is likely to be had from trying to establish links wherever possible.

We shall require the following terminology and notation. Let T denote the duration of the variable of interest; no distinction in notation will be made between the random variable and the values which it takes. Apart from brief exceptions in Sections 6.5 and 6.6, T will be continuous. The probability distribution of T is the main object of interest, and this can be expressed in three equivalent ways as follows:

(a) **The Survivor Function.** This is the probability that an individual survives for length of time T, and is denoted by $G(T)$. The survivor function is the complement of the distribution function, $F(T)$, but $G(T)$ is of more direct relevance in practice.

(b) **The Completed Length of Service (CLS) Density Function.** This is the density function associated with $F(T)$, related to the survivor function by

$$f(T) = -\frac{dG(T)}{dT}.$$

(c) **The Loss Intensity (or Rate).** This is denoted by $\lambda(T)$ and is defined as follows:

$$Pr\{\text{loss in } (T, T + \delta T) | \text{survival to } T\} = \lambda(T)\,\delta T.$$

This function is known by many different names appropriate to different applications. In reliability theory and life-testing it is known as the hazard function, age specific failure rate or, more briefly, as the failure rate. Actuaries call it the force of mortality and in manpower applications it is commonly referred to as the force of separation or the propensity to leave. It appeared in Chapter 5 as the loss intensity and we shall emphasize the continuity by preserving the same terminology here. The relationship between $\lambda(T)$, $G(T)$ and $f(T)$ is easily deduced as follows:

$$\begin{aligned} f(T)\delta T &= Pr\{\text{loss in } (T, T + \delta T)\} \\ &= Pr\{\text{survival to } T\} Pr\{\text{loss in } (T, T + \delta T) | \text{survival to } T\} \\ &= G(T)\lambda(T)\,\delta T. \end{aligned}$$

Hence

$$\lambda(T) = f(T)/G(T) = -\frac{d \log G(T)}{dT}. \tag{6.1}$$

Conversely,

$$G(T) = \exp\left[-\int_0^T \lambda(x)\,dx\right] \tag{6.2}$$

which is equivalent to (5.103) of Chapter 5. The function $\lambda(T)$ provides a concise description of the leaving process, enabling one to identify times in an individual's service when the risk of leaving is particularly high or low.

In practice it is sometimes easier to estimate one of these functions rather than another, but we can easily pass from one to the other using the relations set out above. Most of the subsequent model-building will focus on $f(T)$ or $G(T)$ but the interpretation of the model is often made most easily in terms of $\lambda(T)$. The data from which empirical estimates of these functions have to be made are rarely in the form of a simple random sample. More often they consist of data from a cohort censored at its upper end or of census data relating to the stocks and flows over a short interval of time. Methods of using data of these kinds to estimate survivor functions have been widely discussed; some convenient references are Lane and Andrew (1955), Chiang (1968), Forbes (1971b) and Barlow and coworkers (1972).

6.2 EXPONENTIAL AND MIXED EXPONENTIAL MODELS

Rice, Hill and Trist (1950) published several empirical CLS distributions obtained in their studies at the Glacier Metal Company. They observed that the distributions could be graduated by smooth J-shaped curves of hyperbolic form, but they did not put forward any theoretical model to account for this phenomenon. Silcock (1954) reviewed the literature on turnover up to that date and proposed two models to account for observed CLS distributions. His first model was based on a remark of Rice and coworkers (1950) to the effect that they had found a regularity in the turnover pattern which appeared to be characteristic of the firm and independent of economic and social forces operating outside the firm. Silcock interpreted this as implying a constant loss intensity and observed that this, in turn, would yield an exponential CLS distribution. It is doubtful whether Silcock's interpretation is justified since the loss intensity could certainly depend on factors internal to the firm, including the individual's length of service. Nevertheless, the exponential does provide a testable assumption about the leaving process and it paves the way for more adequate models. The exponential distribution was fitted by Silcock to several CLS distributions and, in every case, the fit was very poor. Two examples are given in Table 6.1, from which it can be seen that the observed distribution is always more highly skew than the fitted exponential. The censoring at 21 months makes it impossible to investigate the form of the upper tail, but it is clear from the data given that the exponential hypothesis is not tenable. Silcock's experience has been borne out in almost all subsequent investigations.

The second model proposed by Silcock (1954) is a generalization of the first. It retains the simple assumption of constant loss rate for individuals but this rate is now supposed to vary in the population from which employees are drawn. This is a very plausible hypothesis. Individuals differ in almost all other aspects of their behaviour and it would be surprising if propensity to leave their jobs was an exception. Denoting the constant loss intensity by λ any individual will have the CLS density

$$f(T) = \lambda \, e^{-\lambda T} \qquad (\lambda > 0; T \geq 0).$$

Suppose now that λ is a random variable with distribution function $H(\lambda)$.

Table 6.1 ***Observed and fitted CLS distributions for two firms***

	Glacier Metal Co. (1944–1947)				J. Bibby & Sons Ltd. (Males, 1950)			
Length of completed service	*Actual number of leavers*	*Exponential fit*	*Type* XI *fit*	*Mixed exponential fit*	*Actual number of leavers*	*Exponential fit*	*Type* XI *fit*	*Mixed exponential fit*
Under 3 months	242	160·2	242·0	242·0[a]	182	103·9	195·4	182·0[a]
3 months	152	138·9	150·3	152·0[a]	103	86·8	87·5	103·0[a]
6 months	104	120·4	103·8	101·4	60	72·4	51·8	60·7
9 months	73	104·5	76·5	72·7	29	60·5	35·0	38·0
12 months	52	90·6	59·2	55·8	31	50·5	25·6	25·5
15 months	47	78·5	47·4	45·7	23	42·1	19·7	18·6
18 months	49	68·1	38·8	39·2	10	35·2	15·8	14·7
21 months and over	487	444·8	488·0	497·2	191	177·6	198·2	186·5
Total	1206	1206·0	1206·0	1206·0	629	629·0	629·0	629·0

[a] These figures agree exactly with those observed because the distribution was fitted by equating percentage points. For further details see Bartholomew (1959).

The CLS distribution for a sample of employees will then have the density function

$$f(T) = \int_0^\infty \lambda^{-\lambda T}\, \mathrm{d}H(\lambda) \qquad (T \geq 0). \tag{6.3}$$

This distribution is always more skew than the exponential distribution with the same mean and it therefore has the main characteristic required by the data. A partial justification of this statement can be seen by considering the ratio

$$r = f(T)/\mu^{-1}\,\mathrm{e}^{-T/\mu} \tag{6.4}$$

where

$$\mu = \int_0^\infty Tf(T)\,\mathrm{d}T = \int_0^\infty \frac{1}{\lambda}\,\mathrm{d}H(\lambda) = E\left(\frac{1}{\lambda}\right).$$

For small T,

$$r \to E(\lambda)/E\left(\frac{1}{\lambda}\right) \geq 1,$$

since the arithmetic mean is always at least as great as the harmonic mean. Equality occurs only in the degenerate case when $H(\lambda)$ places all the probability at one point. For large T, write

$$r = \mu \int_0^\infty \lambda\,\mathrm{e}^{-(\lambda - 1/\mu)T}\,\mathrm{d}H(\lambda).$$

For $\lambda > \mu^{-1}$ the integrand tends to zero; for $\lambda < \mu^{-1}$ it tends to infinity. Therefore, provided that $H(\lambda)$ assigns some probability to values of λ in excess of μ^{-1}, $r \to \infty$. This is bound to be the case except when $H(\lambda)$ degenerates to a single point.

In order to fit (6.3) to data we must specify $H(\lambda)$. Silcock (1954) chose a gamma distribution with density function

$$\frac{\mathrm{d}H(\lambda)}{\mathrm{d}\lambda} = \frac{c^v}{\Gamma(v)}\lambda^{v-1}\,\mathrm{e}^{-c\lambda} \qquad (v > 0; c > 0; \lambda \geq 0)$$

and showed that

$$f(T) = \frac{v}{c}\left(1 + \frac{T}{c}\right)^{-(v+1)} \qquad (T \geq 0). \tag{6.5}$$

The corresponding expressions for the survivor function and the loss intensity are

$$G(T) = (1 + T/c)^{-v} \quad \text{and} \quad \lambda(T) = \frac{v}{c}(1 + T/c)^{-1}. \tag{6.6}$$

This is a J-shaped distribution of the Pearson family, in which it is classified as type XI. The loss intensity has a particularly simple form, indicating a monotonic decline in the propensity to leave with increasing length of service. It may, at first sight, appear paradoxical to have arrived at a decreasing $\lambda(T)$ having started with the assumption that the loss rate was constant for any individual. However, the group of individuals we are observing do not all have the same constant loss rate. The longer members of the group survive the more likely it is that they have low λ's and so we get the apparent decline in λ with time. When Silcock (1954) fitted this distribution to his data the agreement was much improved. Some of Silcock's calculations are given in Table 6.1 under the heading 'Type XI fit'.

The gamma distribution for λ used above is quite flexible, depending as it does on two parameters. Its main advantage in the present context is its mathematical tractability. However, the particular functional form which we adopt does not appear to be critical. An explicit expression for $f(T)$ can be obtained in some cases using an inverse gamma distribution, and an interesting curiosity is provided by the case

$$\frac{\mathrm{d}H(\lambda)}{\mathrm{d}\lambda} = \frac{c^{\rho}}{\Gamma(\rho)\Gamma(1-\rho)}\lambda^{-1}(\lambda - c)^{-\rho} \qquad (\lambda \geq c; 0 < \rho < 1), \tag{6.7}$$

when $f(T)$ itself turns out to be a J-shaped gamma distribution. For some purposes it is simpler to suppose that λ can take only two values, say λ_1 and λ_2. If the associated probabilities are p and $1 - p$ respectively the CLS distribution will be

$$f(T) = p\lambda_1\,\mathrm{e}^{-\lambda_1 T} + (1-p)\lambda_2\,\mathrm{e}^{-\lambda_2 T} \qquad (0 < p < 1, \lambda_1, \lambda_2 > 0;\, T \geq 0). \tag{6.8}$$

This is the two-term mixed exponential distribution used in Chapter 5. It was fitted by Bartholomew (1959) to the data used by Silcock and some of the results are given in Table 6.1. There is very little to choose between the mixed exponential distribution and the type XI in these examples, in spite of the radical difference in the form of $H(\lambda)$ in the two cases.

It cannot be concluded from the foregoing analysis that the 'mixture' models proposed by Silcock (1954) provide a true explanation of the leaving process. There are other models which give equally good agreement with the data and at least one of them gives the same functional form for $f(T)$. If the model does contain any degree of truth then there are obvious implications for the personnel manager. Since, according to the model, there are innate differences between individuals, the way to influence the leaving rate is by screening the intake. If we could identify people with small or big λ's wastage could be decreased or increased by appropriate selection. The model therefore suggests that we ought to seek ways of discriminating between entrants according to their propensity to leave. This is a problem for the psychologist but the success of his methods can certainly be put to a statistical test.

The two-term mixed exponential can clearly be extended to three and more terms, but the number of parameters increases rapidly and the effort is hardly worthwhile unless large amounts of data are available. In Section 6.4 and again in Chapter 8 we shall meet CLS distributions having the form

$$f(T) = \sum_{i=1}^{k} p_i \lambda_i e^{-\lambda_i T} \qquad \left(T \geq 0;\ \sum_{i=1}^{k} p_i = 1 \right). \tag{6.9}$$

If the p_i's are non-negative this belongs to the family (6.3), but this need not necessarily be the case. When these models do lead to a CLS distribution with positive p_i's there is no statistical means of discriminating between mixture and other models using data on length of service alone.

6.3 THE LOGNORMAL MODEL

Another type of distribution which has been successfully used to graduate CLS distributions is the lognormal. The discovery that good fits could be obtained using this distribution appears to have been made by Lane and Andrew (1955) and since then their findings have been confirmed many times. It has been found to apply at all levels of skill and in many countries, and has become a principal tool in manpower planning. One of the defects of the mixed exponential distributions is that their density functions are monotonic decreasing for all T. In practice it is often possible to observe a mode in the CLS distribution near to the origin and the mixed exponential is incapable of reproducing this characteristic. When the data is coarsely grouped this mode may be lost to sight in the first group and for many practical purposes it may be safe to ignore it. Nevertheless, it would be more satisfying to construct a model which would account for this common feature of CLS distributions. The lognormal distribution does predict such a mode and we shall shortly describe models which lead to this distribution.

The lognormal distribution also offers statistical advantages. It implies that $\log_e T$ is normally distributed with mean to be denoted by ω and variance σ^2. Hence many of the problems of inference can be brought within the province of 'normal' theory. It appears in practice that the parameter σ is fairly constant within a particular profession or occupational category (usually in the range (1, 2)). Other factors such as sex, age, place of residence, etc., make their influence felt through the value of ω. This parameter has two simple interpretations. Since ω is the mean of the logarithm of CLS, e^{ω} is the geometric mean of CLS. It is also the median or 'half-life' of the distribution.

Aitchison (1955) pointed out that the lognormal law was a consequence of Kapteyn's law of proportionate effect. This can be applied in the present context to deduce that the form of the distribution of completed length of service in successive jobs in a person's career will tend to lognormality. The argument goes as follows. Let T_j denote the length of time a person spends in his jth job. Then the law states that

$$T_{j+1} = T_j u_{j+1} \qquad (j = 1, 2, 3, \ldots) \tag{6.10}$$

where $\{u_2, u_3, \ldots\}$ is a sequence of random variables with known joint distribution. This may be regarded as an expression of the way that past experience influences future service. A person who has long service in one job will tend to stay a long time in the next. An immediate consequence of equation (6.10) is that

$$\left.\begin{aligned} T_{j+1} &= T_1 \sum_{i=2}^{j+1} u_i \qquad (j \geq 1) \\ &\text{or} \\ \log_e T_{j+1} &= \log_e T_1 + \sum_{i=2}^{j+1} \log_e u_i. \end{aligned}\right\} \tag{6.11}$$

Provided that the joint distribution of the u's is such that the central limit theorem applies to the sum of their logarithms, it follows that $\log_e T_{j+1}$ will tend to normality as j increases. Thus the model predicts that a new entrant to a job who has had several previous jobs will have a CLS distribution which is approximately lognormal.

If the u's have the same joint distribution for all members of the population from which the recruits come an important conclusion can be drawn. Suppose that we classify all recruits according to the number, n, of previous jobs they have had. Then, within these groups the model predicts that σ, the standard deviation of $\log_e T$, should be constant. On the other hand, the parameter ω, which is the expectation of $\log_e T$, depends on the characteristics of the individual through $\log_e T_1$. Thus if we compare groups of recruits with different personal attributes but the same number of previous jobs we would expect all groups to have a similar σ but with different ω's. Even if the groups are not homogeneous with respect to the number of previous jobs the conclusion may still apply, because if the factors u_i have mean near to one then $\log_e u_i$ will have mean near zero so that ω will depend only slightly on n. Provided that n does not vary too much, the shape of the overall distribution is not likely to be seriously distorted by the effect which variation in n will have on σ. This theoretical deduction from the model corresponds with the empirical observations we made above about the factors which appear to affect ω. Examples can be found in Lane and Andrew (1955) and Young (1971). The lognormal model is like Silcock's in that it incorporates a feature allowing for individual differences on the propensity to leave. In the lognormal case these are reflected in the quantities T_1 which will vary between individuals. The theory goes further, however, by supposing that these initial characteristics can be modified by subsequent job experience. Even if this explanation is only partly true it has an important practical consequence. Equation (6.11) shows the important role played by length of service in the first job on subsequent lengths of stay. This suggests *that if particular care is taken to ensure satisfaction in a person's first job then their subsequent rate of turnover will be reduced*. We have already noted that existing data are compatible with this hypothesis but they do not provide enough information to enable a satisfactory test to be made. Further support for the model could be sought by classifying leavers by their number of previous jobs. If the model described here is correct we should find that σ increased with j, the number of jobs held before the present one. If the u's were independent we should find that σ^2 was proportional to j.

The foregoing model explains, in part at least, why lengths of service will tend to be lognormally distributed as an individual's career progresses. It does not explain the occurrence of the lognormal distribution for people in their first job. This has been observed to occur in practice, for example, in jobs filled by new graduates. At best, therefore, the model can only provide a partial explanation. However, Marshall (1971a) showed that the model can be given a different interpretation. Suppose that instead of being the length of service in the first job T_1 is interpreted as the length of time a person *expects* to spend in a given job at the time of joining. During the early stages of his experience in the job he will be subject to the influence of many factors. Some of these will be favourable and incline him to stay longer than he originally expected. Others will be unfavourable and will tend to precipitate leaving. In this way the employee's original expectation will be modified by random perturbations. If we assume that the effect of each factor is to multiply the current expectation of service by a random factor, u, then T_n can be interpreted as the actual length of service for a person subject to n factors. The same stochastic assumption about the u's then ensures the approximate lognormality of the actual length of service.

According to this version of the model, individual differences will still be reflected in T_1, but this is now something which is no longer open to modification in the job since the individual brings it with him. Influence may be brought to bear by trying to change the mean values of the u's in such a way as to increase (or decrease, if necessary) the actual length of service.

Since both versions of the model recognize individual variability we might try to build this feature in more formally. In doing so we shall find additional support for the lognormal hypothesis. Let us introduce a mixed lognormal distribution along the lines of the mixed exponential—and for the same reasons there are now two parameters which may vary, but the foregoing discussion suggests that it would be reasonable to hold σ constant and allow ω to have a distribution. The mixed lognormal density then becomes

$$f(T) = \int_{-\infty}^{+\infty} \frac{1}{\sqrt{2\pi}\sigma T} \exp\left[-\frac{1}{2}\left(\frac{\log_e T - \omega}{\sigma}\right)^2\right] \mathrm{d}H(\omega). \qquad (6.12)$$

Note the range of ω over the whole real line. If we suppose that $\omega \sim N(\mu, \tau^2)$ then it is easy to show that

$$\log_e T \sim N(\mu, \sigma^2 + \tau^2). \qquad (6.13)$$

In other words, the distribution has the same form but will be more skew than the original because of the larger variance of $\log_e T$. Thus, provided that the mixing distribution is reasonably normal (and many human variables do vary in this way), the lognormal form will survive heterogeneity in the scale parameter.

Taken together, the arguments and models set out in this section provide reasons for expecting the lognormal distribution to arise and persist. They do not provide a fully satisfying explanation. For example, even if the first version of the model were correct it offers no explanation of why the relation between T_{j+1} and T_j should have the simple linear form postulated by (6.10). However, the widespread occurrence of the distribution in connexion with length of service suggests that there must be something akin to a central limit effect operating and this, inevitably, conceals the true stochastic nature of the constituent increments whose sum induces the normality.

6.4 MARKOV PROCESS MODELS

An interesting class of models for the leaving process can be developed using a Markov process to represent movements between internal states (of mind) through which an individual passes before leaving. Herbst (1963) pioneered this approach, using what he called a decision process model. Similar models, though with important differences of interpretation, have since been discussed by Hoem (1971) in continuous time and by Young (1971) and Feichtinger (1971) in discrete time. A very similar model was discussed in relation to survival after cancer in Chapter 5, but there the interest was not directly in the survival time. The distinctive thing about Herbst's model is that the intermediate states are hypothetical in the sense that they cannot be directly observed. Herbst gives reasons for believing that the states in his model do correspond to what happens in practice but their existence can at best only be inferred indirectly from the shape of the observed CLS distribution.

The model is most easily explained by reference to Figure 6.1. Each individual is supposed to enter the system in the undecided state and then to

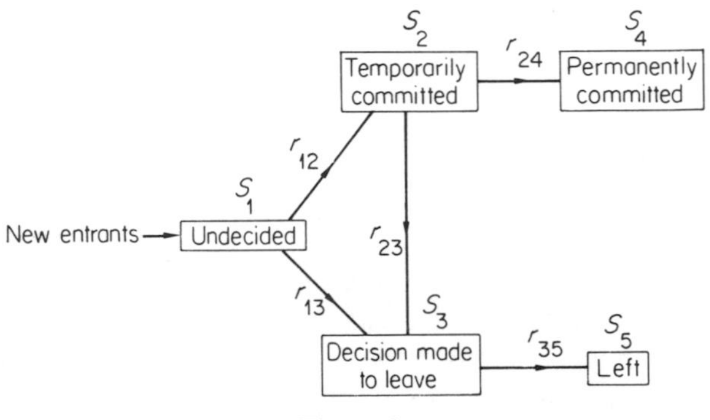

Figure 6.1

pursue a path through the network until he arrives at one or other of the terminal states S_4 and S_5. All that we can observe in practice is the time taken to reach S_5; all else about the system must be inferred from the form of this CLS distribution. Movements through the network are governed by a set of transition rates denoted by r's in the diagram. The system is thus being modelled by a continuous time Markov process of the kind discussed in Chapter 5 with transition matrix

$$\mathbf{R} = \begin{pmatrix} r_{11} & r_{12} & r_{13} & 0 & 0 \\ 0 & r_{22} & r_{23} & r_{24} & 0 \\ 0 & 0 & r_{33} & 0 & r_{35} \\ 0 & 0 & 0 & 0 & 0 \\ 0 & 0 & 0 & 0 & 0 \end{pmatrix} \tag{6.14}$$

where $r_{11} = -r_{12} - r_{13}$, $r_{22} = -r_{23} - r_{24}$, $r_{33} = -r_{35}$. Because the matrix is triangular its eigenvalues are equal to its diagonal elements. The $\bar{n}_i(T)$'s of (5.15) may be interpreted as probabilities of being in S_i at time T

if there is one person in S_1 to start with. The survivor function is then given by

$$1 - \bar{n}_5(T) = G(T) = d_{51} + c_{53}\,e^{r_{11}T} + c_{54}\,e^{r_{22}T} + c_{55}\,e^{r_{33}T}, \qquad (6.15)$$

adopting the notation of (5.23) in the last chapter. Solving (5.17) with the eigenvalues derived from **R** we find

$$\left.\begin{aligned}
d_{11} &= 0\\
c_{13} &= 1\\
c_{14} &= c_{15} = 0\\
d_{21} &= 0\\
c_{23} &= -c_{24} = r_{12}/(r_{11} - r_{22})\\
c_{25} &= 0\\
d_{31} &= 0\\
c_{33} &= \frac{r_{13}}{(r_{11} - r_{33})} + \frac{r_{12}r_{23}}{(r_{11} - r_{22})(r_{11} - r_{33})}\\
c_{34} &= \frac{-r_{12}r_{23}}{(r_{11} - r_{22})(r_{22} - r_{33})}\\
c_{35} &= \frac{-r_{13}}{r_{11} - r_{33}} + \frac{r_{12}r_{23}}{(r_{11} - r_{33})(r_{22} - r_{33})}\\
d_{41} &= \frac{r_{12}r_{24}}{r_{11}r_{22}}\\
c_{43} &= \frac{r_{12}r_{24}}{r_{11}(r_{11} - r_{22})}\\
c_{44} &= \frac{-r_{12}r_{24}}{r_{22}(r_{11} - r_{22})}\\
c_{45} &= 0\\
d_{51} &= 1 - \frac{r_{12}r_{24}}{r_{11}r_{22}}\\
c_{53} &= \frac{-r_{33}}{r_{11}}\,\frac{r_{13}(r_{11} - r_{22}) + r_{12}r_{23}}{(r_{11} - r_{22})(r_{11} - r_{33})}\\
c_{54} &= \frac{r_{33}r_{12}r_{23}}{r_{22}(r_{22} - r_{33})(r_{11} - r_{22})}\\
c_{55} &= \frac{r_{13}}{r_{11} - r_{33}} - \frac{r_{12}r_{23}}{(r_{11} - r_{33})(r_{22} - r_{33})}
\end{aligned}\right\}. \qquad (6.16)$$

Only the last four coefficients are needed for $G(T)$, but the others are included for the benefit of the reader who wishes to follow through the calculations.

Herbst was interested in discovering whether the model provides an adequate description of organizational commitment. He therefore compared (6.15) with length of service distributions from two firms collected by Hedberg (1961). Firm A had a high loss of entrants and firm B a low loss. There are seven parameters to be estimated and this would be overambitious unless the data were very extensive. Fortunately, Herbst had very large samples and the parameters were estimated by equating percentage points. The following estimates were obtained:

Firm A

$$G(T) = 0{\cdot}1493 + 0{\cdot}4544\,\mathrm{e}^{-0{\cdot}4720T} + 0{\cdot}4194\,\mathrm{e}^{-0{\cdot}0966T} - 0{\cdot}0231\,\mathrm{e}^{-10{\cdot}15T} \tag{6.17}$$

Firm B

$$G(T) = 0{\cdot}4700 + 0{\cdot}1274\,\mathrm{e}^{-0{\cdot}3680T} + 0{\cdot}4025\,\mathrm{e}^{-0{\cdot}1142T} \tag{6.18}$$

In the case of firm B, $-r_{33}$ was effectively infinite. The observed and fitted values of $G(T)$ are given in Table 6.2. The closeness of the fit with such large samples is remarkable and seems to provide strong evidence in favour of Herbst's theory. The distributions could have arisen as mixtures of exponential distributions, as described in Section 6.2. (It is true that there is a negative coefficient in firm A's survivor function but its contribution is negligible.) Notice that the first term is a constant; this can be thought of as proportional to the survivor function of a degenerate exponential distribution with infinite mean. It relates to the people who eventually reach the permanently committed category.

Further insight into the nature of the model can be gained by estimating the transition intensities. The procedure is as follows. First, we estimate the seven parameters of the survivor function, d_{51}, c_{53}, c_{54}, c_{55} and r_{11}, r_{22} and r_{33}, as in arriving at (6.17) and (6.18). Secondly, we use the relationship between the rates and the coefficients of (6.16) to estimate the intensities. There are seven coefficients, in this example, and only five intensities. Since r_{11}, r_{22} and r_{33} are essentially sums of pairs of intensities it is convenient to take these three equations together with any two of the last four equations in (6.16). A check on the goodness of fit can then be made by seeing whether the two remaining equations are satisfied by the estimates. Once again the agreement between the prediction and the actual values of the coefficients is extremely good. The estimated rates are given in Figure 6.2. In the high loss firm there is greater 'relative pressure towards leaving' at both the undecided

Table 6.2 Observed and estimated values of $100\{1 - \bar{n}_5(T)\}$ ***for Hedberg's data***

	Firm A 7628 *entrants*		*Firm* B 968 *entrants*	
Month	*Actual*	*Theoretical*	*Actual*	*Theoretical*
0	100·00	100·00	100·0	100·0
1	82·66	82·66	91·7	91·7
2	68·85	68·85	85·1	85·1
3	58·95	58·94	79·8	79·8
4	52·10	51·66	75·6	75·4
5	46·61	46·17	71·7	71·8
6	42·32	41·92	67·7	68·7
7	38·73	38·53	64·9	66·1
8	35·74	35·77	63·0	63·8
9	33·52	33·46	61·3	61·9
10	31·78	31·56	59·5	60·2
11	30·19	29·81	58·5	58·7
12	28·46	28·33	57·2	57·4
13	27·20	27·03	55·9	56·2
14	25·85	25·87	55·4	55·2
15	24·79	24·83	54·8	54·3
18	22·25	22·30	53·1	52·2
21	20·36	20·43	51·4	50·7
24	19·06	19·04	49·7	49·6
27	18·01	18·00	48·8	48·9
30	17·31	17·23	48·2	48·3
36	16·45	16·22	47·7	47·7
42	15·77	15·65	47·4	47·3
48	15·35	15·33	47·2	47·2
54	15·18	15·16	47·1	47·1
60	15·05	15·06	—	—
66	15·00	15·00	—	—

and temporarily committed stages. This tends to strengthen our belief in the consistency and appropriateness of the model.

The statistical problem of estimating the parameters efficiently clearly requires further investigation. Sverdrup (1965) and Hoem (1971) have considered the problem when data are available on individual transitions. In the case of the present example the fit is so good that more efficient procedures would be superfluous.

The method used above to derive the CLS distribution is perfectly general and works with an **R** matrix of any size or structure. With relatively simple

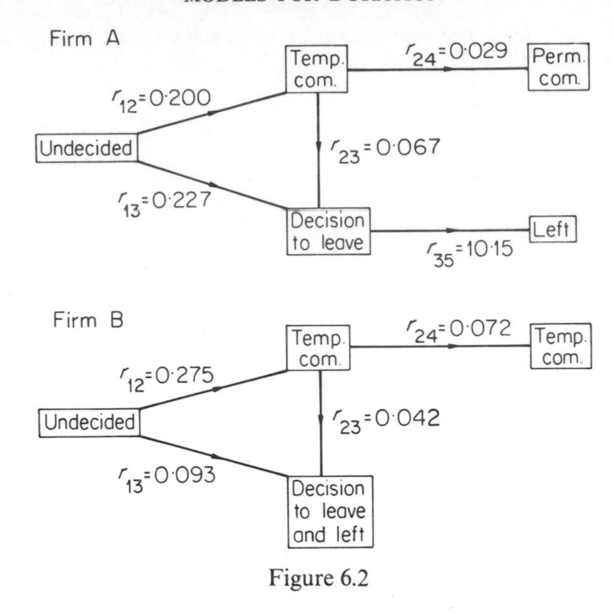

Figure 6.2

networks such as those considered here a simple direct method is available which avoids the necessity of solving the system of simultaneous equations. First, we prove a simple lemma.

Lemma

If, from a state S_1, one-step transitions are possible to any one of the states $S_2, S_3, \ldots, S_k$ with transition intensities $r_{12}, r_{13}, \ldots, r_{1k}$ respectively, then

$$Pr\{S_1 \to S_i \text{ in one step}\} = r_1 / \sum_{j=2}^{k} r_{1j} \qquad (i = 1, 2, \ldots, k).$$

Proof. We first find the joint probability of the required transition and that it takes place in $(T, T + \delta T)$. Thus

$$Pr\{S_1 \to S_i \text{ in one step in } (T, T + \delta T)\}$$
$$= Pr\{\text{survives in } S_1 \text{ to} T\} Pr\{\text{transition takes place in } (T, T + \delta T)\}.$$

In S_1 the total transition intensity is $\sum_{j=2}^{k} r_{1j}$ so the required survival probability is

$$\exp\left(-T \sum_{j=2}^{k} r_{1j}\right).$$

The second factor on the right is $r_{1j}\,\delta T$, by definition. Multiplying these two factors and integrating with respect to T we obtain the required marginal probability as follows:

$$Pr\{S_1 \to S_i \text{ in one step}\} = \int_0^\infty r_{1i} \exp\left[-T\sum_j r_{1j}\right] \mathrm{d}T = r_{1i} \Big/ \sum_{j=2}^{k} r_{1j}$$

$$(i = 2, 3, \ldots, k).$$

Returning to Herbst's example, it is clear from the diagram that there are only two routes by which an individual can reach S_5. These are

(a) $S_1 \to S_2 \to S_3 \to S_5$ and (b) $S_1 \to S_3 \to S_5$.

Successive transitions are independent so we can use the lemma to calculate the probabilities as follows:

$$Pr\{(\mathrm{a})\} = \frac{r_{12}}{r_{12} + r_{13}} \times \frac{r_{13}}{r_{23} + r_{24}} = \frac{r_{12}r_{23}}{r_{11}r_{22}}$$

and

$$Pr\{(\mathrm{b})\} = \frac{r_{13}}{r_{12} + r_{13}} = \frac{-r_{13}}{r_{11}}.$$

The sojourn time in S_1 will be exponential with parameter $-r_{11}$, in S_2 it will be exponential with parameter $-r_{22}$ and in S_3 exponential with parameter $-r_{33}$. If an individual takes route (a) the distribution of total time in the system will be distributed like the sum of three independent exponential variables with the above parameters. If the second route is followed the distribution will be that of the sum of two exponential random variables with parameters $-r_{11}$ and $-r_{33}$. Let us denote the survivor function of these sums by $G(-r_{11}, -r_{22}, -r_{33}; T)$ and $G(-r_{11}, -r_{33}; T)$ respectively. The survivor function for the total sojourn time, regardless of route, will thus be

$$G(T) = \frac{r_{12}r_{23}}{r_{11}r_{22}} G(-r_{11}, -r_{22}, -r_{33}; T) - \frac{r_{13}}{r_{11}} G(-r_{11}, -r_{33}; T) + \frac{r_{12}r_{24}}{r_{11}r_{22}}. \tag{6.19}$$

The last term in (6.19) is the probability of ultimate absorption in S_4. To complete the determination we must find the survivor functions on the right-hand side of (6.19). These are known as generalized Erlang distributions and may be found from the following result.

If we have n independent exponential random variables with parameters $\lambda_1, \lambda_2, \ldots, \lambda_n$ then the survivor function of their sum is

$$G(\lambda_1, \lambda_2, \ldots, \lambda_n; T) = (-1)^{n-1} \sum_{j=1}^{n} \mathrm{e}^{-\lambda_j T} \prod_{\substack{i=1 \\ i \neq j}}^{j} \left(\frac{\lambda_j}{\lambda_j - \lambda_i}\right). \tag{6.20}$$

Taking $n = 3$ with $\lambda_1 = -r_{11}, \lambda_2 = -r_{22}, \lambda_3 = -r_{33}$ and then $n = 2$ with $\lambda_1 = -r_{11}, \lambda_2 = -r_{33}$ and substituting in (6.19) we arrive at the same survivor function as in (6.15) and (6.16). If the network contains a loop there will be infinitely many paths and the method then loses much of its attractiveness.

So far we have concentrated on models in which there is only one exit from the system. In many applications there are several, as, for example, when leavers are classified by reason for leaving. In actuarial terminology this would be described as a multiple decrement problem. If we are interested only in the total time in the system, regardless of the reason for leaving, then all that is required is to give all the terminal states the same label. On the other hand, if we can identify leavers by their reason for leaving we may be interested in the length of service distributions conditional on each reason. The method based on Erlang distributions readily lends itself to the determination of such conditional length of service distributions.

To illustrate the method, suppose that the state S_4 in Herbst's model had been 'declared redundant' instead of 'permanently committed'. Then we might have been interested in the two conditional lengths of service distributions for redundant and voluntary leavers. For the voluntary leavers a simple conditional probability argument gives the survivor function

$$G_L(T) = \left\{G(T) - \frac{r_{12}r_{24}}{r_{11}r_{22}}\right\} \bigg/ \left(1 - \frac{r_{12}r_{24}}{r_{11}r_{22}}\right) \tag{6.21}$$

where $(1 - r_{12}r_{24}/r_{11}r_{22})$ is the probability of ultimate loss in S_5. The corresponding survivor function for the redundant leavers would be

$$G_R(T) = \frac{r_{12}r_{24}}{r_{11}r_{22}} G(-r_{11}, -r_{22}; T) \bigg/ \frac{r_{12}r_{24}}{r_{11}r_{22}} = G(-r_{11}, -r_{22}; T). \tag{6.22}$$

The combined survivor function for the whole system will thus be

$$G(T) - \frac{r_{12}r_{24}}{r_{11}r_{22}} + \frac{r_{12}r_{24}}{r_{11}r_{22}} G(-r_{11}, -r_{22}; T). \tag{6.23}$$

One effect of classifying reasons for leaving will be to provide more data for estimating the parameters of the model and testing its goodness of fit.

6.5 RANDOM WALK MODELS

The model with which we introduce this section was devised by Lancaster (1972) in a study of the distribution of strikes in British industry. The idea on which it is based is capable of development and application elsewhere.

In particular, we shall consider how far it might serve as a model for the leaving process.

Strikes arise as an action of last resort when the two parties to a dispute cannot reach agreement. Let us visualize them at the outset of a strike as separated by a 'distance' d. It is natural to think of this distance as a sum of money if the dispute is over pay, but the issues will usually be more complicated. As the strike progresses the gap between the parties changes as new factors enter the situation and as the hardship and inconvenience caused by the strike makes itself felt. Lancaster's model is a stochastic description of the way in which the gap is closed.

The progress of the negotiations can be represented graphically as shown in Figure 6.3. Points on the graph represent intermediate stages in the negotiations. When the path reaches the horizontal line at d the gap is closed and

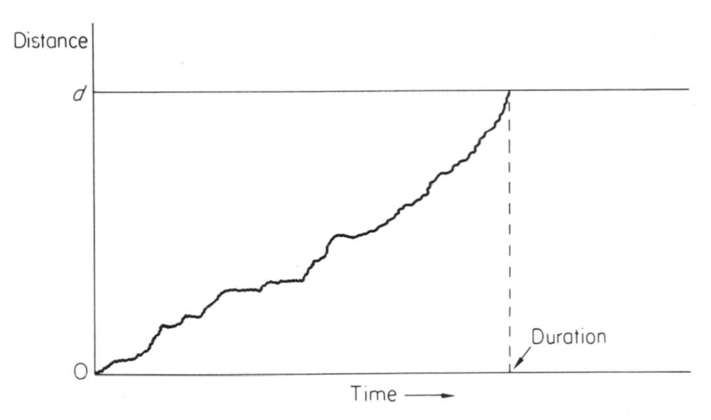

Figure 6.3 Illustration of how the duration of a random walk is related to absorption at a boundary

the duration of the strike is then the horizontal distance from the origin to the point of absorption. Lancaster supposed that the path of the negotiations was a Wiener process. That is, in moving from T to $T + \delta T$ the path moves up or down by an amount which is normally distributed with mean $\mu\,\delta T$ and variance $\sigma^2\,\delta T$ independently of previous increments. The properties of this process are well known; see, for example, Cox and Miller (1965). Absorption is only certain if $\mu \geq 0$, that is if the process has a non-negative 'drift'. The density function of the time to absorption then has the density function

$$f(T) = \frac{d}{\sqrt{2\pi}\sigma T^{\frac{3}{2}}} \exp\left[-\frac{1}{2}\left(\frac{d - T\mu}{\sigma\sqrt{T}}\right)^2\right] \qquad (d, \sigma, \mu > 0). \qquad (6.24)$$

This is sometimes known as the inverse Gaussian distribution and the statistical properties have been studied by Tweedie (1957). Although we have used three parameters in the derivation this is essentially a two-parameter distribution since it can be expressed in terms of d/σ and μ/σ. Lancaster fitted this distribution to British strike data with considerable success, as Table 6.3 shows. The distribution is rather similar to the lognormal. Its density rises to a

Table 6.3 The fit of the inverse Gaussian distribution to the duration of strikes beginning in 1965 in two sectors of British industry (Lancaster, 1972)

	All stoppages in vehicles and cycles		*All stoppages in metal manufacturing*	
Duration (days)	*Observed*	*Fitted*	*Observed*	*Fitted*
2	34	34·0	43	47·1
3	19	18·2	37	30·8
4	10	11·3	21	21·0
5	8	7·7	19	15·3
6	6	5·6	11	11·6
7	5	4·2	8	9·2
8	2	3·3	8	7·4
9	3	2·6	9	6·1
10	2	2·1	3	5·1
11–15	6	6·5	16	16·7
16–20	4	3·1	4	9·1
21–25	4[a]	4·3[a]	4	5·6
26–30	—	—	3	3·7
31–40	—	—	3	4·3
41–50	—	—	5	2·3
> 50	—	—	5	3·7
Total	103	102·9	199	199·0

[a] These frequencies relate to all strikes lasting longer than 20 days. The parameters were estimated by maximum likelihood. The distribution was truncated by the omission of strikes of one days duration (which are not recorded); it is the truncated distribution which is fitted here (see Lancaster, 1972, for details).

mode and then declines slowly with a long upper tail, but the tail dies away more rapidly than that of the lognormal.

It is possible to interpret the model in a way which makes it meaningful in the manpower context. The distance d might be thought of as a critical level of ‘pressure’ on the individual which, when exceeded, would result in his leaving. The random increments would represent the effects of the environmental factors which either increase or decrease this pressure. The positive drift which the model requires would imply an overall tendency for pressure to increase linearly with time. Whether this hypothetical description has any basis in reality is an open question but the similarity of the inverse Gaussian to the lognormal distribution means that it would provide a reasonable fit to data.

The model is clearly capable of many variations according to the stochastic assumptions which we make about the random walk or path by which the gap is closed. Some of the possibilities will be explored in a tentative way below.

Consider a process in discrete time in which increments are added at unit intervals of time. Let the ith increment be of size x_i so that the sample path has coordinates $(\sum_{j=1}^{i} x_j, i)$ for $i = 1, 2, 3, \ldots$. The duration of the walk is the number of steps before $\sum_{j=1}^{i} x_j$ exceeds d for the first time. Let this number be n; then if $G(n)$ denotes the probability that the duration exceeds n it follows that

$$G(n) = Pr\{x_1 < d, x_1 + x_2 < d, \ldots, x_1 + x_2 + \cdots + x_n < d\}. \quad (6.25)$$

In principle this expression provides the means of computing the survivor function for any model of this class, but in practice, it will not usually be easy to compute the probability, especially for large n. A rough approximation when the x's are independently and identically distributed can be obtained by reasoning as follows. Let us suppose the original increments to be grouped into sets of extent j. Then the observation of the grouped process is tantamount to observing every jth point on the original path. The duration of the walk as judged by the grouped process will usually be longer than the original process because the latter may cross the boundary and then return. However, if j is not too large and if the mean increment is small the two processes will be very similar and the distribution of the one could be used as an approximation to the other. In the case of the grouped process the increments will be approximately normal. By a suitable limiting process, to be discussed below, the discrete normal process can be made to approach the Wiener process for which (6.24) applies. (For further information on approximating stochastic processes by simpler known processes see Feller, 1966, Vol. II, page 330).

The discrete time method simplifies greatly if it is assumed that the increments are non-negative. Under these circumstances (6.25) reduces to

$$G(n) = Pr\{x_1 + x_2 + \ldots + x_n < d\}, \quad (6.26)$$

since the last inequality in (6.25) implies all these which precede it. If the

x_i's are 'small' compared with d, $G(n)$ will only differ significantly from one when n is reasonably large. When this is so we may approximate the probability in (6.26) by its normal approximation, obtaining

$$G(n) = \Phi\left(\frac{d - nm}{\sqrt{n}\,S}\right) \tag{6.27}$$

where $m = E(x_i), S^2 = \text{var}\,(x_i)(i = 1, 2, \ldots, n)$. Other assumptions about the stochastic properties of the x's will lead to different distributions. For example, the x's may be assumed to have different distributions such that $\sum x_i$ does not obey the central limit theorem. To be specific suppose that the x's are mutually independent and that x_i is exponentially distributed with mean θ/i. Then

$$Pr\{x_1 + x_2 + \ldots + x_n < d\} = (1 - e^{-d/\theta})^n. \tag{6.28}$$

The duration here has a geometric distribution which is the discrete analogue of the exponential.

It is interesting to investigate the possibility of applying a limiting operation to these discrete processes in the hope of arriving, for example, at a survivor function, like (6.27), but with n replaced by a continuous variable T. The Wiener process can be reached by such a procedure when the x's are identically and normally distributed, but problems arise when the increments are restricted to be positive. To see the difficulty let the time interval between steps be δT and introduce $T = n\,\delta T$. Suppose we now let $n \to \infty$ and $\delta T \to 0$ such that $n\delta T$ remains constant. The size of the increments will also have to be made small and this may be done by letting m and s^2 depend on δT in such a way that the limiting distribution does not degenerate. This can be achieved by letting $m = \mu\delta T$ and $s^2 = \sigma^2\,\delta T$. In the limit we then have

$$G(T) = \lim G(n) = \Phi\left(\frac{d - T\mu}{\sigma\sqrt{T}}\right). \tag{6.29}$$

However, in proceeding to the limit we have invalidated the central limit argument which originally led to the normal approximation. As $\delta T \to 0$ the means of the x's tend to zero faster than their standard deviation. In the case of normal x's this can be done without affecting the form of their distribution but with positive random variables the distribution becomes highly skew as $\delta T \to 0$. Hence the number of increments needed to produce approximate normality will increase and it may be shown that it does so at such a rate that the normal approximation will not be valid for all T. Nevertheless, (6.29) can be regarded as a convenient approximation to the discrete form (6.27). It is also an interesting distribution in its own right, having the principle characteristics of many observed CLS distributions. The density function is

$$f(T) = \frac{1}{2\sqrt{2\pi}\,\sigma}\left(\frac{d}{T^{\frac{3}{2}}} + \frac{\mu}{T^{\frac{1}{2}}}\right) \exp\left[-\frac{1}{2}\left(\frac{d - T\mu}{\sigma\sqrt{T}}\right)^2\right]. \tag{6.30}$$

This expression is very similar to that obtained from the Wiener process in (6.24). In fact, it may be viewed as a mixture, in equal parts, of two densities, one of which is that of T as in (6.24) and the other is the density of $1/T$ with parameters d and μ interchanged.

The possibilities for using these distributions in manpower and other applications has yet to be explored, but the models underlying them seem sufficiently interesting to warrant further investigation. Further generalization is possible by having the time intervals between the addition of increments unequal or by making these intervals random variables.

6.6 MODELS IN DISCRETE TIME

Apart from the brief discussion at the end of the last section the discussion in this chapter has been in terms of continuous time. This is because the various models lead in the main to well-known distributions which are easily handled. Our main interest has been in relating the shapes of survivor functions and loss intensities to the probability models underlying them, and it is easier to visualize a shape in terms of a curve than a discrete set of points. However, there are discrete analogues of all our models and the purpose of this section is to draw attention to some instances in which a discrete formulation offers mathematical advantages or added insights.

The Herbst type of model is very easily expressed in discrete time. In this case changes of state are supposed to take place at intervals of time according to a set of transition probabilities. For Herbst's model the matrix would have the following structure:

$$\mathbf{P} = \left[\begin{array}{ccc|cc} p_{11} & p_{12} & p_{13} & 0 & 0 \\ 0 & p_{22} & p_{23} & p_{24} & 0 \\ 0 & 0 & p_{33} & 0 & p_{35} \\ \hline 0 & 0 & 0 & 1 & 0 \\ 0 & 0 & 0 & 0 & 1 \end{array}\right]. \tag{6.31}$$

The survivor function is simply the complement of the probability of moving from S_1 to S_5 by time T, that is

$$G(T) = 1 - p_{15}(T). \tag{6.32}$$

The element $p_{15}(T)$ is easily calculated by raising **P** to the Tth power and picking out the number in the (1,5)th position. If the transition matrix is given numerically it is therefore very simple to compute the survivor function for any system. If the transition probabilities are unknown and have to be estimated from the data much of the advantage is lost because the multiplication then has to be done algebraically.

The case of multiple decrements is also easy to handle in discrete time. Suppose we label the transient states from 1 to l and the absorbing states from $l + 1$ to k. Then the transition matrix will have the structure and dimensions as shown below:

$$\begin{array}{c|c|c} & \leftarrow l \rightarrow & \leftarrow (k-l) \rightarrow \\ \hline \uparrow\, l \,\downarrow & \mathbf{P}^* & \mathbf{Q} \\ \hline \uparrow\, (k-l) \,\downarrow & \mathbf{O} & \mathbf{I} \end{array}$$

We first show how to find the probabilities of absorption in each of the absorbing states and then use them to find the conditional survivor functions. Let a_{ij} denote the probability that an individual in transient state i is ultimately absorbed in state j. Then an elementary probability argument gives

$$\begin{aligned} a_{ij} &= p_{ij} + \sum_{h=1}^{l} \sum_{m=1}^{\infty} Pr\{\text{individual goes from } i \text{ to } h \text{ in } m \text{ steps}\} p_{hj} \\ &= p_{ij} + \sum_{h=1}^{l} \sum_{m=1}^{\infty} p_{ih}^{(m)} p_{hj} \qquad (i = 1, 2, \ldots, l; j = l+1, l+2, \ldots, k). \end{aligned} \tag{6.33}$$

Reversing the order of summation, $\sum_{m=1}^{\infty} p_{ih}^{(m)} p_{hj}$ is the (i, j)th element of the matrix product $(\mathbf{P}^*)^m\mathbf{Q}$ and p_{ij} is the corresponding element of **Q** so that, in matrix notation (6.33) becomes

$$\mathbf{A} = \sum_{m=0}^{\infty} (\mathbf{P}^*)^m \mathbf{Q} = (\mathbf{I} - \mathbf{P}^*)^{-1}\mathbf{Q} \tag{6.34}$$

where **A** is the $l \times (k - l)$ matrix of elements a_{ij} and **I** is the unit matrix of dimension $l \times l$. The inverse always exists because the row sums of $\mathbf{P}^*$ are less than one.

To find the survivor function conditional upon a particular reason for leaving we replace the original transition matrix by a modified matrix whose elements are probabilities conditional on absorption in the chosen state. Let X_T be a random variable giving the state of the system at time T and let k

be the index of the absorbing state on which we wish to condition. Then

$$\begin{aligned} Pr\{X_{T+1} = j, X_{\infty} = k|X_T = i\} &= Pr\{X_{\infty} = k|X_T = i\} \times \\ &\qquad Pr\{X_{T+1} = j|X_T = i, X_{\infty} = k\} \\ &= a_{ik} \times \tilde{p}_{ij} \end{aligned} \tag{6.35}$$

where $\tilde{p}_{ij}$ is the probability of transition from i to j in one step given ultimate absorption in k. The left-hand side of (6.35) is also equal to

$$Pr\{X_{T+1} = j|X_T = i\}Pr\{X_{\infty} = k|X_{T+1} = j, X_T = i\}.$$

The conditioning event $X_T = i$ in the last probability can be suppressed because of the Markov property and hence this product is

$$p_{ij}\, a_{jk}. \tag{6.36}$$

Equating (6.35) and (6.36) the required conditional probability is

$$\tilde{p}_{ij} = p_{ij}\frac{a_{jk}}{a_{ik}} \qquad (i, j = 1, 2, \ldots, l). \tag{6.37}$$

The modified matrix thus has these elements in place of **P*** and **Q** is replaced with a column vector with elements chosen to make the row sums unity. The submatrix **O**|**I** is replaced by a vector (0 0 . . . 0 1). The application of the technique described at the beginning of this section will then yield the conditional survivor function.

The theory of absorbing Markov chains on which the foregoing discussion is based may be found in Kemeny and Snell (1960). Applications of the theory to demographic processes have been given by Feichtinger (1971).

Finally, we show that the multiplicative model leading to the lognormal distribution can also be expressed as an absorbing Markov chain. We do this not because the discrete formulation offers any analytical advantages but because of the additional insight which it offers and the possible variations which it suggests.

Let the possible lengths of service be $1, 2, 3, \ldots, N$ where N is some large number representing the maximum possible length of service. Similarly, let us number the successive jobs $1, 2, \ldots, n$ and then introduce a Markov chain with states $(i, j)(i = 1, 2, \ldots, N; j = 1, 2, \ldots, n)$. Diagramatically the situation may be set out as in Figure 6.4. From any state in the first row (length of stay in first job) transitions are possible only into the second row (second job). Similarly, in each subsequent row transitions can only be made into the next row. The set of transition probabilities out of any state corresponds to the distribution of $u_{j+1}x_j$ in the continuous model. To make the

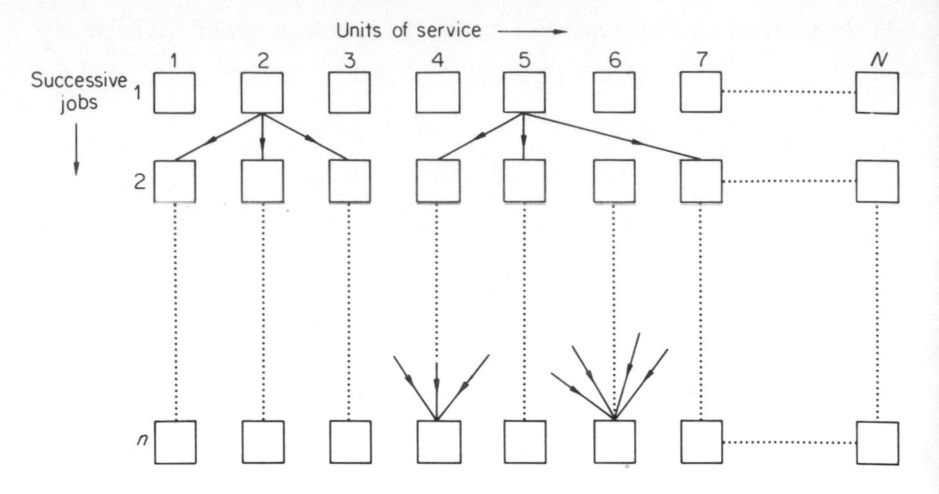

Note: The diagram shows typical sets of transition paths. The model allows transition from all states in each row.

Figure 6.4 The discrete 'lognormal' model

correspondence as near as possible this distribution will have to be such that its dispersion is proportional to the column index of the starting state (x_j). The CLS distribution will be the set of absorption probabilities associated with the states in the last row of Figure 6.4—each state in that row corresponds to a different length of service in the nth job. Perhaps the simplest set of transition probabilities meeting the requirement is

$$\begin{aligned} P_{(i,j),(i,j+1)} &= 1 - p \\ P_{(i,j),(2i,j+1)} &= p \\ &= \text{otherwise}. \end{aligned} \tag{6.38}$$

In effect, this means that in moving from one job to the next length of service either remains the same or doubles (that is $u_{j+1} = 1$ or 2). After n steps the absorption probabilities are

$$\begin{aligned} Pr\{\text{length of service} = 2^i \text{ at time } n\} &= \binom{n}{i} p^i(1-p)^{n-i} \qquad (i = 1, 2, \ldots, n) \\ &= 0 \text{ otherwise}. \end{aligned} \tag{6.39}$$

The logarithm of length of service is thus binomially distributed and it is obvious in what sense this will approximate the normal. This formulation opens up the possibility of using the theory of absorbing Markov chains to study a whole class of models about how experience in one job affects length of service in the next.

We have reviewed a wide range of models for the leaving process without exhausting all the possibilities, even of the models we have formulated. In spite of this, some of the commonest distributions used in reliability theory and life-testing have found no place in our discussion. Among these are the gamma distribution and the various extreme value distributions—in particular the Weibull distribution. These distributions are capable of producing the long upper tail required only for very small values of their shape parameters. We have already noted that a J-shaped gamma distribution can arise as a mixture of exponentials, but the unimodal gamma distribution has a monotonic *increasing* failure rate.

Weibull distributions have been successfully fitted to data on the distribution of wars and strikes in American industry by Horvath (1968). He proposes an extreme value model analogous to Weibull's original model for the strength of materials as follows. During the course of a conflict, whether a war or a strike, there are many obstacles to a settlement. The time taken to reach agreement on any one of these issues may be thought of as a random variable and the conflict is supposed to be ended as soon as the first obstacle is overcome. This is thus a 'weakest point' theory according to which one probes all the potential points of breakthrough until success is achieved at one of them. If there are many such points and if the distribution of the time to breakthrough on each is the same then the extreme value theory shows that the appropriate limiting form of the distribution of duration is the Weibull. The attraction of this model lies in its generality since it is independent of the actual form of the component distribution. However, its plausibility on other grounds may be questioned. In many bargaining situations it is not reasonable to suppose that the conflict will end when one of the obstacles to agreement has been overcome. It may be necessary to reach a settlement on several or, indeed, all of the points at issue. Even if the model is accepted as realistic up to this point it is doubtful whether the time to reach agreement on each issue would have the same distribution or even whether there would be enough issues to make the limit a good approximation. Nevertheless, the model does fit the data and it should clearly be entertained as a possible explanation.

A different kind of model for the duration of a war was proposed by Weiss (1963). His first model relates the chance of ending the conflict at a given time to the number of deaths at that time. In a second model he allows the chance of termination to be a function of both time and the number of deaths. These models are more general than those discussed in this chapter in that they make the loss intensity a function of something other than time (though, in this case, the number of deaths is a monotonic function of time). Such a generalization is clearly a step in the direction of realism but the added complexity which it brings is a serious obstacle. It is difficult enough to distinguish among the simpler models of this chapter and it therefore seems unlikely that further elaboration will be useful until much more detailed data are available.

In later chapters we shall meet further examples of distributions of duration. Chapter 8 deals with multistage renewal processes and we shall find there the distribution of stay in such a system and find that this, too, has the mixed exponential form. In Chapter 9 we shall discusss the distribution of the duration of an epidemic but the interest there will not be in the distribution as such but in using such things as its expectation as a means of identifying critical parameters in the model.

CHAPTER 7

Renewal Theory Models for Recruitment and Wastage

7.1 INTRODUCTION

In the course of developing models for graded systems of fixed total size in Chapters 3 and 5 we had to determine the number of recruits required. This calculation, however, was merely an intermediate step on the way to finding the grade sizes. In this chapter we shall consider the input and output of fixed size systems in more detail as matters of independent interest. The two main factors on which recruitment depends are the propensity of individuals to leave and expansion or contraction of the system. Leaving creates vacancies whose number may then be increased or decreased by planned changes in total size. It is thus clear that the stochastic element in the processes arises from the uncertain nature of the leaving process. In Chapters 3 and 5 we made, for the most part, the simple assumption that the loss probabilities or intensities were constant and, in particular, independent of length of service. Now we shall generalize this aspect by making use of the various models for the leaving process discussed in Chapter 6. In this chapter we shall assume that loss intensities are independent of grade, which enables us to ignore the grade structure of the system; in the following chapter this assumption will be relaxed.

We have already seen that the loss of intensity and the CLS distribution are equivalent ways of expressing the dependence of propensity to leave on length of service. For the purposes of the present chapter the CLS distribution is the more convenient description. It is also possible to develop the theory in discrete or continuous time. We have already noted the advantages of the continuous time formulation in discussing models for the leaving process, and this approach continues to offer advantages here. An indication of the lines on which the discrete time version could be developed is given in Section 5.6 of Bartholomew (1971).

The model which provides the basis for our discussion is a simple renewal process. Renewal, or replacement, theory appears to have arisen in connexion with the study of birth and death in human populations. More recently it has been developed in the context of industrial replacement. There are obvious similarities (and differences) between the replacement of components in such things as computers and motor cars and the replacement of people in a firm. To see the relevance of renewal theory to the turnover of people in an organization consider first a 'one-man' firm. Suppose the first incumbent stays in his job for length of time x_1; he is replaced immediately by a successor who stays for lengths of time x_2. Continuing in this way there will be a succession of completed length of service $x_1, x_2, x_3, \ldots$, and replacements will have to be made at times $x_1, x_1 + x_2, x_1 + x_2 + x_3, \ldots$, and so on. Such a process is an example of a renewal process. Most of renewal theory is concerned with the case when the x's are identically and independently distributed. In the manpower application this assumption requires that all occupants of a given position have the same CLS distribution and that there is no interaction between successive job holders. Neither requirement is likely to be met exactly but both seem to be sufficiently near the truth to justify an investigation of their consequences.

The one-man firm is of little practical interest but it provides the starting point for dealing with N-man systems. To make the necessary extension we shall make the further assumption that the coming and going of people in any one job is independent of that in all others. This assumption enables us to deduce the aggregate behaviour of the system by 'adding up' the component one-man systems. In renewal theory terminology we are interested in the 'pooled' output of a number of independent renewal processes (see Cox, 1962, Chapter 6).

There are two aspects of the renewal process which have particular relevance in manpower applications. The chief of these is the number of replacements which are required in a specified interval of time. Ideally, we would like to know the distribution of this number or, failing that, its mean and variance. The main application is to the prediction of recruitment needs. The second aspect is the 'age' distribution of those in service at any particular time, where age here is used to mean time since joining the system. By comparing observed and expected age distributions it is sometimes possible to diagnose or forecast promotion bottlenecks or recruiting problems.

Following the practice of earlier chapters we shall begin by considering a very simple model in some detail and then generalize it by relaxing one assumption at a time. Thus in Section 7.2 we deal with the case of turnover of employees in an organization of constant size. In the following section we generalize this theory to cover the case of a system in which the form of the CLS distribution is a function of the time at which an individual joins. Finally, we derive the extension of the theory needed to deal with expanding and contracting systems.

The relevance of the theory is not confined to the prediction of flows in and out of an organization. It also throws considerable light on the important practical problem of measuring labour turnover. This problem has a long history surrounded by much confusion. We shall argue that the approach via a stochastic model serves to clarify the issues and provides criteria against which the suitability of indices can be judged.

7.2 RECRUITMENT AND WASTAGE IN ORGANIZATIONS OF CONSTANT SIZE

The simple renewal model

In an organization of constant size each loss must be associated with a gain. The recruitment and wastage processes are therefore essentially the same. In this section we shall use the notation and terminology appropriate to the former but our results will, where necessary, be translated into wastage terms. As we have already remarked, the formulae for predicting recruitment needs can be deduced either from the results of Chapter 5 or by a direct appeal to renewal theory. In order to emphasize the continuity of our development we shall adopt the first course.

In the notation of Chapter 5 the expected number of recruits required in the interval $(0, T)$ is $\bar{n}_{k+1}(T)$. This quantity is obtained by setting $j = k + 1$ in (5.52) of that chapter. Our assumption that the loss intensity is independent of the grade means that $p_{i,k+1}(T)$ is equal to

$$F(T) = \int_0^T f(t)\,dt$$

for all i. Making this change of notation and writing N for the total size of the organization the integral equation for $\bar{n}_{k+1}(T)$ becomes

$$\bar{n}_{k+1}(T) = NF(T) + \int_0^T F(T-x)\frac{d\bar{n}_{k+1}(x)}{dx}\,dx. \tag{7.1}$$

The quantity $\bar{n}_{k+1}(T)$ is related to the more familiar renewal density, which we denote by $h(T)$, and the recruitment rate $R(T)$ by the equations

$$\bar{n}_{k+1}(T) = N\int_0^T h(x)\,dx = \int_0^T R(x)\,dx. \tag{7.2}$$

Differentiating both sides of (7.1) with respect to T and using (7.2) we find

$$\left.\begin{aligned} h(T) &= f(T) + \int_0^T h(x)f(T-x)\,dx \\ &= f(T) + \int_0^T h(T-x)f(x)\,dx \end{aligned}\right\} \tag{7.3}$$

which is the well-known integral equation for the renewal density. Under very general conditions it is known that

$$\lim_{T \to \infty} h(T) = \mu^{-1}$$

where μ is the mean of the CLS distribution.

The direct derivation of (7.3) starts with the interpretation of $h(T)\,\delta T$ as the probability of a renewal in $(T, T + \delta T)$. This event occurs either if the original member of the system leaves in the interval or if the last person to be recruited leaves. The former event has probability $f(T)\,\delta T$; the latter has probability

$$h(x)\,\delta x\, f(T - x)\,\delta T$$

for someone who joined in $(x, x + \delta x)$. Summing these probabilities and cancelling the factor δT yields (7.3)

As a consequence of (7.2), either of the equations (7.1) or (7.3) may be used to predict recruitment needs. We prefer to work with the renewal density rather than the expected number of losses in $(0, T)$ because the former provides the best starting point for the development of the approximations discussed later. It is important to notice that we have assumed that the origin of our time scale corresponds to an initial state of the organization when all employees had zero length of service. This case is of particular importance but a more general assumption will be made in Section 7.3.

The integral equation for the renewal density can sometimes be solved by taking the Laplace transform of both sides. This gives

$$h^*(s) = f^*(s) + h^*(s) f^*(s)$$

whence

$$h^*(s) = f^*(s)/\{1 - f^*(s)\}. \tag{7.4}$$

The usefulness of this technique depends upon whether or not the right-hand side of this equation can be inverted. Unfortunately this does not appear to be possible either for the type XI or lognormal CLS distributions, so we shall have to rely on approximate methods. In the case of the mixed exponential distribution the solution of the equation by the Laplace transform is straightforward and is given below.

Solution of the renewal equation for the mixed exponential CLS distribution

When the CLS distribution has the form

$$f(T) = p\lambda_1\, e^{-\lambda_1 T} + (1 - p)\lambda_2\, e^{-\lambda_2 T}$$

its Laplace transform is

$$f^*(s) = p\left(\frac{\lambda_1}{\lambda_1 + s}\right) + (1 - p)\left(\frac{\lambda_2}{\lambda_2 + s}\right). \tag{7.5}$$

When this expression is substituted in (7.4), $h^*(s)$ may be inverted by standard methods, as in Bartholomew (1959), to give

$$h(T) = \mu^{-1} + \{p\lambda_1 + (1 - p)\lambda_2 - \mu^{-1}\} \exp\{-(p\lambda_2 + (1 - p)\lambda_1 T\} \tag{7.6}$$

where $\mu = p/\lambda_1 + (1 - p)/\lambda_2$ is the mean of $f(T)$. This formula shows that $h(T)$ approaches its limit in an exponential curve. Further, since $p\lambda_1 + (1 - p)\lambda_2 \geq \mu^{-1}$, it follows that the number of recruits required will always be in excess of the number predicted by equilibrium theory. The expected number of recruits needed in any time interval can be found by integrating $h(T)$. Some illustrative calculations are given in Table 7.1, using two of the mixed exponential distributions fitted in Table 6.1.

Table 7.1 Percentage recruitment (wastage) figures in successive quarters for a new firm with mixed exponential CLS distribution

	Recruitment using the CLS distributions fitted to data from:			
Quarter	Glacier Metal Co.	Total	J. Bibby & Sons (males)	Total
1	22·6	69·5	34·6	108·3
2	18·2		28·6	
3	15·3		24·2	
4	13·4		20·9	
5	12·1	44·6	18·5	65·4
6	11·3		16·8	
7	10·8		15·5	
8	10·4		14·6	

We have not carried the calculations beyond the end of the second year because it is doubtful whether the fit of the mixed exponential distribution is adequate in the upper tail. This remark is based on the results of a comparison of the mixed exponential with the lognormal distribution using Lane and Andrew's (1955) data. These give estimates of the CLS distribution over the range 0 to 40 or 50 years. The lognormal distribution is much more successful in graduating the long upper tail than is the mixed exponential.

The figures in Table 7.1 show, in a striking fashion, the rapid decline in recruitment which can occur in a new organization. In the examples we have chosen, the figure is roughly half in the second year what it was in the first. This conclusion has obvious implications both for recruitment planning and the interpretation of wastage figures. It suggests that the high initial wastage might be reduced by careful selection of employees with a view to rejecting those with short service prospects. It must be emphasized that we have assumed that the CLS distribution does not change with time. This would be a questionable assumption under the conditions likely to exist when a new organization is established.

For the purpose of interpreting crude wastage figures† the meaning of Table 7.1 is clear. *A change in the wastage rate does not necessarily indicate a change in those factors which precipitate leaving.* It may simply reflect a change in the length of service structure. This fact makes it quite meaningless, for example, to compare the crude wastage rate of a new firm with that of an old one.

An approximate solution of the renewal equation

We have seen that some empirical CLS distributions can be satisfactorily graduated by a mixed exponential curve—at least in the interval 0 to 21 months. However, we have also pointed out that, over longer periods, the lognormal is more satisfactory because it has a longer upper tail. The same is true of the type XI distribution obtained from Silcock's model. Unfortunately it is not possible to obtain a simple explicit expression for $h(T)$ for either of these distributions. In order to investigate the form of $h(T)$ over longer periods we therefore require an approximate method for obtaining a solution to the renewal equation.

The approximation which we will use was derived by Bartholomew (1963b). It was intended for use when the CLS distribution is extremely skew and has the following form

$$h^0(T) = f(T) + F^2(T) \Bigg/ \int_0^T G(x)\,\mathrm{d}x \tag{7.7}$$

where $G(T) = 1 - F(T)$. The approximation has the following properties in common with the exact solution of the renewal equation.

(a) If $f(T) = \lambda\,\mathrm{e}^{-\lambda T}$ then $h^0(T) = h(T) = \lambda$ for all T.

(b) $\lim_{T\to\infty} h^0 T = \lim_{T\to\infty} h(T) = \mu^{-1}$.

† Crude wastage is defined as the number of leavers in the period divided by the average number of employees during the same period (see Section 7.5).

(c) $h^0(0) = h(0) = f(0)$.

(d) $$\left.\frac{\mathrm{d}^i h^0(T)}{\mathrm{d}T^i}\right|_{T=0} = \left.\frac{\mathrm{d}^i h(T)}{\mathrm{d}T^i}\right|_{T=0} \qquad (i = 1, 2).$$

These properties suggest that the approximation is likely to be good everywhere if $f(T)$ is close to the exponential and will always be good near $T = 0$ and when T is large. The method of derivation used in Bartholomew (1963b), given for a more general equation in Section 7.3, also suggests that the approximation will be good if $f(T)$ has a long upper tail. A further useful property of the approximation is that $h^0(T)$ is an upper bound for $h(T)$ if the loss intensity associated with $f(T)$ is non-increasing. Such distributions have been studied in reliability theory where they are known as DFR (decreasing failure rate) distributions (see, for example, Barlow and coworkers, 1972, Chapter 5). We shall test the adequacy of the approximation in special cases after we have considered two examples.

The simplicity of the approximation is apparent when we consider its form for Silcock's model. In this case

$$f(T) = \frac{\nu}{c}\left(1 + \frac{T}{c}\right)^{-(\nu+1)}, \qquad G(T) = \left(1 + \frac{T}{c}\right)^{-\nu}$$

and

$$\int_0^T G(x)\,\mathrm{d}x = \mu\left\{1 - \left(1 + \frac{T}{c}\right)^{-\nu+1}\right\}.$$

Hence, from (7.7),

$$\left.\begin{aligned} h^0(T) &= \frac{\nu}{c}\left(1 + \frac{T}{c}\right)^{-(\nu+1)} + \frac{\left\{1 - \left(1 + \frac{T}{c}\right)^{-\nu}\right\}^2}{\mu\left\{1 - \left(1 + \frac{T}{c}\right)^{-\nu+1}\right\}} \\ &\sim \frac{1}{\mu}\left\{1 - \left(\frac{c}{T}\right)^{\nu-1}\right\} \quad \text{if } \nu > 1. \end{aligned}\right\} \tag{7.8}$$

When ν is a little greater than one, the approach to equilibrium is very slow. If $0 < \nu < 1$ the mean is infinite and $h(T)$ approaches zero like $T^{\nu-1}$. In the case $\nu = 1$ the zero limit is approached like $(\log T)^{-1}$. The calculations made by Silcock (1954) for eight distributions gave six out of eight values of ν between 0·5 and 1. We would therefore expect the wastage rate to go on declining slowly over all periods likely to be of practical interest.

The approximation also takes a simple form for the lognormal CLS distribution. Thus we have

$$\left.\begin{aligned} f(T) &= \frac{1}{\sqrt{2\pi}\,\sigma T}\exp\left\{-\frac{1}{2}\left(\frac{\log_e T-\omega}{\sigma}\right)^2\right\},\\ F(T) &= \Phi\left(\frac{\log_e T-\omega}{\sigma}\right) \end{aligned}\right\} \tag{7.9}$$

and

$$\int_0^T G(x)\,\mathrm{d}x = T\left\{1-\Phi\left(\frac{\log_e T-\omega}{\sigma}\right)\right\}+\mu\Phi\left(\frac{\log_e T-\omega}{\sigma}-\sigma\right)$$

where $\mu = e^{\omega+\frac{1}{2}\sigma^2}$ and $\Phi(.)$ is the standard normal distribution function. The approximation can therefore be calculated using tables of the normal probability integral. It is clear from (7.9) that equilibrium will not be reached until $X = (\log_e T - \omega)/\sigma$ is large enough for $\Phi(X)$ to be near one. To investigate this point in more detail let us examine the form of $h^0(T)$ for large T. If X is large we may write

$$\Phi(X) \sim 1 - \frac{1}{\sqrt{2\pi}\,X}e^{-\frac{1}{2}X^2}.$$

Straightforward manipulation then gives

$$h^0(T) \sim \mu^{-1}\left\{1-\frac{1}{\sqrt{2\pi}}\frac{\sigma}{X(X-\sigma)}e^{-\frac{1}{2}(X-\sigma)^2}\right\}^{-1}. \tag{7.10}$$

If $\sigma = 2$ and $X = 4$, say, then

$$h^0(T)|_{X=4} = \frac{1.02}{\mu}.$$

When $X = 4$, $T = \mu e^{\sigma} \doteqdot 400\mu$. This means that it would take about 400 times the average CLS to get within 2 per cent of the equilibrium value. Bearing in mind that the lognormal distribution cannot represent the true state of affairs beyond about 40 or 45 years, it is clear that equilibrium behaviour will be of limited practical interest.

The transient behaviour of the renewal model is illustrated in Table 7.2, assuming a lognormal CLS distribution. Calculations have been made for a typical case by taking $\omega = 0$ and $\sigma = 2$. If the unit of time is one year then $\omega = 0$ corresponds to a 'half-life' of one year. For this distribution $\mu = e^2 = 7.389$ so that $h(\infty) = 0.1353$.

***Table 7.2** Approximation to the renewal density for a lognormal CLS distribution with parameters $\omega = 0$ and $\sigma = 2$*

$\log_e T$	0	1	2	3	4	5	10	∞
T	1	2·71	7·39	20·1	54·6	148·4	22,026	∞
$h^0(T)$	0·574	0·424	0·318	0·244	0·194	0·194	0·125	0·135

The figures in this table show the extreme slowness with which $h^0(T)$ approaches its limit. Even after 20 years the renewal density is still roughly twice its equilibrium value. The fact that $h(T)$ for a comparable mixed exponential distribution reaches equilibrium in a matter of a few years serves to emphasize the need for accurate graduation of the tail of the distribution if long-term predictions are required.

We have assumed in the foregoing discussion that conclusions drawn from the behaviour of $h^0(T)$ can be applied to $h(T)$. This assumption can be tested in certain cases as we shall now show. Some comparisons of $h(T)$ and $h^0(T)$ were made in Bartholomew (1963b) for the discrete time version of the renewal equation. The CLS distribution which was used for these calculations was the discrete analogue of the type XI curve. In all cases the agreement was good even for the most skew distributions considered. A comparison for the case of continuous time can be made for the mixed exponential CLS distribution. The results are given in Table 7.3 for the two CLS distributions which were used for the calculations given in Tables 6.1 and 7.2. The agreement in this case is very good for the first six months but less good thereafter. However, $h^0(T)$ is an upper bound for $h(T)$ for all T and so statements about the rate of convergence to the limit can be based upon it.

Butler (1970) compared the exact and approximate renewal densities for a lognormal CLS distribution. He calculated the exact values by solving (7.3) by numerical methods; some of his results are given in Figure 7.1. It appears from these calculations and those of Bartholomew (1963b) that $h^0(T)$ provides quite a useful approximation. Any approximate solution can be improved by using $h^0(T)$ as a starting point for the iterative solution of the integral equation. Thus

$$h^{(1)}(T) = f(T) + \int_0^T h^0(T - x)f(x)\,\mathrm{d}x \tag{7.11}$$

should be a better approximation than $h^0(T)$.

Table 7.3 Comparison of $h(T)$ and $h^0(T)$ for two mixed exponential distributions

			T									
			0	0·2	0·4	0·6	0·8	1·0	1·5	2·0	10·0	∞
(Glacier Metal Co.)	$p = 0{\cdot}6513$ $\lambda_1 = 0{\cdot}2684$ $\lambda_2 = 2{\cdot}4228$	$h(T)$	1·020	0·840	0·712	0·620	0·555	0·508	0·440	0·411	0·389	0·389
		$h^0(T)$	1·020	0·841	0·715	0·628	0·567	0·525	0·466	0·439	0·392	0·389
(J. Bibby & Sons)	$p = 0{\cdot}4363$ $\lambda_1 = 0{\cdot}2339$ $\lambda_2 = 2{\cdot}5335$	$h(T)$	1·530	1·300	1·120	0·979	0·870	0·784	0·643	0·567	0·479	0·479
		$h^0(T)$	1·530	1·301	1·125	0·993	0·894	0·821	0·705	0·643	0·489	0·479

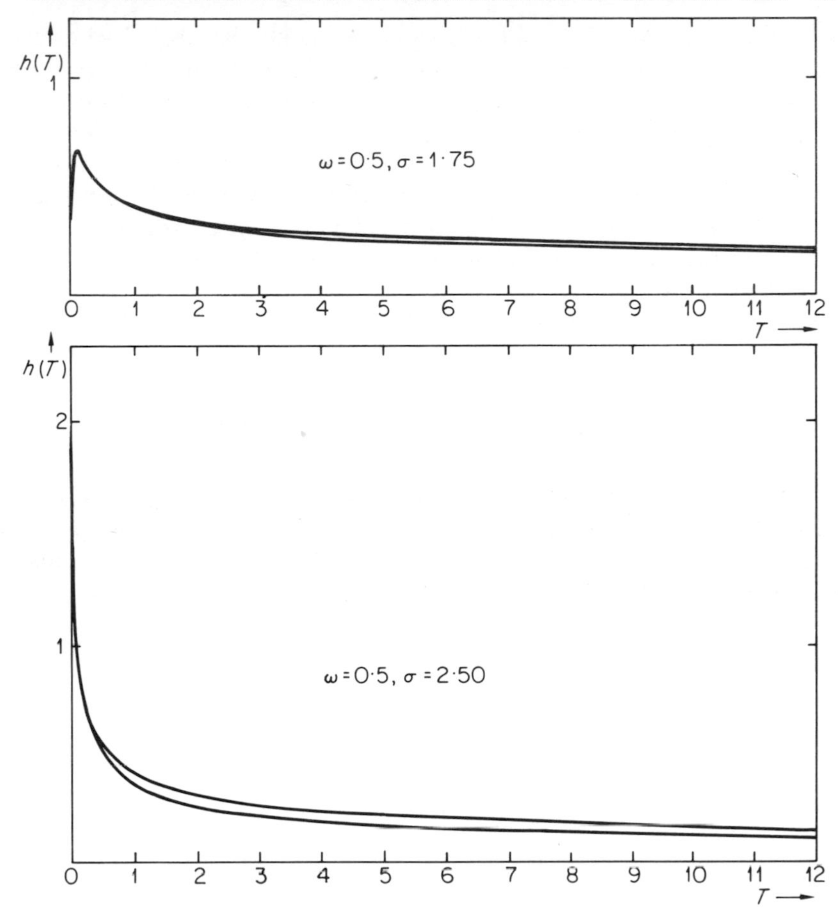

Figure 7.1 Comparisons of $h^0(T)$ (upper curve) and $h(T)$ (lower curve) for a lognormal CLS distribution

Distribution of the number of recruits

Standard renewal theory provides methods for finding the distribution of the number of recruits as well as its expectation. Let us begin by considering an organization with $N = 1$. The expected number of recruits required in the interval $(0, T)$ is then

$$\bar{n}_{k+1}(T) = \int_0^T h(x)\,\mathrm{d}x.$$

(The subscript $k + 1$ is not necessary in the present context so we shall omit it throughout the remainder of this section.) For large T it is known (see Cox,

1962, page 40) that $n(T)$ is approximately normally distributed with mean $T\mu^{-1}$ and variance $V^2T\mu^{-1}$, where V^2 is the square of the coefficient of variation of the CLS distribution. For the lognormal distribution $V^2 = e^{\sigma^2} - 1$ which, for $\sigma = 2$, is equal to 53·6. Long-term predictions of recruiting needs are thus subject to a high degree of uncertainty. However, we have already seen in the case of the mean, that the limit is approached so slowly that asymptotic results are practically useless. The same is true of the variance, so we must use exact results when dealing with this kind of CLS distribution.

The exact value of the variance of $n(T)$ can be found by using the following equation:

$$E\{n^2(T)\} = \bar{n}(T) + 2\int_0^T \bar{n}(T - x)h(x)\,dx \tag{7.12}$$

(see Parzen, 1962, page 179). Since we have the means of finding $\bar{n}(T)$ and $h(T)$ either exactly or approximately for any CLS distribution we can compute $E\{n^2(T)\}$ and hence the variance of $n(T)$. We illustrate the calculations using the mixed exponential distribution for which $h(T)$ was found earlier. If we let

$$a = p\lambda_1 + (1 - p)\lambda_2 - \mu^{-1}$$

$$b = p\lambda_2 + (1 - p)\lambda_1$$

then

$$h(T) = \mu^{-1} + a\,e^{-bT}$$

and

$$\bar{n}(T) = T\mu^{-1} + \frac{a}{b}(1 - e^{-bT}). \tag{7.13}$$

Making the necessary substitutions in (7.12) and subtracting $\bar{n}^2(T)$ we find the following expression for the variance

$$\text{var}\,\{n(T)\} = \frac{T}{\mu}\left\{1 + \frac{2a}{b}\right\} + \frac{a}{b}\left\{1 - \frac{4}{\mu b} + \frac{a}{b}\right\} + \frac{2a}{b}\left\{\frac{1}{\mu} - \frac{a}{b}\right\}T\,e^{-bT}$$
$$- \frac{a}{b}\left\{1 - \frac{4}{\mu b}\right\}e^{-bT} - \frac{a^2}{b^2}e^{-2bT}. \tag{7.14}$$

The corresponding results for an organization of size N are simply obtained by multiplying the mean and variance given above by N. Since $n(T)$ for a large organization can be regarded as the sum of the numbers for single member systems, the central limit theorem ensures its approximate normality. Some numerical values of expectations and standard errors are given in Table 7.4. For the purposes of this calculation we have considered two

***Table 7.4** Expectations and standard errors of numbers of recruits required in various time intervals for two mixed exponential CLS distributions when* $N = 1{,}000$

		Time interval			
		0–3 months	0–6 months	0–12 months	0–24 months
$p = 0{\cdot}6513$	Mean	226	408	695	1,142
$\lambda_1 = 0{\cdot}2684$	S.E.	16·2	21·8	29·0	39·2
$\lambda_2 = 2{\cdot}4228$	Approx. S.E.	18·4	22·6	29·2	39·2
$p = 0{\cdot}5377$	Mean	458	738	1,072	1,508
$\lambda_1 = 0{\cdot}2187$	S.E.	27·0	34·3	41·8	52·5
$\lambda_2 = 4{\cdot}8940$	Approx. S.E.	32·9	36·3	42·4	52·6

hypothetical firms with 1,000 employees. We assume the CLS distribution to be mixed exponential in both cases. In the first case we have used the parameter values obtained by fitting the curve to the Glacier Metal Co. data; in the second we have used a more skew member of the mixed exponential family suggested by data relating to the United Steel Cos. given in Silcock (1954).

The rows labelled 'Approx, S.E.' were obtained using only the first two terms of the variance as given by (7.14). The approximation overestimates the true value but the difference is negligible for all but the shortest time periods. It is clear from the table that predictions for even short periods are subject to considerable uncertainty.

The foregoing results relate to the total number of replacements in $(0, T)$. More often we require to know about the number of replacements in some fairly short interval (T_1, T_2) at a distance from the origin. If this distance is such that the system has reached equilibrium then it is well known that the numbers of renewals in non-overlapping intervals are independent Poisson variables. This result is due to Khintchine (1960, Chapter 5). In view of the slowness which manpower replacement systems approach their equilibrium state it is highly desirable to have some results on the distribution of the number of renewals in the transient state. A limiting result relevant to this requirement was given by Grigelionis (1964). It was re-derived by Butler

(1970) and its adequacy as an approximation was investigated by Bartholomew and Butler (1970).

The result concerns the number of replacements required in a system of size N in an interval of time (T_1, T_2) as $T_2 \to T_1$ and $N \to \infty$, with the expected number of leavers held constant. Let $T_2 - T_1 = \delta T$ be sufficiently small for the chance of two or more replacements in the interval to be negligible. Then

$$\left.\begin{aligned} Pr\{\text{one replacement in the } i\text{th job}\} &= h(T_1)\,\delta T + O(\delta T^2) \\ Pr\{\text{no replacement in the } i\text{th job}\} &= 1 - h(T_1)\,\delta T + O(\delta T^2) \end{aligned}\right\}. \quad (7.15)$$

The expected number of replacements in $(T_1, T_1 + \delta T)$ for the whole system is thus $Nh(T_1)\,\delta T$. For this to remain fixed as $\delta T \to 0$ and $N \to \infty$ we must have $\delta T = K/N$ for some fixed K. The total number of replacements is thus a sum of Bernoulli variables whose probabilities of being non-zero are tending to zero as their number tends to infinity. Under the conditions specified the limiting distribution is Poisson in form with mean $h(T)K$ (see Feller, 1968, page 282). We note in passing that the result remains true if the probability of replacement associated with the ith job depends on i (Feller's result is stated for this more general case). This means that the different jobs could have different CLS distributions or that the time origins for the job streams could be different. We shall make use of this generalization in Chapter 8.

The practical value of limit theorems lies in their usefulness as approximations before the limit is reached. Bartholomew and Butler (1971) investigated the adequacy of the Poisson approximation both theoretically and by simulation, using a mixed exponential CLS distribution. They made exact calculations of the probabilities in (7.15) and simulated a small system with $N = 33$. Their conclusions can be summarized by saying that the approximation in only likely to be good if $(T_2 - T_1)$ is less than about one-tenth of the average CLS. In practice, average CLS's are of the order of a few years so the approximation should be reasonable if the prediction interval is no longer than a few months.

Butler (1971) also tested the adequacy of the theory using data collected for employees in a steel works. He found that the flow numbers in small homogeneous groups over fairly short time periods were approximately distributed in the Poisson form. However, there were a few large flows which could not be accounted for by the Poisson hypothesis and it seemed likely that these arose due to the transfer of groups of individuals between departments. Overall, the agreement with the theory was encouraging.

The derivation of the Poisson distribution given above does not imply that the numbers leaving in non-overlapping intervals will be independent. In general they will not be and Bartholomew and Butler (1971) investigated this point also. Their calculations suggest that the correlation between the numbers of renewals in contiguous intervals is negligible if the intervals are short enough for the Poisson approximation itself to be adequate.

In spite of the qualifications with which the Poisson approximation must be hedged it provides a very useful rule-of-thumb in practice. It means that approximate standard errors can be calculated by simply taking the square root of the expected values. The latter are needed in any case, so the extra labour of finding measures of error is minimal.

The age distribution

A useful way of learning about the past of an organization and identifying possible problems in the future is to look at the age distribution. As already noted, age in this context refers to age in the system and we use the term here instead of the more natural 'length of service' to avoid confusion with CLS. The identification of abnormal characteristics in an age distribution presupposes some knowledge of what is normal. Renewal theory enables us to determine what form the age distribution will take under various assumptions about past history. To illustrate the approach we take the case of an organization of fixed size set up with a full complement at time zero and operating according to the assumptions of the renewal model.

Some founder members may still be in post at time T and any such must obviously have age T. The probability that a randomly selected person has age T is thus the proportion who will be expected to survive to T. That is

$$Pr\{x = T\} = G(T) \tag{7.16}$$

where x denotes age. For values of $x < T$ we have

$$\begin{aligned} &Pr\{\text{individual at } T \text{ has age in } (x, x + \delta x)\} = a(x|T)\,\delta x, \text{ say,} \\ &= Pr\{\text{he joined in } (T - x, T - x + \delta x) \text{ and survived for } x\} \\ &= Pr\{\text{joined in } (T - x, T - x + \delta x)\}\; Pr\{\text{survived for } x\} \end{aligned}$$

since the survival time is independent of the time of joining. Thus

$$a(x|T)\,\delta x = h(T - x)\,\delta x G(x) \qquad (0 \le x < T)$$

or

$$a(x|T) = h(T - x)G(x). \tag{7.17}$$

The age distribution thus depends on the CLS distribution, directly through $G(x)$ and indirectly through the renewal density.

For typical CLS distributions we have seen that $h(T)$ is a decreasing function over most, if not all, its range. Hence $h(T - x)$ is an increasing function of x for fixed T. The survivor function $G(x)$ is always non-decreasing in x. The product of the two functions may therefore increase or decrease, but if T is large and x is small $h(T - x)$ will change only slowly so that $a(x|T)$ will be like $G(x)$ near the origin. In the limit as $T \to \infty$ this correspondence occurs for all x. Under these conditions $G(T) \to 0$ so that the discrete lump of probability at $x = T$ vanishes. Also, for fixed x,

$$\lim_{T \to \infty} h(T - x) = \mu^{-1}.$$

Hence

$$a(x|\infty) = \mu^{-1}G(x) \qquad (0 \leq x < \infty). \tag{7.18}$$

The steady-state age distribution is thus proportional to the survivor function and so is a non-decreasing function of age.

This result provides a means of estimating a survivor function in the absence of any information about leavers. It requires, of course, the steady-state assumption which is unlikely to be exactly satisfied. Nevertheless, there are circumstances where flow data are not available and an estimate of the survivor function, however rough, is well worth having. If, on the other hand, $G(x)$ can be estimated from flow data the estimate can be compared with the age distribution to see whether the system is near its steady state.

Age distributions can be calculated under more complicated assumptions about past history and in this way the shape of the age distribution can be related to this history. We have seen, for example, how a system maintained at a constant size will eventually have an age distribution whose density decreases with increasing age. A 'bump' in one age group of an observed distribution would thus provide clear evidence of a happening in the past giving rise to an excess number in that group.

7.3 RECRUITMENT IN AN ORGANIZATION OF CONSTANT SIZE BUT CHANGING CLS DISTRIBUTION

The assumption that the CLS distribution is the same for all persons irrespective of when they join is unlikely to be true over long periods of time. Changes may occur gradually as a result of trends in economic or educational factors or abruptly because of the introduction of a new recruitment policy. There has been very little research on the effect of such changes so our account will necessarily be incomplete. We shall give some basic formulae which should be useful in a fuller investigation of the subject.

Let the density function of CLS for a person who joins the organization at time x after it was established be $f_x(T)$. We shall assume that this density has the same functional form for all x but that its parameters are continuous functions of x. The individual is thus characterized by the time at which he enters the organization. An integral equation for the renewal density can easily be constructed from first principles. The expected number of leavers in $(T, T + \delta T)$ from among those who joined in $(x, x + \delta x)$ is

$$h(x)\,\delta x\, f_x(T - x)\,\delta T.$$

Integrating with respect to x, adding in a term for the initial members and equating to $h(T)\,\delta T$ we find

$$h(T) = f_0(T) + \int_0^T h(x) f_x(T - x)\,\mathrm{d}x. \tag{7.19}$$

This equation is not amenable to solution by taking Laplace transforms and we have not succeeded in finding a solution for any distributions of present interest. Even if we take $f_x(T)$ to be exponential with its parameter depending on x in a simple way no progress seems possible. The simplest way of investigating the solution would be to solve the discrete time version of (7.19) by numerical methods. In this case we find the following difference equation for $h(T)$:

$$h(T) = f_0(T) + \sum_{i=0}^{T-1} h(i) f_i(T - i) \qquad (T = 1, 2, \ldots) \tag{7.20}$$

where $f_i(T)$ is now a discrete probability distribution and $h(T)$ is the expected number of recruits needed at time T. In spite of the intractability of the generalized renewal equation it is possible to obtain an approximation to the solution similar to that given in (7.7). We shall give the argument in full and, in so doing, will justify the earlier approximation which is a special case.

In addition to satisfying (7.19), $h(T)$ is also the solution of the equation

$$F_0(T) = \int_0^T h(x) G_x(T - x)\,\mathrm{d}x, \tag{7.21}$$

because both sides of this equation are alternative ways of expressing the expected proportion of initial members who have left at time T. It now follows that we may rewrite (7.19) as

$$h(T) = f_0(T) + \frac{F_0(T) \displaystyle\int_0^T h(x) f_x(T - x)\,\mathrm{d}x}{\displaystyle\int_0^T h(x) G_x(T - x)\,\mathrm{d}x}. \tag{7.22}$$

The approximation consists of treating

$$\frac{\int_0^T h(x) f_x(T-x)\,dx}{\int_0^T f_x(T-x)\,dx} \quad \text{and} \quad \frac{\int_0^T h(x) G_x(T-x)\,dx}{\int_0^T G_x(T-x)\,dx}$$

as equal. The resulting approximation is then

$$h^0(T) = f_0(T) + \frac{F_0(T)\int_0^T f_x(T-x)\,dx}{\int_0^T G_x(T-x)\,dx}. \tag{7.23}$$

Each of the fractions following (7.22) is a weighted average of the renewal density. They will be close in value for any T either if the weight functions are similar or if $h(x)$ is nearly constant over the range of x for which the weights have appreciable density. The former situation occurs if $f_x(T)$ is close to the exponential with constant mean and the latter when $f_x(T)$ is highly skewed with slowly changing parameters. An investigation of the accuracy of the approximation has been undertaken by Butler (1970) who showed that it is reasonably good.

So far we have assumed that the parameters of the CLS distribution are continuous functions of time. A case of considerable practical interest arises when there is an abrupt change from one set of parameter values to another. This could be a direct result of new selection methods or a new source of recruitment becoming available. In order to develop the theory for such a system let us take as our time origin the point at which the change in CLS distribution takes place. Let $f_1(T)$ be the density function of the CLS distribution for members recruited before the change and $f_2(T)$ the density function for those recruited afterwards. We shall denote by $d_1(T)$ the density function of the *residual CLS* distribution. This is the distribution of the remaining length of service of a member of the system selected at random at time zero. If we consider the process for $T \geq 0$ we have what Cox (1962, Chapter 2) has called a modified renewal process. The renewal density then satisfies the integral equation

$$h(T) = d_1(T) + \int_0^T h(T-x) f_2(x)\,dx. \tag{7.24}$$

On taking the Laplace transform of each side of (7.24) we find

$$h^*(s) = d_1^*(s)/\{1 - f_2^*(s)\}. \tag{7.25}$$

Whether or not this result is of value in solving the renewal equation will depend on the forms of $f_2(T)$ and $d_1(T)$. This last distribution will depend on the way in which the organization operated before time zero. In order to take the analysis further let us suppose that it started at time X with its members all having the same CLS distribution $f_1(T)$. Under these circumstances the residual CLS distribution at $T = 0$ is the same as the forward recurrence time and is thus given by

$$d_1(T) = f_1(X + T) + \int_0^X h(-x) f_1(x + T)\,\mathrm{d}x \tag{7.26}$$

where $h(-x)$ is the renewal density at time $X - x$ after the start of the initial process. In the limit as $X \to \infty$

$$d_1(T) \to \frac{1}{\mu_1} G_1(T) \tag{7.27}$$

where μ_1 is the mean of $f_1(T)$. In this case

$$d_1^*(s) = \frac{1}{\mu_1 s}\{1 - f_1^*(s)\}$$

so that (7.25) becomes

$$h^*(s) = \frac{1}{\mu_1 s} \frac{\{1 - f_1^*(s)\}}{\{1 - f_2^*(s)\}}. \tag{7.28}$$

As a simple illustration of this result let us suppose that

$$f_i(T) = \lambda_i \mathrm{e}^{-\lambda_i T} \qquad (i = 1, 2).$$

Then

$$h^*(s) = \frac{\lambda_1(\lambda_2 + s)}{s(\lambda_1 + s)} = \frac{\lambda_2}{s} + \frac{\lambda_1 - \lambda_2}{\lambda_1 + s}.$$

This can easily be inverted to give

$$h(T) = \lambda_2 + (\lambda_1 - \lambda_2)\mathrm{e}^{-\lambda_1 T}. \tag{7.29}$$

The new equilibrium is thus approached exponentially at a rate which depends on the initial renewal density λ_1. To take a more realistic example let

$$f_i(T) = p_i \lambda_{1i} \mathrm{e}^{-\lambda_i T} + (1 - p_i)\lambda_{2i} \mathrm{e}^{-\lambda_{2i} T} \qquad (i = 1, 2)$$

for which

$$h^*(s) = \frac{(s + \lambda_{12})(s + \lambda_{22})(s + \nu_1)}{\mu_1 s(s + \lambda_{11})(s + \lambda_{21})(s + \nu_2)} \tag{7.30}$$

where

$$\left.\begin{aligned} \nu_i &= \lambda_{1i}\lambda_{2i}\mu_i \\ \mu_i &= p_i/\lambda_{1i} + (1 - p_i)/\lambda_{2i} \end{aligned}\right\} \qquad (i = 1, 2).$$

By resolving (7.30) into partial fractions it is evident that $h(T)$ has the form

$$h(T) = A + B\,\mathrm{e}^{-\nu_2 T} + C\,\mathrm{e}^{-\lambda_{11} T} + D\,\mathrm{e}^{-\lambda_{21} T}. \tag{7.31}$$

Standard methods yield the following expressions for the coefficients:

$$A = \mu_2^{-1}$$

$$B = \frac{-(\nu_1 - \nu_2)(\lambda_{12} - \nu_2)(\lambda_{22} - \nu_2)}{\mu_1 \nu_2(\lambda_{11} - \nu_2)(\lambda_{21} - \nu_2)}$$

$$C = \frac{-(\lambda_{11} - \nu_1)(\lambda_{12} - \lambda_{11})(\lambda_{22} - \lambda_{11})}{\mu_1 \lambda_{11}(\lambda_{11} - \nu_2)(\lambda_{21} - \lambda_{11})}$$

$$D = \frac{(\lambda_{21} - \nu_1)(\lambda_{22} - \lambda_{11})(\lambda_{22} - \lambda_{21})}{\mu_1 \lambda_{21}(\lambda_{21} - \nu_2)(\lambda_{21} - \lambda_{11})}.$$

The implications of these formulae will be clearer if we consider a numerical example. Let

$$\left.\begin{aligned} \lambda_{11} &= 11{\cdot}0 \\ \lambda_{21} &= 1{\cdot}1 \end{aligned}\right\}, \qquad \left.\begin{aligned} \lambda_{12} &= 5{\cdot}5 \\ \lambda_{22} &= 0{\cdot}55 \end{aligned}\right\} \qquad p_1 = p_2 = \tfrac{1}{2}.$$

Then

$$\mu_1 = 0{\cdot}5, \qquad \mu_2 = 1{\cdot}0, \qquad \nu_1 = 6{\cdot}050 \quad \text{and} \quad \nu_2 = 3{\cdot}025.$$

In this example the expected length of service for members recruited after time zero is twice that of the original members. Substitution of the numerical values in (7.31) gives, for the renewal density,

$$h(T) = 1 - 0{\cdot}798\,\mathrm{e}^{-3{\cdot}025T} + 0{\cdot}655\,\mathrm{e}^{-11T} + 1{\cdot}143\,\mathrm{e}^{-1{\cdot}1T}. \tag{7.32}$$

This function is plotted on Figure 7.2 where it may be compared with the renewal density for the case when $f_1(T)$ and $f_2(T)$ are exponential with means 0·5 and 1·0 respectively. In the mixed exponential case the initial drop in recruitment is greater but the approach to the new equilibrium value is much slower. Over the period shown the total recruitment would need to be higher with the mixed exponential CLS distributions than with the exponential.

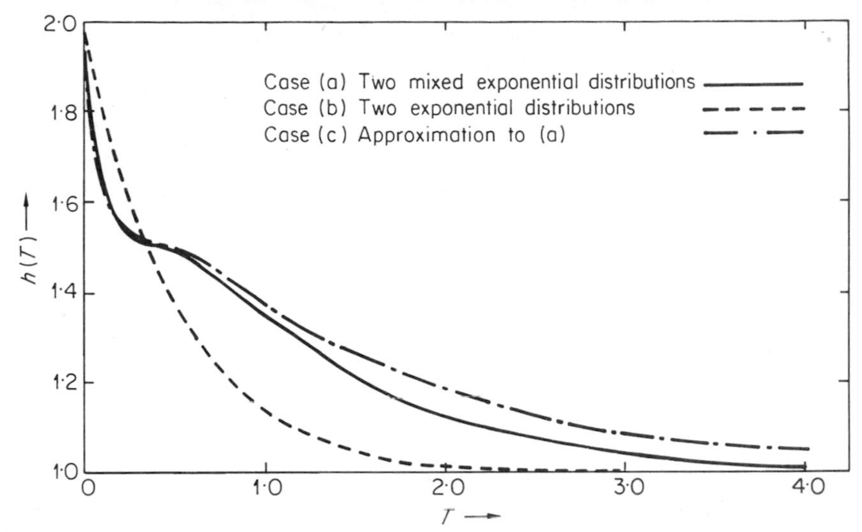

Figure 7.2 The renewal density after a change in the CLS distribution

It is possible to find an approximation for $h(T)$ by the same method as before. By using the argument that led up to (7.21) we find that

$$D_1(T) = \int_0^T G_2(x)h(T-x)\,dx \tag{7.33}$$

where

$$D_1(T) = \int_0^T d_1(x)\,dx.$$

Hence we arrive at the approximation

$$h^0(T) = d_1(T) + \frac{D_1(T)F_2(T)}{\int_0^T G_2(x)\,dx}. \tag{7.34}$$

It may be verified that this expression gives the exact solution when the two distributions are exponential. The accuracy of the approximation for the numerical example discussed above can be judged from Figure 7.2, where $h^0(T)$ is plotted as a broken line. For $T \leq 1$ the approximation is very close but it is less good thereafter. It should provide a useful guide in situations where the renewal equation (7.19) cannot be solved explicitly. An example of such a case arises when the CLS distribution is lognormal. Then we find that the functions

$$F_2(T) \quad \text{and} \quad \int_0^T G_2(x)\,dx$$

of (7.34) can be easily obtained as in Section 7.2. The same would be true of $d_1(T)$ and $D_1(T)$ if it were possible to assume that the system had reached equilibrium by time zero. However, our previous results show that this is a most unreasonable assumption. If X is known we could determine $d_1(T)$ from (7.26), but, in practice, it would be preferable to make a direct empirical estimate. This could be obtained from the length of service structure at $T = 0$ and the known CLS distribution, $f_1(T)$.

7.4 RECRUITMENT AND WASTAGE IN AN ORGANIZATION WITH CHANGING SIZE

An expanding organization

When the size of the organization is changing the recruitment and wastage figures for a given period are not necessarily the same. In the expanding case recruitment will exceed wastage because of the need to fill newly established places. The converse is true in a declining organization and if the contraction is sufficiently rapid it may even be necessary to have a 'negative' recruitment in the form of dismissals. At the beginning of this section we shall avoid the complications introduced by contraction and restrict attention to expanding organizations. The theory for such systems can then be deduced directly from the results of Chapter 5.

Equation (5.52) of Chapter 5 relates the expected number of losses in $(0, T)$ to the recruitment rate throughout the period. Expressed in the simplified notation of this chapter it becomes

$$\bar{n}(T) = N(0)F(T) + \int_0^T F(T - x)R(x)\,\mathrm{d}x \tag{7.35}$$

where

$$R(T) = M(T) + \frac{\mathrm{d}\bar{n}(T)}{\mathrm{d}T}. \tag{7.36}$$

Since $\bar{n}(T)$ is a non-decreasing function of T and $M(T)$ cannot be negative in a continuously expanding organization, it follows that $R(T) \geq 0$ for all T. On substituting for $R(x)$ in (7.35) we obtain an integral equation for $\bar{n}(T)$ which may be solved by the Laplace transform technique. Using the properties of the transform listed in Chapter 5 (Section 5.5) we find

$$\bar{n}^*(s) = N(0)\frac{f^*(s)}{s} + M^*(s)\frac{f^*(s)}{s} + \bar{n}^*(s)f^*(s)$$

whence

$$\bar{n}^*(s) = \frac{f^*(s)N^*(s)}{1 - f^*(s)} = N^*(s)h^*(s). \tag{7.37}$$

In arriving at (7.37) we have used the fact that

$$N(T) = N(0) + \int_0^T M(x)\,dx$$

and hence that

$$N^*(s) = N(0)/s + M^*(s)/s.$$

On inverting each side of (7.37) we have

$$\bar{n}(T) = \int_0^T N(x)h(T - x)\,dx. \tag{7.38}$$

Equations (7.37) and (7.38) show that the renewal density plays a fundamental role in the theory of expanding organizations. The methods and approximations of Section 7.2 can thus be used to find $h(T)$ and then $\bar{n}(T)$ can be determined from (7.38). Having found the expected number of losses, we can easily obtain the expected number of recruits by adding $N(T) - N(0)$ to $\bar{n}(T)$.

The foregoing theory can easily be adapted to cover the case of an organization which remains constant in size for a period and then enters on expansion. Let us transfer the time origin to the point at which the change takes place. Then the only modification required is the replacement of the term $N(0)F(T)$ in (7.35) by the expected number of those in the system at time zero who will have left at time T. If we denote the density function of residual CLS for members of the organization at time zero by $d(T)$, then the Laplace transform of $\bar{n}(T)$ is found to be

$$\bar{n}^*(s) = \left[f^*(s)N^*(s) + \frac{N(0)}{s}\{d^*(s) - f^*(s)\} \right] / \{1 - f^*(s)\}. \tag{7.39}$$

If the system has reached equilibrium before expansion begins

$$d(T) = \frac{1}{\mu}G(T)$$

and

$$d^*(s) = \frac{1}{\mu}G^*(s) = \frac{1}{\mu s}\{1 - f^*(s)\}$$

so that

$$\bar{n}^*(s) = N^*(s)h^*(s) + \frac{N(0)}{\mu s^2} - \frac{N(0)}{s}h^*(s). \tag{7.40}$$

This equation may be inverted term by term to give

$$\bar{n}(T) = \int_0^T N(x)h(T-x)\,\mathrm{d}x + N(0)\left\{\frac{T}{\mu} - \int_0^T h(x)\,\mathrm{d}x\right\}. \tag{7.41}$$

The expected number of losses in this case differs from the earlier result of (7.38) by an amount equal to the last term of (7.41). It is known (see, for example, Cox, 1962, Chapter 4) that

$$\lim_{T\to\infty}\left\{\frac{T}{\mu} - \int_0^T h(x)\,\mathrm{d}x\right\} = -\tfrac{1}{2}(V^2 - 1)$$

where V is the coefficient of variation of the CLS distribution. Hence, for skew distributions with $V > 1$, there will be a smaller total loss when expansion is delayed until the system has reached equilibrium than when it takes place initially.

The statistics of recruitment and wastage are usually expressed not in terms of total change but as rates. Thus, for example, the crude wastage rate in the interval (T_1, T_2) is defined as

$$w(T_1, T_2) = \{n(T_2) - n(T_1)\}\Big/\left\{\frac{1}{(T_2 - T_1)}\int_{T_1}^{T_2} N(x)\,\mathrm{d}x\right\} \tag{7.42}$$

where $n(T_1)$ and $n(T_2)$ are the observed numbers of losses at T_1 and T_2 respectively. If the function $N(T)$ is known only at the end-points of the interval, as in many published figures, the denominator in (7.42) is replaced by $\frac{1}{2}\{N(T_2) + N(T_1)\}$. For theoretical purposes it is useful to define an instantaneous wastage rate obtained by letting $T_2 \to T_1$ in (7.42) and replacing $n(T_1)$ and $n(T_2)$ by their expectations. Thus we let

$$w(T) = \frac{\mathrm{d}\bar{n}(T)}{\mathrm{d}T}\Big/N(T). \tag{7.43}$$

We shall make use of $w(T)$ in the following section but for present purposes we shall need $w(T_1, T_2)$.

In order to illustrate the effect of expansion on crude wastage we shall give two examples. Let us assume that the CLS distribution is mixed exponential; then we have shown that

$$h(T) = \mu^{-1} + a\,\mathrm{e}^{-bT}$$

where

$$a = p\lambda_1 + (1-p)\lambda_2 - \mu^{-1}, \qquad b = p\lambda_2 + (1-p)\lambda_1$$

and

$$\mu = p/\lambda_1 + (1-p)/\lambda_2.$$

In our first example we assume that the system is subject to a linear growth law of the form

$$N(T) = MT,$$

M being the rate of growth. The function $\bar{n}(T)$ can be obtained from (7.38) by straightforward integration. On substituting the result in (7.42) we find that

$$w(T_1, T_2) = \frac{T_2 - T_1}{\mu} + \frac{2a}{b(T_1 + T_2)}\left\{T_2 - T_1 - \left(\frac{e^{-bT_1} - e^{-bT_2}}{b}\right)\right\}. \quad (7.44)$$

Some calculations based on this formula are given in Table 7.5 for three CLS distributions of the mixed exponential family. The parameter values selected were those obtained by fitting the distribution to the data relating to the Glacier Metal Co., J. Bibby & Sons and the United Steel Cos. which have already been used in this chapter.

A comparison of these figures with those in Table 7.1 shows that although the wastage rate still decreases it does not do so as rapidly as before. It can be seen from (7.44) that $w(T_1, T_2)$ does not depend on the value of M and hence the wastage approaches the same equilibrium value regardless of the rate of expansion.

The previous example is based on a continuous growth function. In our second example we shall suppose that $N(T)$ jumps from an initial value of $N(0)$ to a new value of $N(0) + N$ at $T = T_J$. The expected number of losses in this case is clearly

$$\begin{aligned}\bar{n}(T) &= N(0)\int_0^T h(x)\,dx \qquad (T \leq T_J)\\ &= N(0)\int_0^T h(x)\,dx + N\int_0^{T-T_J} h(x)\,dx \qquad (T > T_J).\end{aligned}$$

***Table 7.5** Percentage wastage in successive quarters for a group expanding at a constant rate*

Quarter	Glacier Metal Co.	J. Bibby & Sons	United Steel Cos.
1	23·5	35·8	49·6
2	21·3	32·9	40·5
3	19·5	30·2	33·5
4	18·0	28·0	28·6
5	16·8	26·2	25·2
6	15·9	24·6	22·7
7	15·1	23·3	20·9
8	14·5	22·2	19·4

Some quarterly wastage rates for this kind of growth are given in Table 7.6. Calculations are given for two values each of T_J and N, assuming a CLS distribution like the one for the United Steel Cos.

Table 7.6 Percentage wastage in successive quarters for a firm expanded from $N(0)$ to $N(0) + N$ at $T = T_J$ assuming the United Steel Cos. CLS distribution

	$T_J = \frac{1}{2}$ *year*		$T_J = 1$ *year*	
Quarter	$N = \frac{1}{2}N(0)$	$N = N(0)$	$N = \frac{1}{2}N(0)$	$N = N(0)$
1	45·8	45·8	45·8	45·8
2	28·0	28·0	28·0	28·0
3	27·9	32·4	19·0	19·0
4	18·9	21·2	14·4	14·4
5	14·4	15·6	23·3	29·0
6	12·1	12·7	16·7	19·5
7	11·0	11·3	13·3	14·7
8	10·4	10·6	11·5	12·3

The effect of an abrupt increase in size is to arrest the decreasing wastage rate. With a large increase in size there is a temporary increase in wastage but the effect is short lived and after a year its influence has almost vanished.

A contracting organization

The theory which has been developed above is not restricted in its application to expanding organizations. It applies whenever the recruitment rate $R(T)$ is non-negative for all T. The case of an organization which is being reduced in size may or may not fall into this category. If the desired rate of contraction is small the losses from the system may be more than sufficient to achieve it. New entrants will then be required to prevent the numbers from dropping too rapidly. On the other hand, if a rapid run-down is necessary natural wastage may not be enough to reach the target. In such cases it becomes necessary to introduce involuntary wastage by removing redundant members from the system. One important reason for studying contracting systems is to determine the conditions necessary to avoid redundancy. We shall therefore examine this question first.

The maximum rate of run-down which can be attained without redundancy is clearly that which occurs when no recruits are admitted to the system. This will be called *natural contraction*. We can obtain this by putting

$R(T) = 0$ for all T in (7.35) and (7.36), but a more direct approach is available. Let the residual CLS distribution of all members of the system at $T = 0$, when contraction begins, be $d(T)$. Then the expected proportion remaining at time T is

$$N(T) = N(0)\{1 - D(T)\} \tag{7.45}$$

where $D(T)$ is the distribution function associated with $d(T)$. If the required size exceeds that given by (7.45) for any T then some redundancy will be required. The wastage rate (see 7.43) for an organization undergoing natural contraction is obviously

$$\frac{d(T)}{1 - D(T)}$$

and this will, typically, be a decreasing function of T. Hence observed wastage rates will decline even if there has been no change in the leaving behaviour of the members.

The above result has important practical implications, but two words of caution should be added. We have assumed that the propensity to leave is unaffected by the decision to reduce the size of the organization. In practice the psychological effects of being in a declining firm might very well increase the tendency to leave. This would have the effect of making a faster run-down possible. Secondly, we have treated the organization as a whole but in practice some grades or divisions would have to be run-down more rapidly than others. It is perfectly possible to use the theory for subgroups of a larger system and this should be done in preference to treating the system as a whole.

We shall continue the investigation by means of examples. It is common to specify a desired rate of contraction in terms of a percentage rate of run-down. In continuous time this is tantamount to requiring

$$N(T) = N(0)\,e^{-\alpha T} \quad (0 < \alpha < \infty). \tag{7.46}$$

It is an important practical question to know how large α can be without the need for redundancies. That is, we require the largest α for which

$$N(0)\,e^{-\alpha T} \geq N(0)\{1 - D(T)\} \tag{7.47}$$

for all T.

In the case when the CLS distribution is exponential, $d(T) = \lambda\,e^{-\lambda T}$ whatever the initial age distribution of the population, so the condition becomes

$$e^{-\alpha T} \geq e^{-\lambda T} \quad \text{or} \quad \alpha \leq \lambda. \tag{7.48}$$

Redundancy will thus be avoided if the rate of contraction is less than the wastage rate. Thus, for example, if a firm had an annual wastage rate of 15 per cent its size could be reduced at that rate without causing redundancies. This result seems obvious at first sight and might be supposed to hold whatever the CLS distribution. In fact, as we shall see, it is true only when the CLS distribution is exponential, which is rarely the case in practice. The extent to which this simple rule of thumb may err in practice will be examined by assuming that the CLS distribution is mixed exponential.

If contraction starts at $T = 0$ when all members have zero length of service then

$$d(T) = f(T) = p\lambda_1 \, e^{-\lambda_1 T} + (1 - p)\lambda_2 \, e^{-\lambda_2 T}$$

and the condition on α becomes

$$e^{-\alpha T} \geq p \, e^{-\lambda_1 T} + (1 - p) \, e^{-\lambda_2 T}$$

or

$$1 \geq p \, e^{-(\lambda_1 - \alpha)T} + (1 - p) \, e^{-(\lambda_2 - \alpha)T}. \tag{7.49}$$

This is clearly bounded below one if and only if

$$\alpha \leq \min(\lambda_1, \lambda_2). \tag{7.50}$$

The equilibrium wastage rate in this case is $(p/\lambda_1 + (1 - p)/\lambda_2)^{-1}$, which will always be larger than the value of α given by (7.50) and often much larger, as we shall see. This result is easily generalized to the case of a k term mixed exponential in which case the condition on α is

$$\alpha \leq \min(\lambda_1, \lambda_2, \ldots, \lambda_k). \tag{7.51}$$

For CLS distributions with a survivor function whose tails do not approach zero exponentially fast there will be no α for which (7.47) holds. The Silcock type XI and the lognormal are both examples of such distributions. For these we could never maintain a constant rate of contraction indefinitely without causing redundancy, though in practice the contraction might well come to an end before the redundancy point was reached.

The assumption that all members have zero length of service when contraction starts is rather improbable in practice, but the conditions (7.50) and (7.51) hold regardless of the initial age distribution. Thus, supposing that the distribution is initially $a(x)$ then

$$d(T) = \int_0^\infty a(x) \frac{f(x + T)}{G(x)} \, dx$$

and hence

$$1 - D(T) = \mathrm{e}^{-\lambda_1 T} \int_0^\infty a(x) \frac{p\,\mathrm{e}^{-\lambda_1 x}}{p\,\mathrm{e}^{-\lambda_1 x} + (1-p)\,\mathrm{e}^{-\lambda_2 x}}\,\mathrm{d}x + \mathrm{e}^{-\lambda_2 T}$$

$$\times \int_0^\infty \frac{a(x)(1-p)\,\mathrm{e}^{-\lambda_2 x}}{p\,\mathrm{e}^{-\lambda_1 x} + (1-p)\,\mathrm{e}^{-\lambda_2 x}}\,\mathrm{d}x$$

$$= A\,\mathrm{e}^{-\lambda_1 T} + B\,\mathrm{e}^{-\lambda_2 T}, \text{ say, where } A + B = 1, A,B \geq 0.$$

This leads to a condition of the form of (7.49) and hence to (7.50).

The time at which redundancies first occur (assuming the condition is not satisfied) will, of course, depend on the initial age distribution. We can calculate this critical time by finding when $R(T)$ first becomes negative. From (7.36)

$$R(T) = -\alpha N(0)\,\mathrm{e}^{-\alpha T} + \frac{\mathrm{d}\bar{n}(T)}{\mathrm{d}T} \tag{7.52}$$

where $\bar{n}(T)$, given by (7.38), takes the form

$$\bar{n}(T) = N(0) \int_0^T \mathrm{e}^{-\alpha x} h(T - x)\,\mathrm{d}x.$$

Note that (7.38) is valid for all T up to that for which $R(T)$ first becomes negative. Differentiating and substituting in (7.52) gives the result that the time of the first redundancies will be the root of

$$h(T) = \alpha\,\mathrm{e}^{-\alpha T} + \alpha \int_0^T \mathrm{e}^{-\alpha x} h(T - x)\,\mathrm{d}x. \tag{7.53}$$

For the mixed exponential when all members initially have zero length of service

$$h(T) = \frac{1}{\mu} + a\,\mathrm{e}^{-bT},$$

in which case (7.53) becomes

$$\mathrm{e}^{-bT}\left(\frac{ab}{b-\alpha}\right) = \mathrm{e}^{-\alpha T}\left(\alpha - \frac{1}{\mu} + \frac{a\alpha}{b-\alpha}\right). \tag{7.54}$$

It is easily verified that the condition for this to have no solution is equivalent to (7.50). Otherwise, the single solution is given by

$$T = \frac{\log_e ab - \log_e \{(\alpha - 1/\mu)(b - \alpha) + a\alpha\}}{(b - \alpha)}. \tag{7.55}$$

As an illustration we shall use the parameter estimates for the Glacier Metal Co. data. The values given in Table 7.3 are $p = 0{\cdot}6513$, $\lambda_1 = 0{\cdot}2684$ and $\lambda_2 = 2{\cdot}4228$, whence (with a constant rate of contraction) $a = 1{\cdot}0196$, $b = 1{\cdot}6716$ and $\mu^{-1} = 0{\cdot}3890$. Redundancy will thus be avoided with a constant rate of contraction if $\alpha \leq 0{\cdot}2684$. This limit for α corresponds to a reduction in size of $100(1 - e^{-0{\cdot}2684})$ per cent $= 23{\cdot}54$ per cent per annum. Exponential theory would have predicted an upper limit for α of $\mu^{-1} = 0{\cdot}3890$ with a maximum annual rate of run-down of $100(1 - e^{-0{\cdot}3890})$ per cent $= 32{\cdot}2$ per cent. Substitution of $\alpha = 0{\cdot}3890$ in (7.55) shows that such a policy would lead to redundancies occurring after

$$T = \frac{\log_e 1{\cdot}6716 - \log_e 0{\cdot}3890}{1{\cdot}6716 - 0{\cdot}3890} = \frac{1{\cdot}4580}{1{\cdot}2826} = 1{\cdot}14 \text{ years.}$$

If, on the other hand, the system had reached equilibrium before contraction started

$$\bar{n}(T) = N(0) \int_0^T e^{-\alpha x} h(T - x)\,dx + N(0)\left\{\frac{T}{\mu} - \int_0^T h(x)\,dx\right\}$$

from (7.41). Differentiating as before and substituting in (7.52) gives

$$R(T) = e^{-\alpha T}\left(\alpha - \frac{1}{\mu} + \frac{a\alpha}{b - \alpha}\right) - e^{-bT}\left(\frac{a\alpha}{b - \alpha}\right) = 0. \tag{7.56}$$

If $\alpha = \mu^{-1}$, as in the foregoing numerical example, $R(T)$ will be negative for all T and so redundancies will be required at the outset. This stands in marked contrast to the first example where no redundancies occurred for over a year. The reason for this is intuitively clear. In the former case all members originally had zero length of service and hence a high propensity to leave. A degree of recruitment was thus necessary to replace the exceptionally high initial losses and so stop the size declining too rapidly. In the latter case the initial wastage was already at its equilibrium level, which is not sufficient to effect the desired reduction.

The general conclusion of our analysis so far is *that it will not usually be possible to run down an organization at a rate as high as the observed wastage rate without causing redundancies.* We must next consider the problem of how to estimate recruitment and wastage rates if redundancies occur.

When redundancies occur (7.35) is no longer valid but (7.36) continues to hold if we take $R(T) = 0$ whenever the right-hand side is negative. If this happens $\bar{n}(T)$ must be interpreted as the total expected loss – including dismissals. In order to determine $\bar{n}(T)$ and $R(T)$ in this case we must distinguish between those members who leave of their own choice and those who are dismissed. Let the dismissal rate at time T be $R_D(T)$ so that the expected number of dismissals in $(0, T)$ is

$$\bar{n}_D(T) = \int_0^T R_D(x)\,dx. \tag{7.57}$$

Also let $\bar{n}_L(T)$ denote the expected number of those who leave of their own accord in $(0, T)$. Then

$$\bar{n}(T) = \bar{n}_D(T) + \bar{n}_L(T)$$

and hence

$$\frac{d\bar{n}(T)}{dT} = \frac{d\bar{n}_L(T)}{dT} \text{ if } R_D(T) = 0$$

$$= \frac{d\bar{n}_L(T)}{dT} + R_D(T) \text{ if } R_D(T) > 0$$

for a given value of T. Now $R(T) = 0$ if and only if $R_D(T) > 0$ and, consequently, $R_D(T) = 0$ if and only if $R(T) > 0$. Therefore (7.36) must take one of the following forms:

$$\left.\begin{aligned} R(T) &= M(T) + \frac{d\bar{n}_L(T)}{dT} \\ &\text{or} \\ 0 &= M(T) + \frac{d\bar{n}_L(T)}{dT} + R_D(T). \end{aligned}\right\} \tag{7.58}$$

It is clear that the first member of (7.58) will cover both cases if we define $R(T) = -R_D(T)$ whenever

$$M(T) + \frac{d\bar{n}_L(T)}{dT}$$

is negative. The final stage in the solution is to find an equation for $\bar{n}_L(T)$. This may be expressed as follows:

$$\bar{n}_L(T) = N(0)F(T) + \int_0^T F(T-x)\langle R(x)\rangle\,dx - \int_0^T J(T-x|x)\langle -R(x)\rangle\,dx \tag{7.59}$$

where

$$\langle X \rangle = X \quad \text{if } X > 0$$
$$= 0 \quad \text{if } X \leq 0.$$

In (7.59) the first two terms on the right-hand side give the expected number who would have left even if there had been no dismissals. However, some of these potential leavers will have been dismissed so a term must be subtracted to allow for this. The third term is therefore the expected number of redundant members among the potential leavers. The function $J(T|x)$ is the probability that a person who is declared redundant at time x would have left voluntarily in the interval $(x, x + T)$. In general this probability will depend on how those to be dismissed are chosen. Three rules which suggest themselves for investigation are (a) 'last in first out', (b) 'first in first out' and (c) 'selection at random'. When the CLS distribution is exponential the residual length of service distribution is exponential whichever rule is used for selecting those to be dismissed. We thus have $J(T|x) = F(T) = (1 - e^{-\lambda T})$ for all x and hence (7.59) takes the following form:

$$\bar{n}_L(T) = N(0)F(T) + \int_0^T F(T - x)R(x)\,dx. \tag{7.60}$$

Recalling that the first part of (7.58) can be used regardless of the sign of $R(T)$ we have a pair of equations for finding $\bar{n}_L(T)$ and $R(T)$ which are identical in form to (7.35) and (7.36). With an exponential CLS distribution, and only in this case, it is possible to use the same method whether or not redundancies occur. The results which we obtained for this distribution which are given in (7.47) and (7.48) thus apply for all α and λ provided that $\bar{n}(T)$ is replaced by $\bar{n}_L(T)$ when $\lambda > \alpha$.

In the general case the solution is more difficult, though straightforward in principle. We would first obtain $\bar{n}_L(T)$ by substituting for $R(x)$ in (7.59). Secondly, we would find $R(T)$ from (7.58). When this was known we could calculate the expected number of dismissals from (7.57) and the expected number of recruits could be obtained as

$$\int_0^T \langle R(x) \rangle\,dx.$$

In unpublished work, Casson has investigated each of the above redundancy rules and provided methods of calculating recruitment and redundancy.

7.5 THE MEASUREMENT OF LABOUR TURNOVER

We have pointed out in other contexts the role of stochastic models in constructing meaningful measures of social phenomena. One of the simplest examples of their use in this connexion is provided by the problem of measuring labour turnover. Labour turnover refers to the flow of people into and out of an organization. In the context of the renewal models of the present chapter, for which input is simply related to output, turnover is equivalent to wastage. We shall use the two terms synonymously, and practically all measures which have been proposed are based on the wastage flow alone. Some writers speak of measuring stability, which is essentially the converse of turnover, but the two concepts are equivalent—a high stability implies low turnover and vice versa.

The measurement of labour turnover serves two purposes. One is as an important ingredient of manpower models as, for example, when earlier in this chapter we related the crude wastage rate to the age structure of an organization. The second use is as an index of morale or efficiency. It is with this use that we shall be concerned in this section. A high turnover of employees is usually regarded as a sign of poor morale leading to inefficient operation, and there is an extensive literature devoted to the question of how to measure and interpret it. A bewildering variety of measures have been proposed and a good deal of confusion still surrounds the whole question. Some of the literature was listed in Chapter 6 in the course of the discussion of models for the leaving process. To those given there may be added the papers by Bowey (1969), Bibby (1970), Hyman (1970), Forbes (1971b), Van der Merwe and Miller (1971) and Clowes (1972). We shall here aim to clarify the issues by bringing the results of the earlier stochastic analysis to bear. A measure will not, of course, automatically be a good one just because it is based on a stochastic model. An inappropriate model will lead to misleading conclusions, as Stoikov's (1971) model and the subsequent discussion by Bartholomew (1971) shows.

Possibly the earliest, and certainly one of the most widely used, measures of turnover is the crude rate defined by

$$I = \frac{\text{Number who leave in a given interval}}{\text{Average number employed during the same interval}} \times 100\,\%. \qquad (7.61)$$

The inadequacy of such an index is readily apparent from the analysis of Section 7.2, where we used the index to show the influence on loss of the age of a system. A measure which depends so strongly on age structure cannot adequately reflect other attributes—unless the system has reached its steady state. Thus a difference in the crude rate between two firms may simply be a reflexion of the fact that one has a much younger age structure. Lane and Andrew (1955) provide a striking illustration of this sort of situation by giving data on two firms of which the one with the higher wastage also had the longer expected length of service.

In demographic work an analogous situation arises in the measurement of mortality, where the pitfalls of using crude death rates are well understood. Salubrious seaside resorts have high crude death rates, not because they are unhealthy but because they attract as residents a disproportionate number of elderly people. Demographers and actuaries deal with this problem by calculating standardized mortality rates which are estimates of what the crude rate would be if the population had a chosen standard age distribution. Exactly the same course can be followed when measuring turnover. Suppose we adopt a standard age distribution with density function $s(x)$. (We return below to the question of how $s(x)$ might be chosen.) The expected number of leavers in $(T, T + \delta T)$ from those with age in $(x, x + \delta x)$ will then be

$$N(T)s(x)\,\delta x \lambda(x)\,\delta T$$

where $N(T)$ is the size of the system at time T and $\lambda(x)$ is the force of separation common to all members of the system. The total expected number of leavers in an interval (T_1, T_2) will thus be

$$\int_{T_1}^{T_2} N(T)\,\mathrm{d}T \int_0^{\infty} s(x)\lambda(x)\,\mathrm{d}x \tag{7.62}$$

where, again, we treat $N(T)$ as a continuous variable. The average number present in (T_1, T_2) is

$$\int_{T_1}^{T_2} N(T)\,\mathrm{d}T/(T_2 - T_1),$$

so the crude turnover index for this $s(x)$ is estimated to be

$$I = (T_2 - T_1)\int_0^{\infty} s(x)\lambda(x)\,\mathrm{d}x. \tag{7.63}$$

This expression demonstrates the way in which the crude index depends on both the age structure and the force of separation. By adopting a standard form for the age distribution we eliminate variation from this source and have a measure depending solely on the force of separation.

A natural choice for $s(x)$ is the steady-state age distribution which would be attained in a constant size system as given by (7.18). This will give an index showing what the crude rate would be in equilibrium. Making the substitution

$$I = (T_2 - T_1)\int_0^\infty \frac{1}{\mu} G(x)\lambda(x)\,dx = \frac{T_2 - T_1}{\mu}$$

$$\times \int_0^\infty f(x)\,dx = \frac{T_2 - T_1}{\mu}. \tag{7.64}$$

The standardized index thus turns out to be equivalent to the average CLS—a measure proposed by Lane and Andrew (1955).

The foregoing discussion serves to underline what is otherwise obvious, namely that a measure of wastage should depend only on the propensity to leave. This is expressed by $\lambda(x)$ or either of the equivalent functions $f(x)$ and $G(x)$.

The 'best' way of describing the leaving process is thus by giving an estimate of one or other of these functions. Now that graphical output is commonly available on computers, graphs of these functions can easily be displayed and reproduced. A diagram is easy to assimilate and less open to misinterpretation than any index. However, there may continue to be circumstances in which the information conveyed by $\lambda(x)$ has to be summarized into a single number. The mean of the CLS distribution is an obvious possibility which has already arisen in connexion with the standardized index. The median, or half-life, has similar appeal with the added advantage of being both easy to interpret and easy to calculate. An equally simple measure is obtained by quoting the proportion who would be expected to survive for a specified period.

In Chapter 6 we failed to find any one-parameter distribution which successfully fitted CLS data. Any attempt to convey all the relevant information contained in $\lambda(x)$ by a single number is therefore bound to be inadequate. For example, two distributions can have the same proportions surviving to one year, say, but be very different at other points. There is less risk of concealing important differences if two indices are used, each highlighting different features of the distribution. The two parameters ω and σ of a fitted lognormal distribution contain most of the information about a CLS distribution, though not in an easily interpreted form. Perhaps the simplest pair of measures is $G(x_1)$ and $G(x_2)$ for two suitably chosen lengths of service x_1 and x_2 (x_1 might be the end of the training period and x_2 the time an average person is ready for promotion).

The treatment here and in Chapter 6 has dealt with the dependence of leaving on the length of service to the exclusion of all else. In reality $\lambda(x)$ is a function not only of the length of service but of many other things such as sex, place of residence, skill level, etc. An apparent difference between two firms as judged by any of our measures may therefore be the result of different sex ratios or some other such factors. This serves to emphasize that the length of service based indices can only be meaningfully used if the groups involved are homogeneous with respect to other factors which influence leaving. If it is desired to compare heterogeneous groups the indices must be standardized. This can be done in the same way as described for age.

CHAPTER 8

Renewal Theory Models for Graded Social Systems

8.1 INTRODUCTION

In this chapter we return to the study of graded social systems begun in Chapters 3 and 5. The distinguishing feature of the present models is that the grade sizes do not change with time. The models of the earlier chapters were based on the assumption that the transition intensities, or probabilities in the case of discrete time models, were given and hence that the grade sizes were random variables. In the present case the roles of the constants and random variables are reversed; it is the grade sizes which are fixed and the transition rates which are random variables. If the processes are viewed deterministically the two kinds of specification are equivalent as far as their equilibrium predictions are concerned. The differences emerge when we consider their transient and stochastic behaviour.

The assumption of fixed grade sizes is often more realistic than that of fixed promotion and loss rates. In many organizations there is a fixed establishment in each grade which is determined by the work available for its members to perform. This situation is found in many government organizations, notably the Civil Service, and in British universities where the proportion of senior staff in teaching posts is at present limited to 40 per cent of the total establishment. Our models are designed to describe the operation of such systems and, in particular, to show the effect of grade structure on promotion prospects and wastage rates. We shall restrict the discussion to models in continuous time but Chapter 4 contains some closely related results for a discrete time model.

The basic model has the same elements as before but its mode of operation is different. We suppose that the grades form a hierarchy as illustrated in Figure 5.2 in Section 5.4 of Chapter 5. Every movement connected with the system is assumed to arise from a loss. Suppose, for example, that a member of grade 2 leaves. The vacancy thus created must be filled at once in order to maintain the size of the grade. This may be done either by direct recruitment from outside or by transfer from within the organization. If the vacancy is filled by promotion from grade 1 a new vacancy is created in that grade and this in turn must be filled by outside recruitment or demotion. The operation of such a system depends upon two further factors. One is the stochastic mechanism governing losses from the system; the other is the procedure for filling vacancies. The latter factor includes the decision on whether or not to fill the vacancy from within the organization and, if so, how it is to be done.

The stochastic law governing loss from the system will be specified by a loss intensity, as in Chapter 5. Two basic assumptions will be made and these will form the basis of one of our two main methods of classifying the models. In Sections 8.2 and 8.3 we shall assume that the loss intensity is a function of an individual's total length of service only. It follows from this assumption that other factors, such as the speed with which the individual has risen in the hierarchy, do not influence the likelihood of his leaving. In practice, a man's decision to leave might very well result from his having been passed over for promotion. To accommodate this kind of effect we shall, in Sections 8.4 and 8.5, suppose that the loss intensity is a function only of his length of service in the present grade. This can never be true exactly because as a man approaches retirement his age obviously becomes an important factor, but we shall not incorporate this into the model.

Our second main classification of models is based on the promotion rule. A promotion rule specifies the way in which a person is selected to fill a vacancy from within the system. Again we shall concentrate on two extreme cases. The first rule says that a vacancy shall be filled by the most senior member of the grade below. If several members have the same seniority one of them is selected at random. When this rule is applied consistently, the length of service of the member promoted both within his grade and within the organization is at least as great as that of any other person who is eligible. This rule, or something approximating to it, often operates when experience constitutes the main qualification for promotion. The second rule postulates that a man's chance of promotion shall be independent of his seniority or length of service. This would be appropriate when some kind of innate ability was the determining consideration in assessing suitability for promotion. In probabilistic language this rule says that the member to be promoted should be selected at random from among those eligible.

This fourfold classification of renewal models includes a wide range of possibilities between the extremes which it encompasses. Butler (1970) has considered the effect of promoting the most junior and White (1970a) has a similar model in which 'promotees' are drawn from various grades with specified probabilities. Models used for manpower planning purposes in the British Civil Service have a similar feature. We shall concentrate here on the extreme but simple cases which together provide considerable insight into how graded systems work.

Notation and basic relations

All our models may be regarded as generalizations of the simple renewal process which formed the basis of the analysis in the last chapter. These processes are non-Markovian and their full mathematical analysis presents considerable difficulties. In order to make progress possible we shall impose two restrictions at the outset which were only introduced at a late stage in our previous discussion of hierarchical organizations. These are:

(a) That all recruits enter the lowest grade. Vacancies occurring in grade 1 must therefore be filled from outside and vacancies higher up must be filled internally.

(b) Demotions do not take place and promotions are always into the next highest grade. This means that a vacancy in grade 4, say, sets in train a succession of promotions ending with the entry of a new recruit to grade 1.

In particular cases we shall have to specialize still further by assuming that the grade sizes are large or that the system has reached equilibrium.

As far as possible we shall use the same notation and terminology as in previous chapters but some modifications will be necessary. The grade sizes are no longer functions of T so we shall drop the argument and write them as

$$n_1, n_2, \ldots, n_k \quad \text{with} \quad N = \sum_{i=1}^{k} n_i.$$

To conform with the usage of Chapter 5 we should have to write the loss intensity for a member of grade j as $r_{j,k+1}(T)$. Since we shall be assuming that this intensity is independent of j and to avoid the need for subscripts we shall use the notation $\lambda(T)$ as in the last chapter. An extension of the notation of Chapter 7 will be introduced for the promotion rates between grades and the wastage rates from the grades. Let $n_j h_j(T)\,\delta T$ be the expected number of promotions from grade j to grade $j+1$ in $(T, T+\delta T)(j = 1, 2, \ldots, k-1)$. We shall describe $h_j(T)$ as the *promotion density* from grade j. This is a natural extension of the notation $h(T)$ used for the input to the system as a whole. Similarly we shall write $n_j w_j(T)\,\delta T$ for the expected number of losses from grade j in $(T, T+\delta T)(j = 1, 2, \ldots, k)$. When considering the limiting behaviour of the system we shall write

$$\lim_{T\to\infty} h_j(T) = h_j \quad \text{and} \quad \lim_{T\to\infty} w_j(T) = w_j.$$

The density of $h_j(T)$ is the analogue of the promotion intensity $r_{j,j+1}(T)$ of Chapter 5. In fact, it may be regarded as an 'average' intensity but it would be misleading to use the same notation and terminology.

The wastage and promotion densities are related because the number of entries into any grade during a given time interval must exactly balance the number of losses through promotion or wastage. This statement implies that

$$n_{j-1}h_{j-1}(T) = n_j h_j(T) + n_j w_j(T) \qquad (j = 2, 3, \ldots, k-1). \tag{8.1}$$

When $j = 1$ this equation holds if we replace the left-hand side by $R(T)$, which is the rate of input to the system. It also holds with $j = k$ if we define $h_k(T) \equiv 0$. An equivalent way of expressing the result of (8.1) is to write

$$n_j h_j(T) = \sum_{i=j+1}^{k} n_i w_i(T) \qquad (j = 0, 1, 2, \ldots, k-1) \tag{8.2}$$

where $n_0 h_0(T)$ is read as $R(T)$ here and in the remainder of this chapter. This basic relationship holds for all our models and implies that it will be sufficient to find either the $h_j(T)$'s or the $w_j(T)$'s.

The above equations can be used to make certain simple but important deductions about the relationship between the grade structure and the promotion rates. For example, suppose that we wish to have a system which, in equilibrium, gives equal promotion chances at every level of the organization. Then setting $h_{j-1} = h_j = h$ the equilibrium form of (8.1) may be written

$$(n_{j-1} - n_j)h = n_j w_j \qquad (j = 2, 3, \ldots, k-1). \tag{8.3}$$

Since $n_j w_j$ must be non-negative for all j it follows from (8.3) that

$$n_{j-1} \geq n_j \qquad (j = 2, 3, \ldots, k-1). \tag{8.4}$$

This means that the structure of the organization must, in general, taper towards the top, with the exception that grade k may be larger than grade $k - 1$. In fact, we shall meet examples later in which it has to be very much larger. More generally, if $h_{j-1} \neq h_j$ the inequalities (8.4) must be replaced by

$$h_{j-1}n_{j-1} \geq h_j n_j \qquad (j = 2, 3, \ldots, k). \tag{8.5}$$

Using these inequalities we can deduce that if $h_{j-1} \geq h_j$ $(j = 2, 3, \ldots, k-1)$ then the grade sizes may increase or decrease as we move up the hierarchy.

Further results for the equilibrium state can be obtained by using simple renewal-type arguments. If μ is the mean length of service then the equilibrium recruitment rate will be $R = N\mu^{-1}$, which must also satisfy the equation

$$\sum_{i=1}^{k} n_i w_i = N\mu^{-1}. \tag{8.6}$$

If μ_j denotes the average time spent in grade j then the expected throughput for that grade per unit time will be $n_j\mu_j^{-1}$. This is also equal to $n_{j-1}h_{j-1}$ and hence, from (8.1)

$$\frac{n_j}{\mu_j} = n_j(h_j + w_j) = n_{j-1}h_{j-1} \qquad (j = 2, 3, \ldots, k - 1). \tag{8.7}$$

Although these last equations do not lead to any immediate conclusions about the effect of structure on promotion chances they will be useful in the subsequent analysis.

The distribution of the flows

In Chapter 7 we derived an approximation to the distribution of the number of losses in a short interval from a large 'one-grade' system. The result depended on the chance of the loss of any particular individual being small. We remarked that the conclusion did not depend on the renewal densities involved being the same. It therefore follows that the result applies to any of the flows in the graded systems which we are now considering. Our derivation of expected values in the following sections thus also serves to provide approximate standard errors. Butler (1971) investigated the promotion as well as the wastage flows in his empirical study and found that they, too, were approximately Poisson in form.

8.2 A MODEL WITH LOSS RATE A FUNCTION OF TOTAL LENGTH OF SERVICE AND PROMOTION BY SENIORITY

General theory

Our main objective in this section is to determine the promotion and wastage densities. We have seen from (8.2) that it is sufficient to determine one set of these quantities. In the present instance it happens to be more convenient to work with the wastage densities $\{w_j(T)\}$. The method which we shall use (see Bartholomew, 1963a) enables us to find

$$\sum_{i=1}^{j} n_i w_i(T)\,\delta T,$$

which is the expected number of losses from grades 1 to j inclusive in $(T, T + \delta T)$. The individual wastage densities are then obtained by differencing.

If $\lambda(T)$ denotes the loss intensity for members of the organization then, as we saw in Chapter 6, the CLS distribution has density function

$$f(T) = \lambda(T) \exp\left[-\int_0^T \lambda(x)\,\mathrm{d}x\right]$$

and survivor function (8.8)

$$G(T) = 1 - F(T) = \exp\left[-\int_0^T \lambda(x)\,\mathrm{d}x\right].$$

We shall first calculate the required expectation conditional upon f, the number of original members of the organization who remain at time T. This is a random variable whose distribution depends upon T. Since we are assuming the behaviour of the members to be independent, f will have a binomial distribution with parameters N and $G(T)$.

Consider the probability that the individual whose length of service ranks ith from the bottom leaves the organization in $(T, T + \delta T)$. This probability takes one of two forms. If $i > N - f$ the member in question will be one who joined at the beginning having length of service T. Hence the required probability is $\lambda(T)\,\delta T$. If $i \leq N - f$ the probability of the member ranked i leaving in $(T, T + \delta T)$ may be written as

$$\delta T \int_0^T \lambda(x) A_i(x|T, f)\,\mathrm{d}x \tag{8.9}$$

where $A_i(x|T, f)$ is the probability density function of that particular member's length of service. The determination of the probability of leaving thus depends on our being able to find the form of this distribution. It can be obtained directly because it is the ith order statistic from the length of service distribution of all present members of the organization who have joined since the process began. The density function of length of service was given in Chapter 7 as

$$\left.\begin{aligned} a(t|T) &= h(T - t)G(t) \qquad (0 \leq t < T) \\ Pr\{t = T\} &= G(T) \end{aligned}\right\}. \tag{8.10}$$

The density function for those recruited after $T = 0$ may thus be expressed as

$$a^0(t|T) = h(T - t)G(t)/F(T) \qquad (0 \leq t < T) \tag{8.11}$$

and hence

$$A_i(t|T, f) = (N - f)\binom{N - f - 1}{i - 1} a^0(t|T) \left\{\int_0^t a^0(x|T)\,\mathrm{d}x\right\}^{i-1} \times \left\{\int_\tau^T a^0(x|T)\,\mathrm{d}x\right\}^{N-f-i} \qquad (i = 1, 2, \ldots, N - f). \tag{8.12}$$

The conditional expected number of losses in $(T, T + \delta T)$ from grades 1 to j is now obtained by summing the individual probabilities of leaving for all members in these grades. This is greatly facilitated by the following consequence of the promotion rule. The rule implies that no member of grades 1 to j can have longer service than any member of grades $j + 1$ to k. The summation is thus over values of i from 1 to N_j where

$$N_j = \sum_{i=1}^{j} n_i.$$

Dropping the δT from both sides of the equation we thus obtain

$$\left.\begin{aligned} \sum_{i=1}^{j} n_i w_i(T|f) &= \sum_{i=1}^{N-f} \int_0^T \lambda(x) A_i(x|T, f)\,\mathrm{d}x + (N_j - N + f)\lambda(T) \\ &\qquad\qquad (f \geq N - N_j) \\ &= \sum_{i=1}^{N_j} \int_0^T \lambda(x) A_i(x|T, f)\,\mathrm{d}x \quad (f < N - N_j) \\ &\qquad\qquad (j = 1, 2, \ldots, k) \end{aligned}\right\} \tag{8.13}$$

where $w_i(T|f)$ is the wastage density conditional upon f. Reversing the order of summation and integration in the first part of (8·13) we obtain a binomial series which may be summed to give

$$\sum_{i=1}^{N-f} A_i(x|T, f) = (N - f)a^0(x|T). \tag{8.14}$$

The sum in the second part of (8.13) is an incomplete binomial series which can be expressed, using a well-known result, in terms of the incomplete beta function. Thus

$$\sum_{i-1}^{N_j} A_i(x|T, f) = (N - f)a^0(x|T)I_{y(x|T)}(N - N_j - f, N_j) \tag{8.15}$$

where

$$\left.\begin{aligned} y(x|T) &= \int_x^T a^0(u|T)\,\mathrm{d}u, \\ I_y(a, b) &= \int_0^y u^{a-1}(1 - u)^{b-1}\,\mathrm{d}u/B(a, b) \end{aligned}\right\} \tag{8.16}$$

and

$$B(a, b) = \Gamma(a)\Gamma(b)/\Gamma(a + b).$$

Substituting these results in (8.13) using the expression for $a^0(t|T)$ given in (8.11) and remembering that $\lambda(T)G(T) = f(T)$, we find

$$\left.\begin{aligned}\sum_{i=1}^{j} n_i w_i(T|f) &= \frac{N-f}{F(T)}\int_0^T f(x)h(T-x)\,\mathrm{d}x + (N_j - N + f)\lambda(T) \\ &\qquad\qquad (f \geq N - N_j) \\ &= \frac{N-f}{F(T)}\int_0^T f(x)h(T-x)I_{y(x|T)}(N - N_j - f, N_j)\,\mathrm{d}x \\ &\qquad\qquad (f < N - N_j).\end{aligned}\right\} \quad (8.17)$$

The integral equation for the renewal density enables us to replace the integral in the first part of (8.17) by $h(T) - f(T)$. The final step in the determination is to find the unconditional expectations by averaging over f. This gives

$$\sum_{i=1}^{j} n_i w_i(T) = \sum_{f=0}^{N}\sum_{i=1}^{j} n_i w_i(T|f)\binom{N}{f}\{G(T)\}^f\{F(T)\}^{N-f} \qquad (j = 1, 2, \ldots, k). \quad (8.18)$$

Inspection of the foregoing formulae shows that their evaluation requires the prior determination of the renewal density $h(T)$. This can be found either exactly or approximately by the methods of the previous chapter. There is therefore no obstacle, apart from the magnitude of the task, to finding numerical values for the wastage densities. However, this will seldom be necessary because the exact theory can be used to obtain much simpler approximations, as we shall show later.

Exact solution in special cases

The method of solution is most easily illustrated by assuming a constant loss intensity, λ. In this case

$$f(T) = \lambda\,\mathrm{e}^{-\lambda T}$$

and

$$h(T) = \lambda.$$

Making this substitution in (8.17) we have

$$\left.\begin{aligned}\sum_{i=1}^{j} n_i w_i(T|f) &= \frac{(N-f)}{1-e^{-\lambda T}}\lambda(1-e^{-\lambda T}) + \lambda(N_j - N + f)\\ &= N_j\lambda \qquad (f \geq N - N_j)\\ &= (N-f)\lambda\left(1 - \frac{N - N_j - f}{N-f}\right) = N_j\lambda\\ &\quad (f < N - N_j) \qquad (j = 1, 2, \ldots, k).\end{aligned}\right\} \tag{8.19}$$

These equations show that the wastage densities are independent of f, so that

$$\sum_{i=1}^{j} n_i w_i(T) = N_i\lambda \qquad (j = 1, 2, \ldots, k).$$

Hence, for all j

$$w_j(T) = \lambda. \tag{8.20}$$

As we might have expected the wastage densities are constant in time and the same for all grades. The promotion densities can be obtained from (8.2), which gives

$$h_j(T) = (N - N_j)\lambda/n_j \qquad (j = 1, 2, \ldots, k - 1). \tag{8.21}$$

These formulae show how the promotion prospects in the organization depend on the relative grade sizes. As an illustration of their use we return to a question raised earlier in the chapter and ask what structure would yield equal promotion prospects. In order to make the right-hand side of (8.21) independent of j we must have

$$n_j \propto N - N_j = \sum_{i=j+1}^{h} n_i \qquad (j = 1, 2, \ldots, k - 1). \tag{8.22}$$

This condition implies that the grade sizes must decrease in geometric progression as we move up the hierarchy. The exception to this rule is that the size of the highest grade may be any multiple of the one below it. A similar result was obtained in Section 5.3 of Chapter 5 for a system with fixed input and random grade sizes. There we showed that if the promotion rates were equal the expected grade sizes would form a geometric series. Here we have shown that if the grade sizes are in geometric progression the *expected promotion rates* will be equal. This result illustrates a remark we made at the beginning of the chapter about the equivalence of the two kinds of model when interpreted deterministically.

The practical usefulness of the result for a constant loss rate is severely limited because we have seen in Chapter 6 that it has no empirical support. Nevertheless, the extremely simple forms of the solution in this case provide a convenient base-line from which to measure the consequences of introducing greater realism into the model.

A complete solution is difficult to obtain for any other loss intensity unless we consider special cases. One such case is obtained by taking $k = 2$, $n_1 = n_2 = 1$. This specialization yields a process of some intrinsic interest but of little practical relevance. The reason for introducing it at this stage is that it enables us to judge the adequacy of an approximation discussed in the next section.

If $n_1 = n_2 = 1$

$$Pr\{f = i\} = \binom{2}{i}\{G(T)\}^i\{F(T)\}^{2-i} \qquad (i = 0, 1, 2).$$

The incomplete beta function in (8.17) reduces to $y(x|T)$ when $j = 1$, so that (8.18) yields the following expression for $w_1(T)$:

$$w_1(T) = 2h(T)G(T) - f(T)G(T) + 2\int_0^T f(t)h(T-t)\left\{\int_t^T G(x)h(T-x)\,dx\right\}dt. \tag{8.23}$$

The effect of using a more realistic CLS distribution may be illustrated by taking

$$f(T) = \tfrac{1}{2}\{\lambda_1\,e^{-\lambda_1 T} + \lambda_2\,e^{-\lambda_2 T}\} \qquad (T \geq 0).$$

We have already solved the renewal equation for this distribution and found that

$$h(T) = \mu^{-1} + a\,e^{-bT}$$

where, with $p = \frac{1}{2}$,

$$a = \tfrac{1}{2}(\lambda_1 + \lambda_2) - \mu^{-1}, \qquad b = \tfrac{1}{2}(\lambda_1 + \lambda_2) \quad \text{and} \quad \mu = \tfrac{1}{2}(\lambda_1^{-1} + \lambda_2^{-1}).$$

The evaluation of $w_1(T)$ from (8.23) is tedious but straightforward. Without loss of generality we may take $\mu = 1$ and then we find that

$$\left.\begin{aligned} w_1(T) &= \frac{3}{2} - \frac{1}{2b} + \frac{2a^2}{b}e^{-bT} + \left(a^2T - \frac{3a^2}{2b}\right)e^{-2bT} \\ w_2(T) &= 2h(T) - w_1(T) \end{aligned}\right\}. \tag{8.24}$$

Thus the wastage densities both approach their equilibrium values at a rate governed by the exponential factor e^{-bT}. Some illustrative calculations have been plotted on Figure 8.1, which appears below. For these calculations

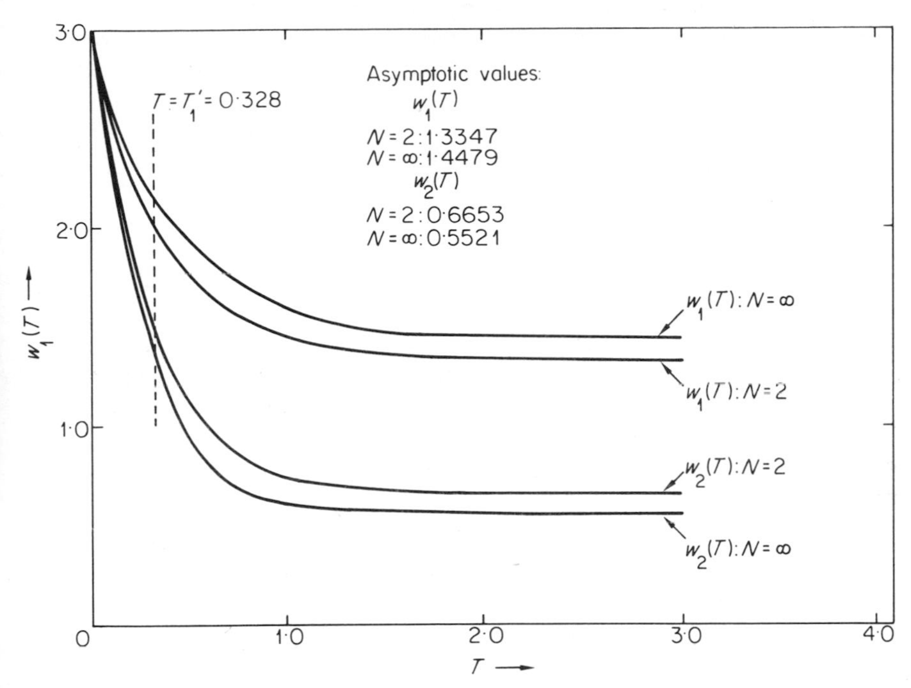

Figure 8.1 Graphs of $w_1(T)$ and $w_2(T)$ for $k = 2$, $n_1 = n_2$ and promotion by seniority

we have taken $\lambda_1 = 10\lambda_2$ which, since $\mu = 1$, implies that

$$\lambda_1 = 5{\cdot}5, \qquad \lambda_2 = 0{\cdot}55, \qquad a = 2{\cdot}025, \qquad b = 3{\cdot}025,$$

and

$$\left.\begin{aligned} w_1(T) &= 1{\cdot}3347 + 2{\cdot}7112\, e^{-3{\cdot}025T} + \{4{\cdot}1006T - 1{\cdot}0209\}\, e^{-6{\cdot}05T} \\ w_2(T) &= 0{\cdot}6653 + 1{\cdot}3388\, e^{-3{\cdot}025T} - \{4{\cdot}1006T - 1{\cdot}0209\}\, e^{-6{\cdot}05T} \end{aligned}\right\}. \tag{8.25}$$

The wastage rate declines more rapidly in grade 2 than in grade 1, but both reach their equilibrium values after approximately the same length of time.

Solution when the grade sizes are large

The general expressions for the wastage densities simplify considerably in two limiting cases which we shall consider in this and the following section.

The first of these relates to the case when the grade sizes are large and the second to the equilibrium state reached as $T \to \infty$. Both cases are of interest in their own right but our more immediate concern is to see whether they will serve as approximations to the exact solution.

We shall examine the effect of allowing n_i $(i = 1, 2, \ldots, k)$ to tend to infinity with the ratios n_i/n_{i+1} $(i = 1, 2, \ldots, k-1)$ held constant. The first step in the derivation is to note that, if the n_i's are large, we may replace

$$\sum_{i=1}^{j} n_i w_i(T) = E_f \sum_{i=1}^{j} n_i w_i(T|f)$$

by

$$\sum_{i=1}^{j} n_i w_i(T|E(f)).$$

The two expressions are asymptotically equal and we shall show that the latter will serve as a good approximation to the former for quite small values of the n_i's. In order to obtain the limiting forms for the wastage densities we must therefore replace f by its expectation $NG(T)$ in (8.17). The range of validity of each part of that equation requires careful consideration because it is defined in terms of f. An obvious way of dealing with the question is to replace f in the inequality $f \geq N - N_j$ by its expectation. This gives

$$NG(T) \geq N - N_j \quad \text{or} \quad NF(T) \leq N_j,$$

and so defines the range of validity of each part of (8.17) in terms of T. Let the critical value of T defined by this inequality be denoted by T'_j. The above step may then be justified by noting that

$$\lim_{\substack{N, N_j \to \infty \\ N_j/N \text{ fixed}}} Pr\{f \geq N - N_j | G(T)\} = 1 \quad \text{if } T < T'_j$$

$$= 0 \quad \text{if } T > T'_j.$$

Asymptotically, therefore, the inequalities in terms of f and T are equivalent. On making the necessary substitutions in (8.17) we find, for $T < T'_j$,

$$\sum_{i=1}^{j} n_i w_i(T) \doteqdot N\{h(T) - f(T)\} + N\lambda(T)\{G(T) - (N - N_j)/N\}$$

$$(j = 1, 2, \ldots, k),$$

which simplifies to give

$$\sum_{i=1}^{j} n_i w_i(T) \doteqdot Nh(T) - (N - N_j)\lambda(T). \tag{8.26}$$

Differencing the equation we finally obtain

$$\left.\begin{aligned} n_1 w_1(T) &\doteqdot Nh(T) - (N - n_1)\lambda(T) \\ w_j(T) &\doteqdot \lambda(T) \qquad (j = 2, 3, \ldots, k) \end{aligned}\right\}. \tag{8.27}$$

For $T > T'_j$ the second part of (8.17) applies. After replacing f by its expectation the incomplete beta function becomes

$$I_{y(x|T)}(NF(T) - N_j, N_j).$$

The limiting form of this function can be found by considering the behaviour of the beta probability density function

$$\frac{1}{B(r, s)} u^{r-1}(1 - u)^{s-1}$$

as r and $s \to \infty$ with their ratio remaining fixed. Under these conditions the density function becomes concentrated at the point $u = r/(r + s)$ and hence it follows that

$$\begin{aligned} I_y(r, s) &\to 0 \quad \text{if } y < \frac{r}{r + s} \\ &\to 1 \quad \text{if } y \geq \frac{r}{r + s}. \end{aligned}$$

If we now apply this result to the second part of (8.17) we obtain

$$\sum_{i=1}^{j} n_i w_i(T) \doteqdot N \int_0^{x_j(T)} f(x)h(T - x)\,\mathrm{d}x \qquad (T > T'_j) \tag{8.28}$$

where $x_j(T)$ satisfies

$$\int_0^{x_j(T)} G(x)h(T - x)\,\mathrm{d}x = N_j/N \qquad (j = 1, 2, \ldots, k - 1). \tag{8.29}$$

The last equation arises from the fact that $x_j(T)$ thus defined is the critical length of service at which the limiting value of the incomplete beta function changes from zero to one.

The limiting form of the solution given by equations (8.27), (8.28) and (8.29) is much simpler than the exact form and again it depends only on the renewal density, $h(T)$. We shall illustrate these results and make some assessment of their suitability as approximations in the finite case by assuming that the CLS distribution is mixed exponential with parameters p, λ_1 and λ_2. For $T < T'_j$ the approximation can be written down at once from (8.27).

When $T > T'_j$ we must use (8.28) and (8.29), remembering that the renewal density has the form

$$h(T) = \mu^{-1} + a\,\mathrm{e}^{-bT}.$$

This gives

$$\sum_{i=1}^{j} n_i w_i(T) \doteqdot N\left(\frac{1}{\mu}\int_0^{x_j(T)} f(x)\,\mathrm{d}x + a\,\mathrm{e}^{-bT}\int_0^{x_j(T)} \mathrm{e}^{bx} f(x)\,\mathrm{d}x\right) \tag{8.30}$$

with $x_j(T)$ given by

$$\frac{1}{\mu}\int_0^{x_j(T)} G(x)\,\mathrm{d}x + a\,\mathrm{e}^{-bT}\int_0^{x_j(T)} \mathrm{e}^{bx} G(x)\,\mathrm{d}x = \frac{N_j}{N} \qquad (j = 1, 2, \ldots, k-1). \tag{8.31}$$

We thus have a pair of parametric equations which can be used to plot the wastage densities as functions of time.

Some calculations have been made for the case $k = 2$, $n_1 = n_2$ using the same parameter values as for the exact solution obtained above. The curves have been plotted on Figure 8.1, where they are labelled '$N = \infty$'. They may be compared with the exact solution for $N = 2$ found earlier which has been drawn on the same figure. In spite of the crudeness of the approximations which have been made it is clear that the magnitude of N has relatively little influence on the wastage densities. Further calculations are needed before this conclusion can be fully justified, but this example suggests that little will be lost by using the formulas derived on the assumption of large grade sizes. Additional support for our conclusion is provided by the fact that the approximation yields the exact solution when the loss rate is constant.

In Sections 8.4 and 8.5 of the present chapter we shall assume that the loss intensity depends on seniority within the grade instead of on total length of service. Under this assumption we shall not be able to obtain the exact theory and shall have to be content with an approximation of the kind given above. Since it will be derived by a deterministic argument it is instructive to see how our present results could have been obtained without recourse to the exact solution.

If N is large the proportion of original members remaining at time T will be approximately equal to $G(T)$. By virtue of the promotion rule these members will occupy the upper part of the hierarchy. In fact, if $G(T) \geq 1 - N_j/N$, grades $j + 1$ to k will consist entirely of original members all with length of service T. These members are all subject to the loss intensity $\lambda(T)$ and hence the expected number of losses in $(T, T + \delta T)$ is $\lambda(T)(N - N_j)\,\delta T$. In consequence, the expected number of losses from grades 1 to j inclusive is the difference between this number and the total number of losses from the whole system in the same interval. Thus we have

$$\sum_{i=1}^{j} n_i w_i(T) \doteqdot Nh(T) - \lambda(T)(N - N_j) \qquad (j = 1, 2, \ldots, k),$$

as in (8.27). This result holds if $T < T'_j$. When $T > T'_j$ we proceed as follows. The length of service distribution for members at time T was given in (8.10) and hence we can easily calculate the proportion of members whose length of service is less than any given value. All members of grades 1 to j must have shorter lengths of service than any member of grades $j + 1$ to k. Hence promotion from grade j to grade $j + 1$ must take place after a length of service $x_j(T)$ which must be such that the proportion of persons whose service does not exceed this value is exactly equal to N_j/N. That is, $x_j(T)$ must satisfy the equations

$$\int_0^{x_j(T)} G(x)h(T - x)\,\mathrm{d}x = \frac{N_j}{N} \qquad (j = 1, 2, \ldots, k - 1),$$

which is the same as (8.29). The expected number of losses from grades 1 to j in $(T, T + \delta T)$ is therefore

$$N\,\delta T \int_0^{x_j(T)} \lambda(x)a(x|T)\,\mathrm{d}x \qquad (j = 1, 2, \ldots, k - 1),$$

which leads directly to (8.28).

The equilibrium solution

We shall now show that the system approaches an equilibrium state as $T \to \infty$. Once more the direct practical usefulness of the results is limited by the fact that the equilibrium is approached very slowly if we make realistic assumptions about the form of the CLS distribution. However, the comparison of the results with those for the exponential CLS distribution gives an indication of the effect of an increase in the dispersion of the CLS distribution. They also serve to show the 'direction' in which the system is moving.

The limiting values of the wastage densities are obtained by finding the limit of the right-hand side of (8.18) as $T \to \infty$. Since

$$\begin{aligned} \lim_{T\to\infty} Pr\{f = i\} &= 1 \quad \text{if } i = 0 \\ &= 0 \quad \text{otherwise} \end{aligned}$$

the term in which $f = 0$ is the only one which has to be considered. As $T \to \infty$ this limit is easily found and we then have

$$\sum_{i=1}^{j} n_i w_i = \frac{N}{\mu} \int_0^{\infty} f(x) I_{y(x)}(N - N_j, N_j)\, dx \tag{8.32}$$

where

$$y(x) = \lim_{T \to \infty} y(x|T) = \frac{1}{\mu} \int_x^{\infty} G(u)\, du.$$

Integration by parts gives the alternative expression

$$\sum_{i=1}^{j} n_i w_i = \frac{N}{\mu^N B(N - N_j, N_j)} \int_0^{\infty} G(x) F(x) \left(\int_x^{\infty} G(u)\, du \right)^{N - N_j - 1} \left(\int_0^{x} G(u)\, du \right)^{N_j - 1} dx. \tag{8.33}$$

These equations show that the equilibrium wastage densities can always be obtained either explicitly or by numerical integration.

An explicit solution can easily be obtained if the CLS distribution has the type XI form. In this case

$$f(T) = \frac{v}{c}\left(1 + \frac{T}{c}\right)^{-(v+1)}$$

$$\mu = c/(v - 1)$$

$$G(T) = (1 + T/c)^{-v}$$

and

$$\int_T^{\infty} G(x)\, dx = \mu \left(1 + \frac{T}{c}\right)^{-v+1}.$$

On substituting these expressions in (8.33) and evaluating the integral we find that

$$\sum_{i=1}^{j} n_i w_i = \frac{N}{\mu} \left\{ 1 - \frac{B\left(N, \dfrac{v}{v-1}\right)}{B\left(N - N_j, \dfrac{v}{v-1}\right)} \right\} \quad (j = 1, 2, \ldots, k - 1). \tag{8.34}$$

If N, N_j and $N - N_j$ are all large, Stirling's approximation to the gamma function gives

$$\sum_{i=1}^{j} n_i w_i \sim \frac{N}{\mu} \left\{ 1 - \left(\frac{N - N_j}{N} \right)^{v/(v-1)} \right\} \quad (j = 1, 2, \ldots, k - 1). \tag{8.35}$$

By allowing v to tend to infinity in either of equations (8.34) or (8.35) we recover the solution for the exponential distribution.

To illustrate the equilibrium theory we have given in Table 8.1 some values of w_1 when $k = 2$ for a type XI CLS distribution with $\mu = 1$. The variance of the distribution is given in the last row of the table; it does not

***Table 8.1** Values of w_1 when $k = 2$ for a type XI CLS distribution with shape jarameter v*

		v				
Structure	N	*Limit as* $v \to 1$	2	3	11	∞
	3	1·50	1·25	1·16	1·04	1.00
$n_1 = 2n_2$	18	1·50	1·32	1·20	1·05	1·00
	∞	1·50	1·33	1·21	1·05	1·00
	3	2·00	1·33	1·20	1·05	1·00
$n_1 = n_2$	18	2·00	1·47	1·28	1·07	1·00
	∞	2·00	1·50	1·29	1·07	1·00
	3	3·00	1·50	1·29	1·06	1·00
$n_1 = 2n_2$	18	3·00	1·63	1·35	1·08	1·00
	∞	3·00	1·67	1·37	1·08	1·00
Variance $\{v/(v-2)\}$		–	–	3·00	1·22	1·00

exist for $v \leq 2$. These calculations suggest two general conclusions. First, *the effect of increasing the dispersion of the CLS distribution is to increase the wastage rate of the lowest grade relative to that of the system as a whole.* This implies a decrease in w_2 and hence a reduction in the promotion rate between grades 1 and 2. The amount of this change depends upon the structure of the organization, being greatest when the higher grade is larger than the lower. Put the other way round this conclusion states that the equilibrium promotion chances would be increased if the variability in completed length of service could be decreased. In a sense this is the converse of the conclusion reached in Chapter 5, Section 5.6. There we found that a decrease in the variability of CLS would tend to reduce the relative size of the lowest grade. The second conclusion is that the total size of the organization is relatively unimportant as far as its effect on w_1 is concerned. It is the relative rather than the absolute values of the grade sizes which are important.

This supports our earlier conclusion based on the transient behaviour of the two-grade system with mixed exponential CLS distribution.

A full investigation of the dependence of the promotion and wastage densities on the structure for larger values of k would require extensive calculations but some further light can be thrown on the question by calculating the structure needed for the promotion rates to be equal. We have already solved this problem for the exponential CLS distributions. For any other distribution the promotion rates change with time and so the question can only be answered for the equilibrium state. We shall assume that the grade sizes are large enough to justify using the approximation given by (8.35). If we express this in terms of the h_j's by means of (8.2) it becomes

$$\frac{n_j}{N} = \frac{1}{\mu h_j}\left(1 - \frac{N_j}{N}\right)^{\nu/(\nu-1)} \qquad (j = 1, 2, \ldots, k-1). \tag{8.36}$$

These equations enable us to determine the relative grade sizes necessary to attain any desired set of promotion densities. They have been solved for the case $k = 5$ and $h_1 = h_2 = h_3 = h_4$ and the results are given in Table 8.2 for two values of ν. For the purposes of this calculation we have supposed that the common promotion density is $2/\mu$.

***Table 8.2** Relative grade sizes necessary to give equal promotion densities, in equilibrium, for $k = 5$, $\mu h = 2$ and a type XI CLS distribution*

	j				
ν	1	2	3	4	5
2	0·268	0·162	0·107	0·075	0·388
∞ (exponential)	0·333	0·222	0·148	0·099	0·198

The distribution with $\nu = 2$ has a much greater degree of skewness than the exponential distribution which has $\nu = \infty$. These calculations show that the structure has to be even more top-heavy when $\nu = 2$ than when $\nu = \infty$ in order to offer equal promotion chances at every level. The values of ν which Silcock (1954) obtained by fitting the type XI curve to his data were all smaller than 2. *It thus appears that the attainment of equal promotion prospects is an impossible goal in a pyramidical structure with typical wastage patterns.*

The limiting solution given by (8.32) is valid for all grade sizes. An equilibrium solution valid when the grade sizes are large can be found in two ways. One is to find the limiting form of (8.30) and (8.31) as $T \to \infty$. The other is to apply a limiting operation to (8.32) in which the grade sizes become large. In each one we reach the following approximate solution:

$$\sum_{i=1}^{j} n_i w_i \doteqdot \frac{N}{\mu} \int_0^{x_j} f(x)\,\mathrm{d}x = \frac{N}{\mu} F(x_j) \tag{8.37}$$

where x_j satisfies

$$\frac{1}{\mu} \int_0^{x_j} G(x)\,\mathrm{d}x = N_j/N \qquad (j = 1, 2, \ldots, k - 1). \tag{8.38}$$

When the CLS distribution has the type XI form these equations lead to the same approximation as given in (8.35). Very similar arguments have been used by Keyfitz in unpublished work for an organization expanding at a constant rate and by Morgan, Keenay and Ray (1973).

8.3 A MODEL WITH LOSS RATE A FUNCTION OF TOTAL LENGTH OF SERVICE AND PROMOTION AT RANDOM

General theory

Promotion is said to be 'at random' whenever the choice is made without regard to seniority. Strictly speaking we should use the description: 'at random with respect to length of service' since selection might well be on the basis of some other observable characteristic. However, provided that any such characteristic is independent of the individual's propensity to leave, our theory is applicable. Since we shall be concerned almost entirely with wastage and promotion densities we shall have no need to specify the promotion rule more precisely. If we were interested in such things as the distribution of 'ability' throughout the organization we should have to postulate how promotion depended on ability. For present purposes the promotion rule states that there is a probability equal to n_j^{-1} that any given member of grade j will be promoted when a vacancy occurs in grade $j + 1$ $(j = 1, 2, \ldots, k - 1)$.

The general theory for this promotion rule is more difficult than it was for promotion by seniority. Nevertheless, it is possible to develop a general theory which can be made the basis for deriving usable approximations. In order to formulate the theory we require the following definitions.

(a) Let $f_j(t|T)$ $(j = 1, 2, \ldots, k-1)$ be the probability density function of the remaining length of service of a person selected at random from grade j when the system is aged T. We may also define $f_0(t|T) \equiv f(t)$. The distribution function is denoted by $F_j(t|T)$ and its complement by $G_j(t|T)$.

(b) Let $Q_j(t|T)$ $(j = 1, 2, \ldots, k-1)$ be the probability that a member of grade j is not promoted in the interval $(T, T+t)$, given that he has not left in the same interval.

(c) Let $m_j(t|T)$ $(j = 1, 2, \ldots, k-1)$ be the number of promotions from grade j to grade $j+1$ in the interval $(T, T+t)$. This number is, of course, a random variable.

The first step in the analysis is to express the expected number of losses from grade j in $(T, T+\delta T)$ in terms of the functions defined above. To do this let us classify the losses according to the time at which they entered the grade. The contribution from the initial members who were in grade j at $T = 0$ is clearly

$$n_j f_0(T|0) Q_j(T|0)\,\delta T.$$

For those who entered between $T - x$ and $T - x + \delta x$ the corresponding expected number is

$$n_{j-1} h_{j-1}(T-x) f_{j-1}(x|T-x) Q_j(x|T-x)\,\delta x\,\delta T.$$

Summing these contributions we have that

$$n_j w_j(T) = n_i f_0(T|0) Q_j(T|0) + n_{j-1} \int_0^T h_{j-1}(T-x) f_{j-1}(x|T-x) Q_j(x|T-x)\,\mathrm{d}x$$

$$(j = 1, 2, \ldots, k). \qquad (8.39)$$

As they stand, these equations involve two unknown functions. The next step in the argument is to express these in terms of known quantities.

First let us consider the probability $Q_j(t|T)$. If we calculate the probability conditional upon the event $m_j(t|T) = i$, say, then it follows at once from the promotion rule that

$$Q_j(t|T, m_j(t|T) = i) = \left(1 - \frac{1}{n_j}\right)^i. \qquad (8.40)$$

The unconditional probability may thus be expressed in the form

$$Q_j(t|T) = \sum_{i=0}^{\infty} Pr\{m_j(t|T) = i\} \left(1 - \frac{1}{n_j}\right)^i$$

$$= E\left(1 - \frac{1}{n_j}\right)^{m_j(t|T)} \qquad (j = 1, 2, \ldots, k-1). \qquad (8.41)$$

In order to complete the determination we require the distribution of $m_j(t|T)$. This is not easy to obtain for any CLS distribution other than the exponential. In that case the promotion process from any grade is a Poisson process and hence $m_j(t|T)$ has a Poisson distribution. This result may be established as follows. The residual length of service distribution at any stage of a member's career always has the same exponential distribution. Hence grade k behaves like a simple renewal process. A standard result of renewal theory (see Cox, 1962, Section 3.1) states that the input to such a system is a Poisson process. Now grade $k - 1$ is subject to two kinds of removal, each being independent Poisson processes. Hence the total output of that grade, which is identical to the input, must be a Poisson process. By repeating the argument for each grade in turn the result is established. The mean number of promotions from grade j in $(T, T + t)$ is clearly $n_j h_j t$ so that

$$Q_j(t|T) = \sum_{i=0}^{\infty} \frac{(n_j h_j t)^i}{i!} e^{-n_j h_j t} \left(1 - \frac{1}{n_j}\right)^i$$

$$= e^{-h_j t} \qquad (j = 1, 2, \ldots, k - 1). \tag{8.42}$$

When the CLS distribution is not exponential we have only been able to make further progress by assuming that the grade sizes are large. Under these circumstances we can use the fact that $m_j(t|T)$ will be approximately distributed in the Poisson form with

$$Em_j(t|T) = n_j \int_T^{T+t} h_j(x)\, dx.$$

(This result follows from a simple extension of the argument used in Chapter 7, Section 7.2, for wastage.) Hence, substituting in (8.41) and taking the limit as $n_j \to \infty$, we find

$$Q_j(t|T) \sim \exp\left[-\int_T^{T+1} h_j(x)\, dx\right] \qquad (j = 1, 2, \ldots, k - 1). \tag{8.43}$$

The second function which we have to find in order to determine $n_j w_j(T)$ from (8.39) is $f_j(t|T)$. If we were told that a member of grade j had been in that grade for time x before his promotion to grade $j + 1$ at time T then it would follow that

$$f_j(t|T, x) = f_{j-1}(t + x|T - x)/G_{j-1}(x|T - x) \qquad (j = 1, 2, \ldots, k - 1) \tag{8.44}$$

where $f_j(t|T, x)$ is the conditional distribution for given x. The distribution of x can be obtained as follows. The probability that we require is

$$Pr\{\text{length of service in grade } j \text{ is in } (x, x+\delta x) \text{ when promoted to grade } j+1 \text{ at time } T\}$$

$$= Pr\{\text{promotion from grade } j \text{ with length of service in } (x, x+\delta x) \text{ at time } T\}/Pr\{\text{stay in grade } j \text{ ended by promotion}\}.$$

Of the two probabilities appearing on the right-hand side of this equation only the numerator depends on x. It may be expressed as

$$Pr\{\text{no loss in } (T-x, T)\}Pr\{\text{no promotion in } (T-x, T)\} \times Pr\{\text{promotion in } T, T+\delta x\} = G_{j-1}(x|T-x)Q_j(x|T-x)h_j(T)\,\delta x. \tag{8.45}$$

Combining this probability with that given by (8.44) we find that

$$f_j(t|T) = C\int_0^T f_{j-1}(x+t|T-x)Q_j(x|T-x)\,\mathrm{d}x \qquad (j = 1, 2, \ldots, k-1). \tag{8.46}$$

The constant C is obviously given by

$$C^{-1} = \int_0^T G_{j-1}(x|T-x)Q_j(x|T-x)\,\mathrm{d}x. \tag{8.47}$$

The wastage densities, $\{w_j(T)\}$, can be determined in principle by the following method. First, solve (8.39), with $j = 1$. The equation then depends only on $f(T)$ and $h(T)$; $f(T)$ is given and $h(T)$ is found by solving the renewal equation. Secondly, knowing $w_1(T)$ and hence $h_1(T)$, determine $f_1(t|T)$ from (8.46). When this is known $w_2(T)$ can be found by putting $j = 2$ in (8.39). This is used to find $f_2(t|T)$ which, in turn, is needed to obtain $w_3(T)$, and so on. These steps can be carried out by numerical methods for any CLS distribution but the work involved is very heavy, especially if k is large. The only CLS distribution for which a simple solution is available is the exponential. If

$$f(T) = \lambda\,\mathrm{e}^{-\lambda T} \qquad (T \geq 0)$$

we have already noted that

$$f_j(t|T) = \lambda\,\mathrm{e}^{-\lambda t} \qquad (T \geq 0)$$

for all j and T. Also, we have shown in this case that

$$Q_j(t|T) = \mathrm{e}^{-h_j t} \qquad (j = 1, 2, \ldots, k-1).$$

Making the appropriate substitutions in (8.39) it is easy to show that

$$w_j(T) = \lambda$$

and

$$h_j(T) = (N - N_j)\lambda/n_j \qquad (j = 1, 2, \ldots, k - 1).$$

These values are identical with those obtained by assuming that promotions were made on the basis of seniority. The reason for this coincidence is that loss is independent of length of service for this distribution. Hence a rule which selects on the basis of length of service will be no different in its effect from one which selects at random. In no other case has it been found possible to obtain the full transient solution, but some further progress can be made in the limiting case when $T \to \infty$.

The equilibrium solution

The direct method of investigating the limiting behaviour of the system would be to let T tend to infinity in (8.39). An alternative and somewhat simpler method is as follows. Let us treat grades $j + 1$ to k as a single grade. This composite grade will behave like a simple renewal process with a CLS distribution having the density function given by

$$\lim_{T \to \infty} f_j(t|T) = f_j(t), \text{ say.}$$

In the limit the renewal density for this process is known to be

$$(N - N_j)\Big/\int_0^\infty tf_j(t)\,\mathrm{d}t \qquad (j = 1, 2, \ldots, k - 1).$$

By definition it is also equal to $n_j h_j$. We shall now show that the mean of the residual CLS distribution depends upon h_j and h_{j-1} so that the equations

$$n_j h_j = (N - N_j)\Big/\int_0^\infty tf_j(t)\,\mathrm{d}t \qquad (j = 1, 2, \ldots, k - 1) \qquad (8.48)$$

provide recurrence relations between h_j and h_{j-1}. Since the distribution $f_j(t)$ is of interest in its own right we shall first show how it can be found and then determine its mean. Letting $T \to \infty$ in (8.46) and using the approximate form of $Q_j(t|T)$ we find

$$f_j(t) = \int_0^\infty f_{j-1}(x + t)\,\mathrm{e}^{-h_j x}\,\mathrm{d}x/G^*_{j-1}(h_j) \qquad (8.49)$$

where

$$G_{j-1}^*(h_j) = \int_0^\infty G_{j-1}(x)\,\mathrm{e}^{-h_j x}\,\mathrm{d}x \qquad (j = 1, 2, \ldots, k-1). \tag{8.50}$$

It will be clear from the form of this equation why we have adopted the notation used earlier (Chapter 5, Section 5.5) for the Laplace transform. Once the limiting promotion densities have been found (8.49) provides the means of determining the residual CLS distributions recursively, beginning with $f_0(t) \equiv f(t)$.

Equation (8.49) can be expressed in terms of Laplace transforms. On transforming each side we obtain

$$f_j^*(s) = \frac{1}{G_{j-1}^*(h_j)}\left(\frac{f_{j-1}^*(s) - f_{j-1}^*(h_j)}{h_j - s}\right) \qquad (j = 1, 2, \ldots, k-1). \tag{8.51}$$

Differentiating each side of this equation and setting $s = 0$ gives

$$\left.\frac{\mathrm{d}f_j^*(s)}{\mathrm{d}s}\right|_{s=0} = \frac{1}{G_{j-1}^*(h_j)}\left(\frac{\left.\frac{\mathrm{d}f_{j-1}^*(s)}{\mathrm{d}s}\right|_{s=0}}{h_j} + \frac{1 - f_{j-1}^*(h_j)}{h_j^2}\right). \tag{8.52}$$

But

$$\left.\frac{\mathrm{d}f_j^*(s)}{\mathrm{d}s}\right|_{s=0} = -\int_0^\infty t f_j(t)\,\mathrm{d}t$$
$$= -(N - N_j)/n_j h_j,$$

from (8.48). Making this substitution in (8.52) and simplifying we have

$$G_{j-1}^*(h_j) = n_j/n_{j-1}h_{j-1} \qquad (j = 1, 2, \ldots, k) \tag{8.53}$$

where $n_0 h_0$ is read as $N\mu^{-1}$ and

$$G_0^*(h_1) = \int_0^\infty G(x)\,\mathrm{e}^{-h_1 x}\,\mathrm{d}x.$$

Thus using (8.53) in conjunction with (8.51) the complete set of promotion densities, and hence the wastage densities, can be found. It can easily be verified that this procedure yields the same solution for the exponential distribution as we obtained in the last section.

An example of practical interest for which a complete equilibrium solution can be obtained is provided by the mixed exponential distribution. The recurrence relations for the residual length of service density functions can be simplified by using (8.53) to give

$$\left.\begin{aligned} f_j(t) &= \frac{n_{j-1}h_{j-1}}{n_j}\int_0^\infty f_{j-1}(x+t)\,\mathrm{e}^{-h_j x}\,\mathrm{d}x \qquad (j = 2, 3, \ldots, k-1) \\ f_1(t) &= \frac{N}{n_1\mu}\int_0^\infty f(x+t)\,\mathrm{e}^{-h_1 x}\,\mathrm{d}x \end{aligned}\right\} \tag{8.54}$$

where

$$f(t) = p\lambda_1\,\mathrm{e}^{-\lambda_1 t} + (1-p)\lambda_2\,\mathrm{e}^{-\lambda_2 t} \qquad (t \geq 0).$$

Repeated application of (8.54) easily yields

$$f_j(t) = \frac{N}{n_j\mu}\prod_{i=1}^{j} h_{i-1}\left(\frac{p\lambda_1\,\mathrm{e}^{-\lambda_1 t}}{\prod_{i=1}^{j}(\lambda_1 + h_i)} + \frac{(1-p)\lambda_2\,\mathrm{e}^{-\lambda_2 t}}{\prod_{i=1}^{j}(\lambda_2 + h_i)}\right)$$

$$(j = 1, 2, \ldots, k-1) \tag{8.55}$$

where $h_0 = 1$. We thus have the very useful result that the successive residual life distributions all have the mixed exponential form. A similar result is arrived at if we take a mixed exponential distribution with more than two components. This fact makes it easy to find $G_j(t)$ and hence $G_j^*(s)$ for all j. On substituting the result in (8.53) the recurrence relation becomes

$$\prod_{i=1}^{j} h_{i-1}\left(\frac{p}{\prod_{i=1}^{j}(\lambda_1 + h_i)} + \frac{1-p}{\prod_{i-1}^{j}(\lambda_2 + h_i)}\right) = \frac{n_j\mu}{N} \qquad (j = 1, 2, \ldots, k) \tag{8.56}$$

where $h_0 = 1$ and $h_k = 0$. When $j = 1$ this equation gives a quadratic equation for h_1. The solution of the equation may then be substituted in (8.56) with $j = 2$ to give a quadratic for h_2. This procedure may be repeated for each grade in turn until the complete solution has been obtained. The wastage densities can then be determined from (8.1). As a simple illustration we have given some values of h_1 in Table 8.3 for different values of the ratio λ_1/λ_2 when $p = \frac{1}{2}$ and $\mu = 1$.

It is clear from the table that the chances of promotion from grade 1 decrease as the dispersion of the CLS distribution increases. The corresponding value of h_1 for promotion by seniority when $\lambda_1/\lambda_2 = 10$ can be calculated from the results plotted in Figure 8.1. We take the curve for $N = \infty$ because our present theory has been derived on the assumption of

***Table 8.3** Values of h_1 when the CLS distribution is mixed exponential with $p = \frac{1}{2}$, $\mu = 1$ and $n_1 = \frac{1}{2}N$*

λ_1/λ_2	1	2	5	10	∞
h_1	1·00	0·94	0·76	0·64	0·50

large grade sizes and find that $h_1 = 2 - w_1 = 0{\cdot}55$. *The random promotion rule thus offers better promotion prospects from grade* 1 *than does the seniority rule.*

A second comparison of the effects of the two rules can be made by considering the structure necessary to achieve equal promotion densities. An exact comparison is not possible for $k > 2$ because our results for the seniority rule were for the type XI distribution and those in the present case are for the mixed exponential. However, the two distributions are similar in shape so we shall choose parameter values for the mixed exponential distribution which enable us to make a qualitative comparison with the figures given in Table 8.2.

If the h_j's are to have a common value h it is clear from (8.56) that the n_j's must satisfy the equations

$$\frac{ph^{j-1}}{(\lambda_1 + h)^j} + \frac{(1-p)h^{j-1}}{(\lambda_2 + h)^j} = \frac{\mu n_j}{N} \qquad (j = 1, 2, \ldots, k-1), \tag{8.57}$$

since $h_k = 0$. The grade sizes are thus expressible as a mixture of two geometric series. There is a close link here with the result for the simple exponential CLS distribution. In fact, the latter can be obtained as a special case by putting $\lambda_1 = \lambda_2$ or $p = 0$ or 1. Some numerical values of the relative grade sizes are given in Table 8.4 for $k = 5$.

In broad terms the conclusion to be drawn from this table is the same as that from Table 8.2, which related to the rule of promotion by seniority.

***Table 8.4** Relative grade sizes required to give equal equilibrium promotion densities for a mixed exponential CLS distribution with $p = \frac{1}{2}$, $\mu = 1$ and $h = 2$ when promotion is at random*

	Grade				
	1	2	3	4	5
$\lambda_1/\lambda_2 = 10$	0·263	0·172	0·125	0·096	0·345
$\lambda_1/\lambda_2 = 1$ (exponential)	0·333	0·222	0·148	0·099	0·198

The structure is slightly less top-heavy when promotion is at random, but it is clearly the form of the CLS distribution rather than the promotion rule which is the main determinant of the grade sizes required. This analysis serves to underline our earlier conclusion that equal promotion opportunities and a tapering structure are incompatible in practice. Modifying the promotion rule can, at best, produce only a marginal improvement.

The determination of the full solution for other CLS distributions is a formidable task but it is relatively easy to obtain h_1 for any distribution. Thus, from (8.53), h_1 satisfies

$$\int_0^\infty G(x)\,\mathrm{e}^{-h_1 x}\,\mathrm{d}x = n_1\mu/N, \tag{8.58}$$

and this equation can always be solved numerically. An upper bound for h_1 is found by observing that the left-hand side of (8.58) is maximized for *given* h_1 by taking

$$\begin{aligned} G(x) &= 1 \quad \text{for } 0 \leq x \leq \mu \\ &= 0 \quad \text{for } x > \mu. \end{aligned}$$

($G(x)$ must be a non-increasing function with

$$\int_0^\infty G(x)\,\mathrm{d}x = \mu$$

and $G(x) \leq 1$ for all x.) Hence, if (8.58) is satisfied when $G(x)$ has this form, then any change in its form will require a reduction in h_1 if the equation is still to hold. Therefore, the solution of

$$\int_0^\mu \mathrm{e}^{-h'_1 x}\,\mathrm{d}x = \frac{1-\mathrm{e}^{-h'_1\mu}}{h'_1} = \frac{n_1\mu}{N}$$

is an upper bound for h_1. Tables of the solution of this equation are given in Barton, David and Merrington (1960). If $\mu = 1$ and $n_1 = \frac{1}{2}N$ we find that $h'_1 = 1{\cdot}59$, which may be compared with the values in Table 8.3. The bound is not attainable since the extremal form of $G(x)$ implies that the CLS is constant, in which case the renewal density does not exist.

Adequacy of the assumption of large grade sizes

The whole of the foregoing theory for random promotion rests on the assumption that the grade sizes are large enough to warrant the replacement of $Q_j(t|T)$ as given by (8.41) by the approximation of (8.43). We shall make a partial test of this assumption by comparing our results with the exact values in a particularly unfavourable case. This occurs when $n_i = 1$ for all i. In this case all promotion rules are equivalent because there is no scope for choice. We can therefore use the exact theory developed for promotion by seniority to make our comparison. Some calculations for the case $k = 2$ are presented in Table 8.5. We have again assumed a mixed exponential CLS distribution

Table 8.5 Comparison of exact and approximate values of h_1 for random promotion when $n_1 = n_2 = 1$

λ_1/λ_2	1	2	5	10	∞
Approximate ($N = \infty$)	1·00	0·94	0·76	0·64	0·50
Exact ($N = 2$)	1·00	0·94	0·78	0·67	0·50

with $p = \frac{1}{2}$ and $\mu = 1$. The approximation slightly underestimates the true promotion density and hence slightly overestimates the wastage density, but the agreement is remarkably good.

8.4 A MODEL WITH LOSS RATE DEPENDING ON SENIORITY WITHIN THE GRADE AND PROMOTION BY SENIORITY

Introduction of the general theory

In this and the following section we shall suppose that the loss intensity for a given individual is a function only of his seniority in his current grade. The course which we shall follow runs parallel to that followed in the earlier part of this chapter. First we shall assume that promotion is by seniority and secondly, in Section 8.5, that it is at random. The symbol τ will be used in the remainder of this chapter to denote length of service within a given grade. Otherwise, we shall use the same notation as before although the roles which the various quantities play will be somewhat different.

The models of this section imply that it is a man's experience in his current job rather than his experience with the firm which influences his propensity to leave. Thus an ambitious person might become increasingly dissatisfied with waiting for promotion and so become more likely to leave with the passage of time. Alternatively, increasing achievement or stronger personal ties might make the member less likely to leave as his seniority increases. In practice it seems probable that length of service both within the grade and within the firm will exert their influence. A general model embodying both features would be intractable so we shall adopt our usual strategy of concentrating on extreme but relatively simple cases.

The wastage and promotion densities will continue to be our main concern but, in addition, we shall meet a new feature of considerable interest. Previously the CLS distribution has been part of the data of the problem when it has sometimes been specified in terms of its associated loss intensity. In the present case it is not given but must be deduced from the loss intensities within the grades and the structure of the system. As a byproduct of the analysis we shall therefore have a new class of models for the leaving process to add to those in Chapter 5. It is of interest to see whether they have the same characteristics as those we have to be typical in practice.

The methods used in Section 8.2 for obtaining the transient behaviour of the system do not carry over to the present case. Neither has any alternative approach been found apart from the special cases noted below. For the most part, therefore, we shall have to rely on equilibrium results valid when the grade sizes are large. The principal exception to this remark is provided by the highest grade. Since each member 'begins again' on promotion to grade k this behaves exactly like a simple renewal process. The CLS distribution appropriate to this grade has a known loss intensity which we shall denote by $\lambda_k(\tau)$. This distribution plays the same role as the residual length of service distribution which appeared in the last section, so its density function will be denoted by $f_{k-1}(\tau)$. Corresponding functions for other grades have the same notation with an appropriate change of subscript. The expected number of entrants to grade k in $(T, T + \delta T)$ is $n_{k-1}h_{k-1}(T)$ and this must clearly satisfy the following integral equation:

$$n_{k-1}h_{k-1}(T) = n_k f_{k-1}(T) + n_{k-1} \int_0^T h_{k-1}(T-\tau) f_{k-1}(\tau)\, d\tau. \tag{8.59}$$

This equation may be solved for $n_{k-1}h_{k-1}(T)/n_k$ by the methods of Chapter 7 and so $h_{k-1}(T)$ and $w_{k-1}(T)$ can both be found. No such equation holds for the lower grades because removals from those grades are partly due to natural loss and partly to promotions to the next higher grade. The general transient solution can be obtained, however, when the loss intensities $\{\lambda_j\}$ are constants; we shall justify this remark in the next section.

Equilibrium theory when the grade sizes are large

Let $\mu_{(j)}$ be the average time spent in grade j $(j = 1, 2, \ldots, k)$ regardless of whether the stay there ends in promotion or wastage. When the system has reached equilibrium the total removal density from that grade must be

$n_j/\mu_{(j)}$. Since the grade sizes must remain fixed, this density can be equated to the input density of that grade. Thus

$$\left.\begin{aligned} n_{j-1}h_{j-1} &= n_j/\mu_{(j)} \qquad (j = 2, 3, \ldots, k) \\ R &= n_1/\mu_{(1)} \end{aligned}\right\}. \tag{8.60}$$

To make further progress we must assume that the grade sizes are large. In Section 8.2 we showed that this restriction was of little practical consequence and we anticipate that the same will be true here. Under this assumption the seniority at which promotions occur from grade j will approach a limiting value which we shall denote by τ_j ($j = 1, 2, \ldots, k - 1$). Of those who enter grade j a proportion

$$F_{j-1}(\tau_j)$$

will leave before reaching the seniority in that grade at which they would be promoted. The input to grade j must therefore be sufficient to allow for this wastage and provide just enough candidates to fill the vacancies occurring in grade $j + 1$. This requires that

$$\left.\begin{aligned} n_{j-1}h_{j-1}G_{j-1}(\tau_j) &= n_j h_j \qquad (j = 2, 3, \ldots, k - 1) \\ RG_0(\tau_1) &= n_1 h_1 \end{aligned}\right\}. \tag{8.61}$$

The final step in the argument is to express $\mu_{(j)}$ in terms of τ_j. We shall then be able to eliminate τ_j between (8.60) and (8.61), and so obtain a recurrence relation between h_j and h_{j-1}.

In view of the fact that everyone leaves grade j not later than τ_j after entering it we have

$$\begin{aligned} \mu_{(j)} &= \int_0^{\tau_j} \tau f_{j-1}(\tau)\,d\tau + \tau_j G_{j-1}(\tau_j) \\ &= \int_0^{\tau_j} G_{j-1}(\tau)\,d\tau \qquad (j = 1, 2, \ldots, k;\ \tau_k = \infty). \end{aligned} \tag{8.62}$$

Substituting for $\mu_{(j)}$ in (8.60) gives

$$\left.\begin{aligned} \frac{n_j}{n_{j-1}h_{j-1}} &= \int_0^{\tau_j} G_{j-1}(\tau)\,d\tau \qquad (j = 2, 3, \ldots, k;\ \tau_k = \infty) \\ \frac{n_1}{R} &= \int_0^{\tau_1} G_0(\tau)\,d\tau \end{aligned}\right\}. \tag{8.63}$$

If we solve these equations for $\{\tau_j\}$ and substitute the solution in (8.61) we obtain a recurrence relation from which the h_j's can be calculated. Equation

(8.63) gives h_{k-1} directly and the remaining h_j's are found in decreasing order of their subscripts.

As a first example suppose that the loss intensity is constant but has a different value in each grade. Thus we may write $\lambda_j(\tau) = \lambda_j$ $(j = 1, 2, \ldots, k)$. We know of no direct empirical evidence in support of this form but we shall show later that it can lead to a CLS distribution similar to those observed in practice. Under this assumption

$$f_{j-1}(\tau) = \lambda_j \mathrm{e}^{-\lambda_j \tau}, \qquad G_{j-1}(\tau) = \mathrm{e}^{-\lambda_j \tau}$$

and

$$\int_0^\tau G_{j-1}(x)\,\mathrm{d}x = (1 - \mathrm{e}^{-\lambda_j \tau})/\lambda_j \qquad (j = 1, 2, \ldots, k).$$

On substituting these expressions in (8.61) and (8.63) and eliminating

$$\mathrm{e}^{-\lambda_j \tau_j}$$

between them we find

$$n_{j-1}h_{j-1} = n_j h_j + n_j \lambda_j$$

and hence

$$\left.\begin{aligned} h_j &= \sum_{i=j+1}^{k} n_i \lambda_i / n_j \qquad (j = 1, 2, \ldots, k-1) \\ R &= \sum_{i=1}^{k} n_i \lambda_i \end{aligned}\right\}. \tag{8.64}$$

When $\lambda_j = \lambda$, for all j, these expressions reduce to those given in (8.20) and (8.21). In this case the differences between the various models vanish and the same solution is obtained. The effect of increasing (decreasing) the loss intensity in grade i is to increase (decrease) the promotion density h_j $(j < i)$ by an amount proportional to n_i/n_j. Equation (8.64) can also be used to determine the grade structure necessary to achieve any desired set of promotion densities.

For our second example we suppose that the loss intensity decreases monotonically with

$$\lambda_j(\tau) = \left(\frac{\nu_j}{\kappa_j + \tau}\right) \qquad (j = 1, 2, \ldots, k). \tag{8.65}$$

In this case

$$f_{j-1}(\tau) = \frac{\nu_j}{\kappa_j}\left(1 + \frac{\tau}{\kappa_j}\right)^{-(\nu_j + 1)}, \qquad G_{j-1}(\tau) = \left(1 + \frac{\tau}{\kappa_j}\right)^{-\nu_j}$$

and

$$\int_0^{\tau} G_{j-1}(x)\,dx = \mu_j\left\{1 - \left(1 + \frac{\tau}{\kappa_j}\right)^{-\nu_j+1}\right\} \qquad (j = 1, 2, \ldots, k)$$

where $\mu_j = \kappa_j/(\nu_j - 1)$. Following the same procedure as in the previous example the recurrence relation for the h_j's is found to be

$$\mu_j n_{j-1} h_{j-1} = n_j \Big/ \left\{1 - \left(\frac{n_j h_j}{n_{j-1} h_{j-1}}\right)^{(\nu_j-1)/\nu_j}\right\} \qquad (j = 1, 2, \ldots, k) \tag{8.66}$$

where $h_k = 0$.

The grade structure needed to give equal promotion densities is easily obtained in this case. Setting $h_j = h$ for $j = 1, 2, \ldots, k-1$ in the last equation we see that the n_j's must satisfy the equations

$$\left.\begin{aligned} \frac{n_j}{n_{j-1}} &= \mu_j h\left\{1 - \left(\frac{n_j}{n_{j-1}}\right)^{(\nu_j-1)/\nu_j}\right\} \qquad (j = 2, 3, \ldots, k-1) \\ \frac{n_k}{n_{k-1}} &= h\mu_k \end{aligned}\right\}. \tag{8.67}$$

The solution of these equations gives n_j/n_{j-1} as a function of $\mu_{(j)}$, h and ν_j. If $\nu_j = \nu$ and $\kappa_j = \kappa$ for all j we reach the interesting conclusion that the grade sizes, except for grade k, must form a geometric series. It is a little surprising to find here the same result as in the case of a constant loss intensity. In the example of Section 8.2 the loss intensity had the same form but was a function of total length of service instead of seniority within the grade. In that case the equilibrium structure giving equal h_j's was not geometric (see equation 8.36 and Table 8.2). A numerical comparison of the two cases is given in Table 8.6 for $k = 5$, $h = 2$ and $\mu = 2$. The parameter κ is a scale parameter and does not enter into the comparison. In the last column of the table we have given the limiting case as $\nu \to \infty$; both models then coincide, having a constant loss intensity.

The familiar pattern of a tapering structure with a very large top grade occurs in both cases. The model of Section 8.2 shows the greater divergence from the limiting case in all grades except the first. In the case of this model *the effect of a decreasing loss rate is to increase the relative size of the highest grade.*

Our third example is almost identical mathematically with the second but its practical implications are very different. Suppose that the loss intensities are given by

$$\lambda_j(\tau) = \frac{\nu_j}{\kappa_j - \tau} \qquad (0 \le \tau < \kappa_j; j = 1, 2, \ldots, k). \tag{8.68}$$

Table 8.6 Relative grade sizes, n_j/N, required to give equal promotion densities for the models of Sections 8.2 and 8.4, when promotion is according to seniority

	$\nu = 2$		$\nu = \infty$
j	*Model of Section* 8.2	*Model of Section* 8.4	*Both models*
1	0·333	0·437	0·333
2	0·167	0·235	0·222
3	0·100	0·126	0·148
4	0·067	0·067	0·099
5	0·333	0·135	0·198

In this case the propensity to leave *increases* with seniority until, after a time κ_j, the person leaves automatically. We considered a similar function in Chapter 5 (equation 5.120), but there the loss intensity was a function of total length of service. Our choice of $\lambda_j(\tau)$ for this example would describe the behaviour of someone who became increasingly dissatisfied until a point was reached at which he felt compelled to leave.

The recurrence formula for the promotion densities holds for this example also if we replace μ_j by $\kappa_j/(\nu_j + 1)$ and $(\nu_j - 1)/\nu_j$ by $(\nu_j + 1)/\nu_j$.

The CLS distribution

As we remarked at the beginning of this section, the CLS distribution is no longer given but must be found from the loss intensities and the grade sizes. When promotion is by seniority the distribution is easily obtained as follows. Let $\lambda(T)$ be the loss intensity for a person who has been in the organization for length of time T. Then it is clear that, under the assumption of large grade sizes and in equilibrium,

$$\left.\begin{aligned} \lambda(T) &= \lambda_1(T) && (0 \leq T < \tau_1) \\ &= \lambda_2(T - \tau_1) && (\tau_1 \leq T < \tau_2) \\ & && \dots \\ &= \lambda_k(T - \tau_{k-1}) && (\tau_{k-1} \leq T < \infty) \end{aligned}\right\} \tag{8.69}$$

where $\tau_1, \tau_2, \ldots, \tau_{k-1}$ are the seniorities at which promotions take place. The CLS distribution is then given by

$$f(T) = \lambda(T) \exp\left[-\int_0^T \lambda(x)\,\mathrm{d}x\right] \qquad (T \geq 0).$$

The discontinuities in $\lambda(T)$ will lead to discontinuities in $f(T)$. A full investigation of the shapes to which this model can give rise has not been undertaken but a simple example will illustrate some of the possibilities.

Let $k = 2$ and $\lambda_j(\tau) = \lambda_j$ $(j = 1, 2)$. Then

$$\left.\begin{aligned} f(T) &= \lambda_1 \mathrm{e}^{-\lambda_1 T} && (0 \leq T < \tau_1) \\ &= \lambda_2 \mathrm{e}^{-\lambda_1 \tau_1 - \lambda_2 (T - \tau_1)} && (\tau_1 \leq T < \infty) \end{aligned}\right\}. \tag{8.70}$$

The point of discontinuity τ_1 is found from (8.63), which becomes in this case

$$\mathrm{e}^{-\lambda_1 \tau_1} = n_2 \lambda_2 / N(\lambda_1 + \lambda_2). \tag{8.71}$$

The density function $f(T)$ thus has the form of an exponential density with parameter λ_1 over the first part of its range and the same form with parameter λ_2 over the second part. Its most obvious characteristic is the discontinuity at $T = \tau_1$ and we would be able to recognize this in practice if the model applied. However, whether or not this simple form of the model is realistic, we may note that it possesses one feature which has been observed in CLS distributions. If $\lambda_1 > \lambda_2$ there will be an excess of frequency at the lower and upper ends of the range if we compare the distribution with the exponential curve having the same mean. A similar result holds if we consider an organization with more than two grades provided that the loss intensities decrease as we move up the hierarchy. These facts lead us to expect that models in which loss rates are constant within the grades may be capable of describing observed distributions. This point will be taken up again in the next section, where it will be seen that the form of $f(T)$ is mixed exponential when promotion is at random.

8.5 A MODEL WITH LOSS RATE DEPENDING ON SENIORITY WITHIN THE GRADE AND PROMOTION AT RANDOM

Exact theory in the transient case

When promotion is at random it is relatively easy to obtain integral equations for the promotion and wastage densities in the transient case. Unfortunately, the solutions to these equations are not always easy to find. In this section we shall formulate the basic theory and obtain an explicit solution in one case. It will again be necessary to assume that the grade sizes are large. Later we shall use the exact theory to obtain the equilibrium solution by letting T tend to infinity.

We begin by recalling that the behaviour of grade k does not depend upon the promotion rule and hence that $h_{k-1}(T)$ can be obtained as in the last section from (8.59). Let us now consider grade j and attempt to treat it as a simple renewal process. This would be possible if the length of time spent in that grade by any individual had a distribution independent of T. However, this is not so because there are two intensities of removal acting on members of this grade. There is the loss intensity $\lambda_j(\tau)$ depending only on seniority within the grade and there is also the promotion intensity which is a function of T only. When promotion is at random the probability that any given member of grade j will be promoted in $(T, T + \delta T)$ is $h_j(T)\,\delta T$. Hence, since promotion and loss are independent, the total force of removal acting on a member of grade j with seniority τ at time T is

$$\lambda_j(\tau) + h_j(T) \qquad (j = 1, 2, \ldots, k - 1). \tag{8.72}$$

The sojourn time distribution for grade j thus depends on the time of joining the grade and on the length of service. Renewal processes of this kind were discussed in the last chapter. In order to use the results given there we must find the distribution of the sojourn time.

Let

$$\lambda_{(j)}(\tau|X) = \lambda_j(\tau) + h_j(X + \tau)$$

denote the intensity of removal for a member of grade j who entered grade j at time X and has a length of service τ in that grade. It is tempting to write down the density function of sojourn times as

$$f_{(j)}(\tau|X) = \lambda_{(j)}(\tau|X) \exp\left[-\int_0^T \lambda_{(j)}(x|X)\,dx\right]. \tag{8.73}$$

However, this step requires the assumption that the probability of removal from the grade with seniority in $(\tau, \tau + \delta\tau)$ at time $X + \tau$ is independent of events prior to that time. This is certainly true for the loss probability but it does not hold in general for the promotion process. The probability of promotion in any small increment of time depends in general on the time of the last promotion. The exceptional case occurs when the promotion process is also a Poisson process. This happens if all the loss rates are constant since then the losses from any grade constitute a Poisson process and hence so must the promotions. It also happens asymptotically as the grade sizes increase, as may be deduced from a result of Khintchine (1960, Chapter 5). In order to make use of (8.73) in the general case we shall therefore have to assume that the grade sizes are large. Under these conditions it follows from (7.19) of Chapter 7 that the promotion densities are given by

$$n_{n-1}h_{j-1}(T) = n_j f_{(j)}(T|0) + n_{j-1}\int_0^T f_{(j)}(\tau|T - \tau)h_{j-1}(T - \tau)\,d\tau$$

$$(j = 1, 2, \ldots, k). \tag{8.74}$$

Substituting for $f_{(j)}(\tau|T-\tau)$ from (8.73) we arrive at equations connecting $h_{j-1}(T)$ with $h_j(T)$. We therefore begin by finding $h_{k-1}(T)$ in the way described above and then use (8.74) to compute the remaining densities recursively.

Although it is easy to construct these equations it is difficult to find analytical solutions. One exception to this rule occurs when the loss intensities are constant within the grade. Thus if $\lambda_j(\tau) = \lambda_j$ $(j = 1, 2, \ldots, k)$ the solution proceeds as follows. Putting $j = k$ in (8.73) gives

$$f_{(k)}(\tau|T-\tau) = \lambda_k \, e^{-\lambda_k \tau} \qquad (\tau \geq 0)$$

and hence, by solving (8.74) with $j = k$, we have

$$\frac{n_{k-1}h_{k-1}(T)}{n_k} = \lambda_k \quad \text{and hence} \quad h_{k-1}(T) = \frac{n_k\lambda_k}{n_{k-1}}. \tag{8.75}$$

This result follows from the fact that, in this case, (8.74) reduces to the standard integral equation of renewal theory. Next we put $j = k - 1$ in (8.73) and so find

$$f_{(k-1)}(\tau|T-\tau) = \left(\lambda_{k-1} + \frac{n_k\lambda_k}{n_{k-1}}\right) \exp\left[-\lambda_{k-1}\tau - \frac{n_k\lambda_k}{n_{k-1}}\tau\right], \tag{8.76}$$

which is an exponential distribution with parameter $(n_{k-1}\lambda_{k-1} + n_k\lambda_k)/n_{k-1}$. Substitution of this result in (8.74) with $j = k - 1$ yields another integral equation of the same form, from which we find that

$$\frac{n_{k-2}h_{k-2}(T)}{n_{k-1}} = \frac{n_{k-1}\lambda_{k-1} + n_k\lambda_k}{n_{k-1}}$$

and hence

$$h_{k-2}(T) = \frac{n_{k-1}\lambda_{k-1} + n_k\lambda_k}{n_{k-2}}. \tag{8.77}$$

This procedure can be repeated for each grade to give

$$h_j(T) = \frac{\sum_{i=j+1}^{k} n_i\lambda_i}{n_j}, \qquad R = \sum_{i=1}^{k} n_i\lambda_i \qquad (j = 1, 2, \ldots, k-1), \tag{8.78}$$

which agrees with the equilibrium solution for the case of promotion by seniority given in (8.64). Under the assumption of constant loss rates the two promotion rules are identical in the way in which they affect the promotion or wastage densities. The exact theory derived above thus applies equally to both rules and so establishes that the equilibrium result given in (8.64) is, in fact, the transient solution also.

It would be possible to solve (8.74) by numerical methods for any other kind of loss intensity. A practical question of more immediate concern for the application of the theory is to find whether the assumption of constant loss rates is valid in practice. This could be tested by seeing whether the promotion and wastage rates remain constant over time. Here we have a good example of how the analysis of a theoretical model can guide empirical research.

Equilibrium theory

The equilibrium theory can be readily deduced from equations (8.73) and (8.74) by taking limits as $T \to \infty$. Beginning with $h_{k-1}(T)$ we know from standard renewal theory that

$$\frac{n_{k-1}h_{k-1}}{n_k} = \frac{1}{\mu_{(k)}}$$

where $\mu_{(k)}$ is the mean of the sojourn time distribution for the kth grade. A closely related result also states that

$$\lim_{T \to \infty} \int_{T-\tau}^{T} h_j(x)\, dx = \tau h_j \qquad (j = 1, 2, \ldots, k-1)$$

for fixed τ. The limiting form of the sojourn density function for finite τ may thus be written

$$f_{(j)}(\tau) = \{\lambda_j(\tau) + h_j\} \exp\left[-\int_0^\tau \lambda_j(x)\, dx - h_j\tau\right] \qquad (j = 1, 2, \ldots, k;\, h_k = 0). \tag{8.79}$$

In the limit, therefore, grade j may be treated as a simple renewal process with a renewal distribution given by this equation. It then follows that

$$\left.\begin{aligned} n_{j-1}h_{j-1} &= n_j/\mu_{(j)} \qquad (j = 1, 2, \ldots, k-1) \\ R &= n_1/\mu_{(1)} \end{aligned}\right\}. \tag{8.80}$$

Now

$$\mu_{(j)} = \int_0^\infty \tau f_{(j)}(\tau)\, d\tau = \int_0^\infty G_{(j)}(\tau)\, d\tau$$

where

$$G_{(j)}(\tau) = \int_\tau^\infty f_{(j)}(x)\, dx = \exp\left[-\int_0^\tau \lambda_j(x)\, dx - h_j\tau\right].$$

Therefore

$$\mu_{(j)} = \int_0^\infty G_j(\tau)\,\mathrm{e}^{-h_j\tau}\,\mathrm{d}\tau = G_j^*(h_j) \qquad (j = 1, 2, \ldots, k). \qquad (8.81)$$

On substituting this result of (8.80) we obtain the following recurrence relation for the h_j's:

$$\left.\begin{aligned} n_{j-1}h_{j-1} &= n_j/G_j^*(h_j) \qquad (j = 1, 2, \ldots, k-1) \\ R &= n_1/G_1^*(h_1) \end{aligned}\right\}. \qquad (8.82)$$

These equations are identical with those obtained in Section 8.3 for the same promotion rule when the loss intensity depended on total length of service. They may, in fact, be derived by the same argument, the only difference being that the known and unknown quantities are now interchanged. Thus in the former case we had to calculate h_j from h_{j-1}; now we have to find h_{j-1} from h_j. This is very much easier because the unknown no longer appears in the integrand of a Laplace transform.

To illustrate these results we shall use a parametric form for $\lambda_j(\tau)$ which includes both increasing and decreasing functions of τ. This is achieved if $f_j(\tau)$ has the form of a gamma density function. We therefore suppose that

$$f_j(\tau) = \frac{\lambda_j^{\nu_j}}{\Gamma(\nu_j)}\tau^{\nu_j - 1}\,\mathrm{e}^{-\lambda_j\tau} \qquad (\tau \geq 0, \nu_j > 0, \lambda_j > 0; j = 1, 2, \ldots, k). \qquad (8.83)$$

The loss intensity associated with this distribution is always a monotonic function unless $\nu_j = 1$ when $\lambda_j(\tau) = \lambda_j$. If $\nu_j > 1$, $w_j(\tau)$ increases with τ and approaches a horizontal asymptote. If $0 < \nu_j < 1$, $\lambda_j(\tau)$ is a decreasing function of τ and has an infinite ordinate at $\tau = 0$. By varying ν_j we can thus change the form of dependence of the loss rate on τ.

The Laplace transform of $f_j(\tau)$ is

$$f_j^*(s) = \left(\frac{\lambda_j}{\lambda_j + s}\right)^{\nu_j} \qquad (j = 1, 2, \ldots, k).$$

Since $G_j^*(s) = \{1 - f_j^*(s)\}/s$ and $\mu_j = \nu_j/\lambda_j$, (8.82) becomes

$$n_{j-1}h_{j-1} = n_j h_j\left\{1 - \left(1 + \frac{h_j\mu_j}{\nu_j}\right)^{-\nu_j}\right\}^{-1} \qquad (j = 1, 2, \ldots, k; h_k = 0). \quad (8.84)$$

This equation can be used to investigate the dependence of the limiting promotion densities on the grade structure. In order to make a comparison with our earlier results we shall use it to find the grade structure required to make $h_1 = h_2 = \ldots = h_{k-1} = h$. For this to be so we must have

$$\left.\begin{aligned} \frac{n_j}{n_{j-1}} &= 1 - \left(1 + \frac{h\mu_j}{\nu_j}\right)^{-\nu_j} \qquad (j = 2, 3, \ldots, k-1) \\ \frac{n_k}{n_{k-1}} &= h\mu_k \end{aligned}\right\}. \tag{8.85}$$

If ν_j and λ_j are independent of j for all grades we note that the grade sizes, except for the highest, will form a geometric series. We have found the same result for two of our previous models. Thus it holds when $\lambda_j(T) = \lambda_j(\tau) = \lambda$ for all grades and both promotion rules and also when

$$\lambda_j(\tau) = \frac{\nu}{\kappa + \tau}$$

for all j is promotion is by seniority. Further, when the loss intensity was supposed to depend on the total length of service, we met an example (see Table 8.2) with a structure very close to the geometric form. This is particularly interesting in view of the fact that many hierarchical organizations conform approximately to this pattern. The reason for this is that each man at any given level is often responsible for the work of x people in the level below him. According to Simon (1957a), this number seldom varies very much within a firm; at executive levels x usually lies between 3 and 10. However, we would not expect to find equal h_j's in such organizations because the large grade k's which our theory requires rarely, if ever, occur in practice.

The conclusion about geometric structure does not depend on whether $\lambda_j(\tau)$ is an increasing or decreasing function of τ. The value of ν does, of course, help to determine the constant ratio of the grade sizes. Since the right-hand side of (8.85) is an increasing function of ν the structure tapers more rapidly if ν is large.

The equilibrium CLS distribution

The CLS distribution for a recruit to the organization will be a function of the time at which he joins. No general theory is available but the equilibrium distribution can be found when the grade sizes are large. The method can be most easily understood by visualizing the situation as shown in Figure 8.2.

On this diagram we have plotted a man's grade as a function of his length of service. The vertical jumps correspond to promotions and the horizontal segments to the lengths of service in the various grades. We shall find the probability that a new entrant remains in the organization for a length of

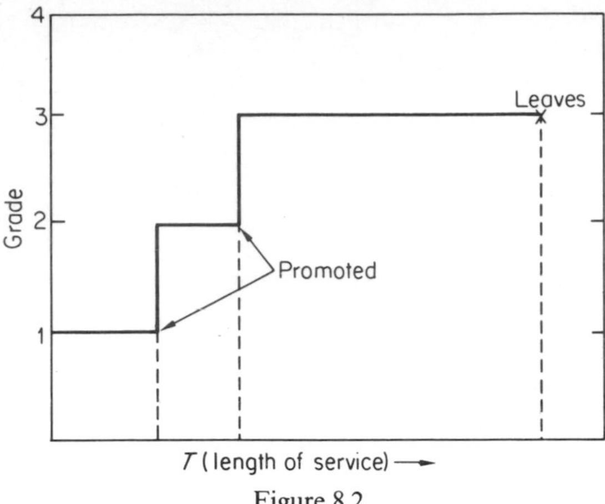

Figure 8.2

time exceeding T. If $f(T)$ is the equilibrium CLS density function then the probability which we shall determine is $G(T)$. To calculate this probability we shall express it in the form

$$G(T) = \sum_{i=0}^{k-1} Pr\{\text{man promoted } i \text{ times } \textit{and} \text{ does not leave in } (0, T)\}. \quad (8.86)$$

First consider the case $i = 0$. We have already seen that, in equilibrium, the promotions from any grade constitute a Poisson process. Therefore

$$Pr\{\text{no promotion in } (0, T)\} = e^{-h_1 T}.$$

Further, since the loss intensity in grade 1 is $\lambda_1(T)$ (T and τ are equivalent in this grade),

$$Pr\{\text{no loss in } (0, T)\} = G_1(T).$$

The promotion and loss processes are independent and so the first term in (8.86) is

$$G_1(T)\, e^{-h_1 T}.$$

When $i = 1$ the probability which we need is obtained as follows. First suppose that promotion takes place from grade 1 to grade 2 in the interval $(T_1, T_1 + \delta T_1)$. The probability of this event is

$$h_1\, e^{-h_1 T_1}\, \delta T_1.$$

Given T_1, the probability of no loss occurring is

$$G_1(T_1) G_2(T - T_1)$$

and the probability of no promotion occurring while the member is in grade 2 is

$$e^{-h_2(T-T_1)}.$$

Therefore the total probability required for the second term on the right-hand side of (8.86) is

$$\int_0^T h_1 \, e^{-h_1 T_1} \, e^{-h_2(T-T_1)} G_1(T_1) G_2(T-T_1) \, dT.$$

The above argument extends immediately to the general case. If the promotions take place at times $T_1, T_2, \ldots, T_i$ then

$Pr\{$man promoted i times and does not leave in $(0, T)\}$

$$= \int_{0 \le T_1 < \ldots} \cdots \int_{< T_i \le T} \left\{ \prod_{j=1}^{i} h_j \, e^{-h_j(T_j - T_{j-1})} G_j(T_j - T_{j-1}) \right\} e^{-h_{i+1}(T-T_i)}$$
$$G_{i+1}(T - T_i) \, dT_1 \, dT_2 \ldots dT_i \qquad (i = 1, 2, \ldots, k-1) \qquad (8.87)$$

where $T_0 \equiv 0$ and $h_k = 0$.

When we sum over i and add in the term for $i = 0$ obtained above we arrive at the expression for $G(T)$. This apparently cumbersome result is easily handled by means of the Laplace transform. The expression on the right-hand side of (8.87) has the form of a convolution integral. Hence, apart from a constant, its Laplace transform is the product of the transforms of the functions

$$G_j(T) \, e^{-h_j T} \qquad (j = 1, 2, \ldots, i+1).$$

Thus we find

$$G^*(s) = \sum_{i=1}^{k} \prod_{j=1}^{i} h_{i-1} G_j^*(s + h_j) \qquad (h_0 = 1). \qquad (8.88)$$

In general, this equation will not be easy to invert unless there is some simplifying feature such as equal promotion densities or constant loss rates. When we considered promotion by seniority we found that the assumption of constant but unequal loss rates could give a CLS distribution with greater dispersion than the exponential. We shall therefore examine the same case for the random promotion rule by setting $\lambda_j(\tau) = \lambda_j \, (j = 1, 2, \ldots, k)$. This gives

$$G_j(\tau) = e^{-\lambda_j \tau}$$

and

$$G_j^*(s + h_j) = \frac{1}{\lambda_j + s + h_j} \qquad (j = 1, 2, \ldots, k; h_k = 0). \tag{8.89}$$

Substituting in (8.88) and using the fact that

$$f^*(s) = 1 - sG^*(s)$$

we obtain

$$f^*(s) = \frac{\lambda_1}{\lambda_1 + h_1 + s} + \sum_{i=2}^{k} \left(\frac{\lambda_i}{\lambda_i + h_i + s}\right) \sum_{j=1}^{i-1} \left(\frac{h_j}{\lambda_j + h_j + s}\right) \qquad (h_k = 0). \tag{8.90}$$

As a check it may be verified that this transform reduces to $\lambda/(\lambda + s)$ if $\lambda_j = \lambda$ for all j and to 1 if $s = 0$. The expression on the right-hand side of (8.90) can be resolved into partial fractions and inverted term by term to give a density function of the form

$$f(T) = \sum_{i=1}^{k} p_i(\lambda_i + h_i)\,\mathrm{e}^{-(\lambda_i + h_i)T} \tag{8.91}$$

where $h_k = 0$ and

$$\sum_{i=1}^{k} p_i = 1.$$

This is the familiar mixed exponential distribution, although the p_i's need not all be positive as we shall see below.

In the simplest case when $k = 2$

$$\begin{aligned} p_1 &= (\lambda_1 - \lambda_2)/(\lambda_1 - \lambda_2 + h_1) \\ &= n_1(\lambda_1 - \lambda_2)/\{n_1(\lambda_1 - \lambda_2) + n_2\lambda_2\}, \end{aligned}$$

which satisfies $0 \leq p_1 \leq 1$ if $\lambda_1 \geq \lambda_2$. Thus, whenever the loss intensity is higher in the first grade than in the second, we shall have a mixed exponential distribution of the kind we have met before. This is the third probability model we have discussed which gives rise to this particular distribution: the other two are discussed in Chapter 6. When fitting this distribution we found that fairly typical parameter values were $p_1 = \frac{1}{2}$ and $(\lambda_1 + h_1)/\lambda_2 = 10$. To achieve these values with the present model we should need $\lambda_1/\lambda_2 = 5{\cdot}5$ and $n_2/n_1 = 4{\cdot}5$. If this model represented the true state of affairs we should not be able to distinguish it from the others if we only observed the total length of service. To do so it would be necessary to identify the two grades, test whether their loss intensities were constant and, if so, to see whether or not their values agreed with those predicted by the fitted distribution.

For $k > 2$ the CLS distribution can take on a variety of shapes. In order to illustrate some of the possibilities we shall list four examples below for the case $k = 3$.

(a) $n_1 = n_2 = n_3,\ \lambda_j = \lambda/j \qquad (j = 1, 2, 3)$

$$f(T) = \frac{\lambda}{324}(209\,\mathrm{e}^{-11\lambda T/6} + 75\,\mathrm{e}^{-5\lambda T/6} + 40\,\mathrm{e}^{-\lambda T/3}).$$

(b) $n_1 = n_2 = n_3,\ \lambda_j = \lambda/(4 - j) \qquad (j = 1, 2, 3)$

$$f(T) = \frac{\lambda}{60}(209\,\mathrm{e}^{-11\lambda T/6} - 405\,\mathrm{e}^{-3\lambda T/2} + 216\,\mathrm{e}^{-\lambda T}).$$

(c) $n_1 = 2n_2 = 3n_3,\ \lambda_j = \lambda/j \qquad (j = 1, 2, 3)$

$$f(T) = \frac{\lambda}{18}(14\,\mathrm{e}^{-4\lambda T/3} + 3\,\mathrm{e}^{-2\lambda T/3} + \mathrm{e}^{-\lambda T/3}).$$

(d) $n_1 = 2n_2 = 4n_3,\ \lambda_j = \lambda/(4 - j) \qquad (j = 1, 2, 3)$

$$f(T) = \frac{\lambda}{6}(20\,\mathrm{e}^{-4\lambda T/3} - (18 - 9T)\,\mathrm{e}^{-\lambda T}).$$

The term involving the factor T in case (d) arises because, in that example, $\lambda_2 + h_2 = \lambda_3$ so that the inversion leading to (8.91) breaks down. The difficulty can be overcome by finding the limit of $f(T)$ as given in that equation when $\lambda_3 \to \lambda_2 + h_2$.

In those cases where the loss intensities decrease as we move up the hierarchy we have a mixed exponential distribution with positive p_i's for both structures considered. In the other two cases, when the loss intensities increase with j, the distribution is less skew and rises to a mode. It is the former kind of distribution which occurs in practice and so our present model can only provide a satisfactory explanation if there is a decreasing sequence of loss intensities. This is what we might expect to find in practice but empirical evidence on the question is lacking.

8.6 AGE STRUCTURE OF THE GRADES

Theory

The whole emphasis in the preceding part of this chapter has been on the determination of the wastage and promotion densities. These quantities do not provide a complete picture of the system and we shall now remedy one of the most important omissions. The efficient operation of an organization may depend critically on there being enough experienced persons available at each level. It may, therefore, be more important from the firm's point of view to achieve an adequate distribution of experienced staff than to ensure good promotion prospects. The basic mathematical requirement for such studies is the distribution of age (measured from entry) for serving members. This distribution can be obtained quite easily in the equilibrium state for both promotion rules. The formulae which we shall obtain depend only on the CLS distribution $f(T)$; they therefore apply whether the loss intensity depends upon seniority within the grade or within the organization. In the former case this distribution must be found by the methods of Section 8.5.

Let $a_j(t)$ denote the probability density function of age measured from entry to the system for serving members of grade j $(j = 1, 2, \ldots, k)$ when the system has reached equilibrium. We shall continue to use the notation $a(t)$ for the corresponding density function for the system when viewed as a whole. From these definitions it follows that

$$Na(t) = \sum_{j=1}^{k} n_j a_j(t). \tag{8.92}$$

When promotion is according to seniority the solution of our problem is almost immediate. We saw in Section 8.2 that, under this promotion rule, no member of grade j can be older than any member of grade $j + 1$. This means that the grades divide the total membership according to age. Let the critical age at which promotion takes place from grade j to grade $j + 1$ be t_j $(j = 1, 2, \ldots, k - 1)$. If the model of Section 8.4 applies, then t_j will be given by

$$t_j = \sum_{i=1}^{j} \tau_i \qquad (j = 1, 2, \ldots, k - 1).$$

It thus follows that

$$\begin{aligned} a_j(t) &= a(t) \Big/ \int_{t_{j-1}}^{t_j} a(x)\,\mathrm{d}x \qquad (t_{j-1} \leq t < t_j) \\ &= \frac{Na(t)}{n_j} \qquad (j = 1, 2, \ldots, k) \end{aligned} \tag{8.93}$$

where $a(t) = \mu^{-1} G(t)$, $t_0 = 0$ and $t_k = \infty$.

When promotion is at random the derivation starts with the observation that

$$a_j(t) \propto a(t)Pr\{\text{member having age } t \text{ is in grade } j\}. \tag{8.94}$$

A person will be in grade j at time t if, and only if, he has experienced exactly $j - 1$ promotions in $(0, t)$. The probability of this event can be determined if it can be assumed that the promotions between adjacent pairs of grades constitute a Poisson process. For this to be so, either the loss intensities must be constant or the system must have large grade sizes and have reached its equilibrium. Under these conditions

$$Pr\{\text{promotion from grade } j \text{ after total service in } (t, t + \delta t)\} = h_j\,\delta t \qquad (j = 1, 2, \ldots, k - 1).$$

The probability of $j - 1$ promotions in $(0, t)$, given that there is no loss in this interval, follows at once from the theory of the time-homogeneous birth process (see also Chapter 6, equation 6.20, and Bartlett, 1955, Section 3.2). In the present notation this probability is

$$P(j - 1|t) = \prod_{i=1}^{j-1} h_i \sum_{i=1}^{j} \mathrm{e}^{-h_i T} \prod_{\substack{l=1 \\ l \neq i}}^{j} \frac{1}{(h_l - h_i)} \qquad (j = 1, 2, \ldots, k - 1). \tag{8.95}$$

If some of the h_i's are equal the approximate form of the probabilities can be found by a limiting operation on this equation. In particular, if all of the promotion intensities have a common value h the distribution of the number of promotions is Poisson for $j < k - 1$ with

$$\left.\begin{aligned} P(j - 1|t) &= \frac{(ht)^{j-1}}{(j - 1)!}\mathrm{e}^{-ht} \qquad (j = 1, 2, \ldots, k - 2) \\ P(k - 1|t) &= 1 - \sum_{i=0}^{k-2} P(i|t). \end{aligned}\right. \tag{8.96}$$

Returning to (8.94) we may therefore write the density function of age in the jth grade as

$$\left.\begin{aligned} a_j(t) &= a(t)P(j - 1|t) \Big/ \int_0^\infty a(x)P(j - 1|x)\,\mathrm{d}x \qquad (j = 1, 2, \ldots, k - 1) \\ &\text{with} \\ a_k(t) &= \frac{1}{n_k}\left\{Na(t) - \sum_{i=1}^{k-1} n_i a_i(t)\right\}. \end{aligned}\right\} \tag{8.97}$$

Examples

We shall illustrate the foregoing theory by supposing that the CLS distribution is mixed exponential with density function

$$f(T) = p\lambda_1 e^{-\lambda_1 T} + (1-p)\lambda_2 e^{-\lambda_2 T} \qquad (T \geq 0).$$

It has been shown in Section 8.5 that this form can arise in those models in which the loss rate depends on the grade. However, our discussion will be primarily in terms of the models of Sections 8.2 and 8.3. In those cases we saw that the mixed exponential distribution gave a satisfactory description of observed CLS distributions except, perhaps, in the region of the upper tail.

When promotion is by seniority the age distribution for the jth grade is, from (8.93),

$$a_j(t) = \left(\frac{N}{n_j}\right)\frac{p e^{-\lambda_1 t} + (1-p) e^{-\lambda_2 t}}{p/\lambda_1 + (1-p)/\lambda_2} \qquad (t_{j-1} \leq t < t_j), \qquad (8.98)$$

where the $\{t_j\}$ satisfy

$$\int_0^{t_j} \{p e^{-\lambda_1 t} + (1-p) e^{-\lambda_2 t}\}\, dt = \frac{N_j}{N}\left(\frac{p}{\lambda_1} + \frac{1-p}{\lambda_2}\right) \qquad (j = 1, 2, \ldots, k-1). \qquad (8.99)$$

If promotion is at random it is clear from (8.94) and (8.95) that $a_j(t)$ is a linear combination of exponential terms. For simplicity we consider the case where the grade structure is such that the promotion densities are equal. In this case (8.96) applies and we find

$$\left.\begin{aligned} a_j(t) &= \frac{p t^{j-1} e^{-(\lambda_1 + h)t} + (1-p) t^{j-1} e^{-(\lambda_2 + h)t}}{(j-1)!\left\{p\left(\dfrac{1}{\lambda_1 + h}\right)^j + (1-p)\left(\dfrac{1}{\lambda_2 + h}\right)^j\right\}} \quad (t \geq 0) \\ & \qquad\qquad (j = 1, 2, \ldots, k-1) \\ a_k(t) &= \left\{N a(t) - \sum_{i=1}^{k-1} n_i a_i(t)\right\}/n_k \end{aligned}\right\}. \qquad (8.100)$$

Inspection of the first part of this equation shows that it is a weighted average of two gamma densities each with index $j-1$. This fact enables us to write down the moments of the distribution and, in particular, the mean, which is

$$\left.\begin{aligned} \xi_j &= \frac{j\left\{\dfrac{p}{(\lambda_1 + h)^{j+1}} + \dfrac{1-p}{(\lambda_2 + h)^{j+1}}\right\}}{\left\{\dfrac{p}{(\lambda_1 + h)^{j}} + \dfrac{1-p}{(\lambda_2 + h)^{j}}\right\}} \qquad (j = 1, 2, \ldots, k-1) \\ \xi_k &= \left\{N\mu - \sum_{i=1}^{k-1} n_i \xi_i\right\} / n_k \end{aligned}\right\}. \qquad (8.101)$$

These expressions yield simple bounds on ξ_j. Without loss of generality we may assume that $\lambda_1 > \lambda_2$, in which case

$$\frac{j}{\lambda_1 + h} < \xi_j < \frac{j}{\lambda_2 + h} \qquad (j = 1, 2, \ldots, k-1).$$

The bounds cannot be attained unless either $p = 1$, $p = 0$ or $\lambda_1 = \lambda_2$. In the latter case the CLS distribution reduces to the exponential and the bounds coincide.

In order to give a clearer idea of the practical implications of these formulae we shall make numerical calculations for two special cases. For the first example we consider a hierarchy with only two grades; for the second we assume an exponential CLS distribution (the case $\lambda_1 = \lambda_2$). One of our principal objects will be to compare the effect of the two promotion rules. Before embarking on this investigation we may remark that there is one obvious difference. If promotion is by seniority the variation of age in grade j, for example, is restricted to the interval (t_{j-1}, t_j); under random promotion the range is unlimited. An organization with random promotion will thus be characterized by a greater spread of experience in each grade.

First suppose that $k = 2$ with N large and $n_1 = 2n_2$. Let us use the same parameter values as in several of our earlier examples and take $p = \frac{1}{2}$, $\lambda_1 = 5{\cdot}5$, $\lambda_2 = 0{\cdot}55$ with $\mu = 1$. When promotion is by seniority t_1, the age at promotion, satisfies (8.99) with $j = 1$,

$$\frac{e^{-5\cdot5t_1}}{5\cdot5} + \frac{e^{-0\cdot55t_1}}{0\cdot55} = \frac{2}{3}.$$

The solution of this equation is $t_1 = 1{\cdot}8240$. This implies that all members of grade 1 have age less than 1·8240; those in grade 2 have a greater age. A useful way of comparing the seniority structure for two promotion rules is to compute the average seniority for each grade. This can easily be found from (8.98) for grade 1. The means for grade 2 can be obtained from the last member of (8.101), which holds for all promotion rules. Numerical values for the above parameter values are given in Table 8.7.

When promotion is at random $a_1(t)$ is given by (8.100) with $j = 1$. For the system we are considering h_1, the renewal density, satisfies

$$\frac{1}{5{\cdot}5 + h_1} + \frac{1}{0{\cdot}55 + h_1} = \frac{4}{3}$$

(see equation 8.58). Thus $h_1 = 0{\cdot}3111$ and on substituting this value in (8.100) we find

$$a_1(t) = 0{\cdot}7500(e^{-0{\cdot}8611t} + e^{-5{\cdot}8111t}) \qquad (t \geq 0). \qquad (8.102)$$

For grade 2, $a_2(t)$ can be found from (8.92) and it, too, is mixed exponential in form. Numerical values of the average seniority have been calculated from these distributions and the results are given in Table 8.7.

Table 8.7 Average seniority in grades 1 and 2 for the two promotion rules and a mixed exponential CLS distribution with $p = \frac{1}{2}, \lambda_1 = 5{\cdot}5, \lambda_2 = 0{\cdot}55$

Grade	*Seniority rule*	*Random rule*
1	0·683	1·041
2	3·642	2·927
Whole organization	1·669	1·669

It may seem surprising that the average age for the whole organization, 1·669, is greater than the average CLS, which is 1. This happens because the serving members contain a disproportionate number who will have 'greater than average' lengths of completed service. As we might have expected, random promotion gives a higher average seniority in grade 1 but a lower average in grade 2 than does promotion by seniority. An apparently paradoxical situation is revealed if we look at the average seniority *within each grade* instead of within the organization as a whole. Consider grade 2 from this point of view. Under promotion by seniority we showed that promotion takes place at seniority 1·824. The average time spent by members of grade 2 in that grade is thus 3·642 − 1·824 = 1·818. If promotion is at random the average age of members when they are promoted must be equal to their average seniority in grade 1. In this example the figure is 1·041: the average age in grade 2 is therefore 2·927 − 1·014 = 1·886. Thus random promotion gives greater average seniority within *both* grades. This conclusion takes no account, of course, of the greater variability which will occur in the random case.

Second, we shall consider the case of general k when the CLS distribution is exponential. Let us assume that the grade sizes satisfy

$$n_{k-1} = n_k, \qquad n_j = n_{k-1}2^{k-j-1} \qquad (j = 1, 2, \ldots, k-2). \qquad (8.103)$$

In such an organization each grade is half the size of the one below it, except for grade k which is equal to grade $k - 1$. This structure has the property that the promotion and wastage densities are all equal to λ, the parameter of the CLS distribution (see Section 8.2).

When promotion is by seniority t_j satisfies

$$\int_0^{t_j} e^{-\lambda x}\,dx = \frac{N_j}{N\lambda} \qquad (j = 1, 2, \ldots, k-1),$$

whence

$$t_j = -\frac{1}{\lambda}\log_e (N - N_j)/N.$$

For the structure given by (8.103) this equation simplifies to

$$t_j = \frac{j}{\lambda}\log_e 2 = \frac{0{\cdot}6932j}{\lambda}.$$

The times of successive promotions thus increase in arithmetic progression. The average ages for the members of each grade can easily be calculated. Thus we find

$$\left.\begin{aligned} \xi_j &= \frac{\lambda N}{n_j}\int_{((j-1)\log_e 2)/\lambda}^{(j\log_e 2)/\lambda} t\,e^{-\lambda t}\,dt = \frac{1}{\lambda}\{1 + (j-2)\log_e 2\} \\ &\qquad\qquad (j = 1, 2, \ldots, k-1) \\ \xi_k &= \frac{1}{\lambda}\{1 + (k-1)\log_e 2\} \end{aligned}\right\}. \qquad (8.104)$$

The average length of time which a member has spent in a given grade is

$$\xi_j - t_{j-1} = \frac{1 - \log_e 2}{\lambda} = \frac{0{\cdot}3068}{\lambda} \qquad (j = 1, 2, \ldots, k-1;\ t_0 = 0)$$

$$\xi_k - t_{k-1} = \frac{1}{\lambda}.$$

This system has the rather interesting property that the average experience within a particular grade is the same for all grades except the highest.

When promotion is at random the mean ages in the grades may be obtained from (8.101) because h_j is independent of j for the above structure. Setting $\lambda_1 = \lambda_2 = \lambda$ and $h = \lambda$ we find

$$\left.\begin{aligned} \xi_j &= j/2\lambda \qquad (j = 1, 2, \ldots, k-1) \\ \xi_k &= (k+1)/2\lambda \end{aligned}\right\}. \tag{8.105}$$

On comparing the results given in this equation with those of (8.104) we see that ξ_2 is the same in both cases. For $j > 2$ the average age is greater when promotion is by seniority but when $j = 1$ is smaller. In the random case the average age in each grade is given by

$$\left.\begin{aligned} \xi_j - \xi_{j-1} &= 1/2\lambda \qquad (j = 1, 2, \ldots, k-1;\ \xi_0 = 0) \\ \xi_k - \xi_{k-1} &= 1/\lambda \end{aligned}\right\}. \tag{8.106}$$

As in the previous example we find that random promotion leads to greater average experience within each grade.

These two examples suggest that *there may be practical advantages in adopting promotion policies which are not too rigidly tied to such things as seniority*. However, much more detailed calculation of the kind illustrated here would be needed to establish this as a firm conclusion.

CHAPTER 9

The Simple Epidemic Model for the Diffusion of News and Rumours

9.1 INTRODUCTION

The diffusion of information in a social group is a phenomenon of considerable interest and importance. A large amount of research has been devoted to the subject and some of this has involved the construction of mathematical or stochastic models. A brief review of the earlier published work on such models is given by Coleman (1964b, in Chapter 17). Our object in this chapter is to describe some stochastic models for diffusion of information and to use them to gain understanding of the phenomenon. As in the earlier part of the book the treatment is theoretical but we have been guided in our choice of assumptions by the limited amount of experimental evidence which is available. A brief review of some of this work is given in the course of this and the next chapter.

The system which we shall study may be described as follows. There is a population of N units which we shall usually describe as people but which may be groups of people as, for example, families. Information is transmitted to members of the group from a source either at an initial point in time or continuously. For example, the source may be a television commercial, a newspaper, a roadside advertisement hoarding or a group of people introduced into the population from outside. Persons who receive the information may become 'spreaders' themselves by transmitting the information to others whom they meet. The process of diffusion continues until all have heard the news or until transmission ceases.

In many cases a stochastic model will be required to provide an adequate description of the process. The chance element enters at two points. Whether or not a given person hears the information will depend on (a) his coming into contact with the source or a spreader and (b) on the information being transmitted when contact is established. In social systems such as the armed forces where there are well-defined channels of communication the chance factor is negligible. In less rigidly organized systems neither (a) nor (b) is a certain event and so the development of the process is unpredictable. Hence it can only be described stochastically.

In spite of the obvious stochastic nature of the process much of the existing theory is deterministic. The reason for this is that the mathematics of the full stochastic version of many of the models is so intractable. We shall often find that we have to fall back on the deterministic version of a model to make progress. However, the deterministic model will be regarded as an approximation and models will always be formulated stochastically in the first instance.

The progress of the spread of information may be described in a variety of ways. Most work, empirical and theoretical, has centred on the growth of the number of people who have received the information at any time—called hearers. There are two random variables relevant to this particular aspect of the process which we shall study. The first is the number of hearers at time T which will be denoted by $n(T)$ or, where there is no risk of confusion, by n. The second way of describing the growth in the number of hearers is by the time taken for n people to hear. This time is denoted by T_n. The two random variables, T_n and $n(T)$, stand in an inverse relation to one another and for many practical purposes either will serve. Our choice between them will be governed chiefly by the mathematical advantages which each offers in a particular circumstance.

Most of the empirical studies of diffusion have concentrated on observing $n(T)$ at intervals of time in the course of the diffusion process. Some of the data resulting from these studies will be found in Pemberton (1936), Ryan and Gross (1943), Dodd (1955), Griliches (1957, 1960) and Hägerstrand (1967). In most cases, when $n(T)$ was plotted against T the curve was found to be sigmoid in form indicating an accelerating rate of adoption in the first phase of the spread followed by a decelerating rate. One of the principal objects of theoretical research has been to construct models which provide a satisfactory explanation of this phenomenon. We shall show how this characteristic of diffusion curves can be generated by two simple—and rather different—models.

Much of the empirical work relates to the diffusion of innovations. Here it is important to distinguish between the time at which a potential innovator hears of the innovation and the time at which he adopts it. In many cases adoption may be almost immediate so that the distinction is of no practical importance. In other cases this is not so. The time to adoption must be treated as the sum of the time to hearing and the subsequent delay before adoption. Ryan and Gross (1943) give data on both aspects of a study of the spread of hybrid corn in two Iowa farming communities. In this and the following chapter we shall concentrate on the hearing aspect of the total process.

A second class of random variables relates to the number of intermediaries between the original source of information and the hearer. Someone who receives the news direct from the source is called a first generation hearer. Those who first hear from a member of the first generation are second generation hearers, and so on. The number of hearers in the gth generation at time T will be denoted by $n_g(T)$, where g takes integer values from 1 to N. We shall be mainly interested in the ultimate number of hearers in each generation for which we shall use the notation n_g $(=n_g(\infty))$. These random variables have a particular interest because news which spreads from person to person is liable to distortion. The distribution of the n_g's therefore gives us some idea of the extent to which a message may be distorted in the course of its diffusion through a social group.

In the case of a population consisting of a school or club it may be realistic to think of all the members as located at a single point since diffusion will tend to take place when the members of the population are together. Similarly, if the means of communication is by telephone or radio, location may be irrelevant over small areas at least. In many other situations, however, the spatial spread of the information will be of prime interest. This is well illustrated by Hägerstrand's (1967) pioneering study of the diffusion of innovations in Swedish communities and with Morrill's (1965) study of the growth of negro ghettos in U.S. cities. Two aspects of the process are of interest in the study of spatial diffusion. One is the rate at which the news spreads throughout the area. That is, we are still interested in the random variable $n(T)$ but we shall now wish to regard it as a function of location as well as of time. The other aspect is the pattern of the state of knowledge at any given point in the course of the diffusion. Pattern is not an easy concept to quantify in this context but it refers to regularities in the location of hearers in relation to the original source. For example, a clustering of hearers in a north–south direction or along a main road would constitute a pattern indicative of the manner in which the diffusion had taken place.

The development of stochastic models for these different aspects of the diffusion process has been somewhat uneven. Our treatment will reflect this unevenness which arises partly from the mathematical intractability of some of the problems and partly from the fact that so much of the theory was first developed with epidemics in mind. There are obvious similarities between epidemics and social diffusion but there are also important differences. For example, the random variables $\{n_g\}$ have no particular relevance in epidemic theory and it is only recently that they have received any attention at all. The word 'epidemic' provides a convenient description of either process and we shall continue to use it without meaning to imply a medical application. A comprehensive account of epidemic theory up to the middle 1950's is given in Bailey (1957).

The plausibility of the various assumptions that we shall have to make depends partly on the kind of information being transmitted. The term 'information' is being used here in a neutral sense to cover such diverse things as rumours, news, advertising material and public announcements. Our models should not therefore be regarded as of universal application but rather as pointers to what might happen under various specified sets of conditions.

Models for diffusion can be conveniently classified according to whether or not they include any mechanism for stopping the act of spreading the information. If everyone who acquires the information continues to pass it on indefinitely, then it will usually be the case that the whole population will ultimately be informed. In these circumstances the main object of the theory is to study the approach to that limiting state. The remainder of this chapter is devoted to such models which, following a well-established usage in epidemic theory, we call *simple epidemic models*. In practice, however, there are many reasons why spreaders may cease this activity, in which case it is possible for the process to die out before the whole population has been informed. This possibility raises a range of questions about the extent to which any significant diffusion will take place at all. Models incorporating this feature will be called *general epidemic models* and they form the subject of Chapter 10.

9.2 THE BASIC MODEL

Description of the model

Our basic model is a special case of the pure birth process. Let E_S denote the transmission of the information from a source to any given member of the population. This is assumed to be a random event with

$$Pr\{E_S \text{ in } (T, T + \delta T)\} = \alpha \, \delta T \qquad (\alpha > 0) \tag{9.1}$$

where α is described as the *intensity of transmission of the source*. In this model we treat contact with the source and reception of information from the source as a single event. The above assumption may thus be expressed by saying that all members are equally exposed to the source. It would be a plausible assumption if the source were a television commercial and if the population consisted of regular viewers. It would not be realistic if the population also included people who rarely view the programme. Our model can be generalized to include variable exposure. However, we shall show that, in most circumstances, α plays a minor role in the development of the process; a simple assumption will therefore suffice.

Let us denote the transmission of news between any given pair of individuals by E_I. Our second assumption about the process is that

$$Pr\{E_I \text{ in } (T, T + \delta T)\} = \beta\, \delta T \qquad (\beta > 0) \tag{9.2}$$

where β is the *intensity of transmission between individuals.* We assume that this probability is the same for all pairs of individuals. This, in turn, implies that we have a homogeneously mixing population. In such a population any uninformed member is equally likely to receive the news from any of the n persons who are active spreaders. The assumption of homogeneous mixing seems plausible only in very small groups and experimental evidence supports this view. Nevertheless, there are advantages in studying the simple model first and then introducing greater realism by way of appropriate generalizations. Finally, we assume that all transmissions, whether from the source or between pairs of individuals, are independent of each other.

We are now in a position to relate the process that we have described to the pure birth process. When exactly n people have received the information we shall say that the system is in state n. A stochastic process is a time-homogeneous birth process if the probability of a transition from state n to state $n + 1$ is given by

$$Pr\{n \to n + 1 \text{ in } (T, T + \delta T)\} = \lambda_n\, \delta T \qquad (\lambda_n \geq 0) \tag{9.3}$$

and if no other types of transition (for example, $n \to n - 1$) are possible. It is obvious that n can only increase and the identification of the two processes will be complete when we have expressed λ_n in terms of the parameters of our model. The number who have heard can be increased in one of two ways. Either the next person to hear will receive the information from the source or from another person. As there are $N - n$ persons who have not heard, the total contribution to λ_n from the source is $(N - n)\alpha\, \delta T$. The contribution from communication between persons is obtained as follows. Of all the possible pairs which could be formed there are $n(N - n)$ which consist of one 'knower' and one 'ignorant'. These are the only pairs which can give rise to the transition $n \to n + 1$ and the total probability associated with them is $n(N - n)\beta\, \delta T$. Combining these results we have

$$\lambda_n = (N - n)(\alpha + \beta n) \qquad (n = 0, 1, \ldots, N - 1). \tag{9.4}$$

The theory associated with our model can thus be developed from that for the birth process with quadratic birth rate, given by (9.4).

The model which we have described was proposed by Taga and Isii (1959) but it is almost identical with the simple epidemic model discussed, for example, in Bailey (1957). In epidemic theory the source consists of one or more persons who introduce the infection to the group. Thus if one person starts the epidemic we have to put $\alpha = \beta$ when we find

$$\lambda_n = \beta(n + 1)(N - n) \qquad (n = 0, 1, \ldots, N - 1). \tag{9.5}$$

This particular case has received the greatest attention and we shall return to it later. In the application to diffusion of news it is not necessary that $\alpha = \beta$ or that β should be a multiple of α as in epidemic theory. It is also worth drawing attention to the fact that the assumptions of the model seem more reasonable when it is applied to the diffusion of news. For example, the application to the epidemic requires that we ignore the incubation period of the disease and that each infected individual remains infectious until the epidemic is over. Both assumptions are unrealistic for many infectious diseases but are quite reasonable for the diffusion of some kinds of information. They then require that the information is transmitted instantaneously and that it is not forgotten.

Analysis of the model

We shall now use the birth process model to make deductions about the development of the diffusion process in time. In view of the fact that our model is a pure birth process it is natural to begin by studying the distribution of $n(T)$. Historically this was the course followed and we shall begin by briefly describing some of the results which have been obtained. However, it is now clear that the approach via the random variables $\{T_n\}$ is capable of yielding more information about the process in a much simpler fashion.

The expression for the distribution of $n(T)$ may be found in Bartlett (1955, Section 3.2), and it has already appeared in another guise in (6.20). It is given by

$$Pr\{n(T) = 0\} = e^{-\lambda_0 T},$$

$$Pr\{n(T) = n\} = \prod_{i=0}^{n-1} \lambda_i \sum_{i=0}^{n} \frac{e^{-\lambda_i T}}{\prod_{\substack{j=0 \\ j \neq i}}^{n} (\lambda_j - \lambda_i)}. \tag{9.6}$$

Since λ_i can be found from (9.5) in terms of α, β and N the problem is solved in principle. Even for small values of N the computation of the distribution is formidable; it is given by Bailey (1957, Table 5.1) for $N = 10$ and $\alpha = \beta$. For large values of N the task is not practicable. The feature of the distribution which is of greatest interest is the mean, $\bar{n}(T)$. When plotted as a function of T it gives a visual representation of the expected development of the process. If we are primarily interested in the rate at which the news is spreading at T we would wish to plot the derivative of $\bar{n}(T)$. This latter curve is often called the 'epidemic curve' and provides a clearer picture of the growth and subsequent decline of the epidemic. The expressions for $\bar{n}(T)$ and its derivative were obtained by Haskey (1954) and are given by Bailey (1957). The formulae are rather complicated but computations have been carried out by Haskey (1954), Bailey (1957) and Mansfield and Hensley (1960) for $N \leq 40$. Two epidemic curves, plotted from their calculations, are given in Figure 9.1 for the case $\alpha = \beta$. Bailey (1963) gave a complete solution in terms of known functions and later, Bailey (1968b), an approximation, valid in large populations, based on a perturbation of the deterministic theory to be discussed later.

The abscissa on Figure 9.1 is plotted in units of β^{-1}. This is the expected time taken for any given pair of people to meet—a fact which follows directly

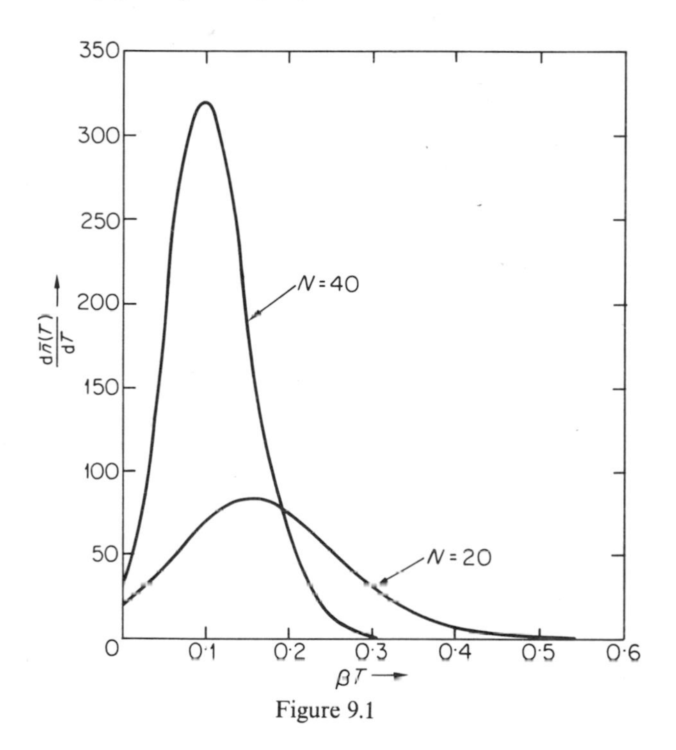

Figure 9.1

from (9.2). Thus, for example, it is clear that the diffusion is completed in the case $N = 40$ in about one-third of the average time taken for two given persons to meet. It will be noted that the spread is more rapid in the larger of the two populations shown in the figure. We shall encounter this phenomenon again below, where the reason for it will be made clear.

The investigation of the form of the epidemic curve for larger N is facilitated by the following observation. Let T' denote the time taken for the news to reach a specified member of the population and let $F(T')$ be its distribution function. Then it is clear that

$$\bar{n}(T) = NF(T) \tag{9.7}$$

and

$$\frac{\mathrm{d}\bar{n}(T)}{\mathrm{d}T} = Nf(T). \tag{9.8}$$

Equation (9.8) shows that the epidemic curve is proportional to the probability density function of the time taken to inform a given member of the population. Williams (1965) exploited this relationship to deduce the moments and asymptotic form of the epidemic curve. We shall derive similar results by considering the random variables $\{T_n\}$.

The time at which the nth person receives the information can be represented as a sum of random variables as follows:

$$T_n = \sum_{i=1}^{n} \tau_i \tag{9.9}$$

where τ_i is the time interval between the reception of the information by the $(i-1)$th and ith persons. It now follows from the Markov property of the birth process that the random variables $\{\tau_i\}$ are independently and exponentially distributed with parameters λ_{i-1} $(i = 1, 2, \ldots, N-1)$. Using this fact the exact distribution can, in principle, be found. It was investigated by Kendall (1957) for $n = N$ who found that when $\alpha = \beta$,

$$W = (N+1)T_N - 2\log_e N$$

has the limiting density function

$$f(W) = 2K_0(2\,\mathrm{e}^{-\frac{1}{2}W})\,\mathrm{e}^{-W} \qquad (-\infty < W < \infty) \tag{9.10}$$

where $K_0(\cdot)$ is the modified Bessel function of the second kind and zero order. This is a unimodal density with some positive skewness. Note that the limiting distribution is not normal because the τ_i's do not satisfy the conditions of the Lindberg–Feller central limit theorem for non-identically distributed random variables. (For these conditions see Feller, 1966, Theorem 3, page 256.) The result has recently been rediscovered by McNeil (1972). For our purposes we shall find it possible to obtain all the information we require from the cumulants of T_n.

The rth cumulant of τ_i is

$$\kappa_r = (r-1)!/\lambda_{i-1}^r.$$

Since the τ_i's are independent the rth cumulant of their sum is

$$\kappa_r(T_n) = (r-1)! \sum_{i=1}^{n} \lambda_{i-1}^{-r} \tag{9.11}$$

where λ_{i-1} is given by equation (9.4). Our main interest is in the expectation of T_n, which is obtained by putting $r = 1$ in (9.11). This gives

$$\begin{aligned} E(T_n) = \kappa_1(T_n) &= \sum_{i=0}^{n-1} \frac{1}{(N-i)(\alpha+\beta i)} \\ &= \frac{1}{\alpha+\beta N}\sum_{i=0}^{n-1}\frac{1}{N-i} + \frac{\beta}{\alpha+\beta N}\sum_{i=0}^{n-1}\frac{1}{\alpha+\beta i} \\ &= \frac{1}{\beta(N+\omega)}\{\phi(N) - \phi(N-n) - \phi(\omega-1) + \phi(n+\omega-1)\} \end{aligned} \tag{9.12}$$

where $\omega = \alpha/\beta$ and

$$\phi(x) = \sum_{i=1}^{\infty} \frac{x}{i(i+x)} = \sum_{i=1}^{\infty}\left(\frac{1}{i} - \frac{1}{i+x}\right)$$

is the digamma function. This function is tabulated in the British Association Mathematical Tables, Volume I (1951) for $x = 0{\cdot}0(0{\cdot}01)1{\cdot}0$ and $10{\cdot}0(0{\cdot}1)60{\cdot}0$. For large x

$$\phi(x) \sim \log_e x + \gamma$$

where $\gamma = 0{\cdot}5772\ldots$ is Euler's constant.

We can now use (9.12) to study the expected development of the diffusion process. Nothing essential will be lost if we suppose that N is large. The limiting behaviour of $E(T_n)$ depends on whether or not n is near to zero or N. Initially let us suppose that n/N is fixed and denoted by p. Then if N is large and $p \neq 0$ or 1

$$E(T_{Np}) \sim \frac{1}{\beta(N+\omega)}\left\{\log_e \frac{pN+\omega-1}{1-p} + \gamma - \phi(\omega-1)\right\}. \tag{9.13}$$

If ω is fixed we have approximately that

$$E(T_{Np}) \sim \frac{\log_e N}{N\beta} + \frac{1}{N\beta}\log_e \frac{p}{1-p}. \tag{9.14}$$

Two important conclusions follow from these formulae. First, it is clear that the parameter β is much more important than ω, and hence α, in determining the rate of diffusion. Secondly, a proportion p will be reached faster in a large than in a small population. Both of these phenomena can be explained by reference to the form of λ_n. Except near the start and finish, the coefficient of β in λ_n is an order of magnitude larger than that of α. Also λ_n is an increasing function of N.

The above formulae do not hold if p is zero or one because, in deriving them, we had to assume that N, $N - n$ and n were large. The case $p = 1$, or $n = N$, is of particular interest because T_N is then the *duration* of the epidemic. Proceeding to the limit we find the expression comparable to (9.13) to be

$$E(T_N) \sim \frac{2 \log_e N - \phi(\omega - 1) + \gamma}{\beta(N + \omega)}. \tag{9.15}$$

Thus (9.14) shows that half of the population will have heard in time $(\log_e N)/N\beta$ and (9.15) that all will have heard in twice that time.

Equation (9.13) also enables us to find the asymptotic form of the function $\bar{n}(T)$. This follows from the fact that, when n and N are both large with $0 < p < 1$,

$$\frac{n}{N} = p \doteqdot F(T)|_{T = E(T_n)}. \tag{9.16}$$

Hence the function on the right-hand side of (9.13) is $F^{-1}(p)$. It follows that, for large N and $\bar{n}(T)$ not near 0 or N,

$$\bar{n}(T) \sim N\left(\frac{e^X - \omega + 1}{e^X + N}\right) \tag{9.17}$$

where $X = \{\beta(N + \omega)T - \gamma + \phi(\omega - 1)\}$. For small values of ω/N this is an S-shaped function, similar to the normal ogive. Under the conditions discussed below, $\bar{n}(T)/N$ becomes the same as the probability integral of a logistic distribution. The epidemic curve is easily obtained by differentiating (9.17). It is a unimodal curve with some degree of positive skewness. An illustrative diagram is given in Williams (1965).

We have already found an expression for the median of the epidemic curve. It is a simple matter to find its mean, which is the time that a randomly selected individual can expect to wait before hearing the news. If we write this quantity as $E(T')$ it is obvious that

$$E(T') = \frac{1}{N} \sum_{n=1}^{N} E(T_n).$$

The summation over n on the right-hand side of (9.12) involves the manipulation of double sums but readily yields

$$E(T') = \frac{1}{\beta N}\{\phi(N + \omega - 1) - \phi(\omega - 1)\}. \tag{9.18}$$

This equation is a slight generalization of equation (21) in Williams (1965), whose analysis was restricted to the case when ω is an integer. As we should expect from the result for the median, the mean of the epidemic curve is of order $(\log N)/N$.

The consideration of mean values does not tell us how far an actual realization of the process may depart from our expectation. To investigate this we may consider the distribution of the random variables $\{T_n\}$. We already have a general expression for the rth cumulant of T_n in (9.11). This result will now be used to show that the time taken for n people to hear is subject to considerable variation. Using the identity

$$\frac{1}{x^r y^r} \equiv \frac{1}{(x + y)^{2r}} \sum_{i=1}^{r} \binom{2r - i - 1}{r - 1} (x + y)^i \left(\frac{1}{x^i} + \frac{1}{y^i}\right)$$

the cumulants given by (9.11) in conjunction with (9.4) may be written

$$\kappa_r(T_n) = \frac{(r - 1)!}{\beta^r} \sum_{i=1}^{r} \binom{2r - i - 1}{r - 1} \frac{1}{(N + \omega)^{2r-i}} \left\{ \sum_{j=1}^{n} \left(\frac{1}{(N + 1 - j)^r} + \frac{1}{(j + \omega - 1)^r} \right) \right\} \quad (r > 1). \tag{9.19}$$

Introducing the polygamma functions defined by

$$\phi^{(s)}(x) = (-1)^s (s - 1)! \sum_{i=1}^{\infty} \frac{1}{(i + x)^s} \qquad (s > 1)$$

the cumulants may be written

$$\kappa_r(T_n) = \frac{(r - 1)!}{\beta^r} \sum_{i=1}^{r} \binom{2r - i - 1}{r - 1} \frac{(-1)^{i-1}}{(i - 1)!(N + \omega)^{2r-i}} \times \{\phi^{(i)}(N - n) - \phi^{(i)}(N) + \phi^{(i)}(\omega - 1) - \phi^{(i)}(\omega + n - 1)\}. \tag{9.20}$$

In the limit as $N \to \infty$, with $n/N = p$ held fixed with $0 < p < 1$,

$$\kappa_r(T_n) \sim \frac{(-1)^r}{\beta^r (N + \omega)^r} \phi^{(r)}(\omega - 1) \qquad (r > 1). \tag{9.21}$$

On the other hand, if $(N - n)$ is fixed so that $p \to 1$ as $N \to \infty$ a different limiting form is obtained. For example, in the case $n = N$

$$\kappa_r(T_N) \sim \frac{(-1)^r}{\beta^r(N+\omega)^r}\{\phi^{(r)}(\omega-1)+\phi^{(r)}(0)\} \qquad (r > 1). \qquad (9.22)$$

A notable feature of (9.21) is that it does not depend on n. In particular, the asymptotic variance of T_n is independent of n. The polygamma functions are tabulated in the British Association Tables referred to above so that numerical values for the cumulants are readily obtainable. It follows from (9.21) that the variance of T_n is a decreasing function of ω, which means that its variability goes down as the 'strength' of the source is increased. The extent of this reduction in variance can be gauged from the fact that $\phi^{(2)}(0) = 1{\cdot}6449$ and $\phi^{(2)}(1) = 0{\cdot}6449$. Even when n is large the distribution of T_n may have considerable skewness and kurtosis. Some illustrative calculations are given in Table 9.1.

Table 9.1 Asymptotic values of the skewness, $\sqrt{\beta_1}$, and the kurtosis, β_2, of T_n

		ω				
		1	2	5	10	∞
$0 < p < 1$	$\sqrt{\beta_1}$	1·14	0·73	0·47	0·32	0·00
	β_2	5·40	4·19	3·44	3·21	3·00
$p = 1$	$\sqrt{\beta_1}$	0·81	0·81	0·97	1·04	1.14
	β_2	4·20	4·33	4·90	5·12	5·40

The covariance of any pair of T_n's can be found at once from the fact that

$$\operatorname{cov}(T_n, T_{n+j}) = \operatorname{var}(T_n) \qquad (j \geq 0). \qquad (9.23)$$

The deterministic approximation

With the simple birth process model there is no need to have recourse to deterministic methods. All that we required to know about the expected behaviour of the process can be determined from the stochastic model. Nevertheless, it is instructive to consider a deterministic version of the model as a preparation for the analysis of the more intractable models which occur later. We treat $n(T)$ as a continuous function. According to (9.4) the expected amount by which it will increase in $(T, T + \delta T)$ is $(N - n)(\alpha + \beta n)\,\delta T$. In the deterministic treatment we suppose that it increases by exactly this amount in each small increment of time. This implies that $n(T)$ satisfies the differential equation

$$\frac{\mathrm{d}n(T)}{\mathrm{d}T} = (N - n(T))(\alpha + \beta n(T)). \tag{9.24}$$

Solving this with the boundary condition $n(0) = 0$ we have for $T > 0$,

$$n(T) = N\frac{\exp[\beta(N + \omega)T - 1]}{\exp[\beta(N + \omega)T] + N/\omega}. \tag{9.25}$$

A comparison of this with (9.17) shows that the stochastic solution and the deterministic approximation are not the same. The deterministic curve lags behind the stochastic curve by an amount which varies with both N and ω. The position for small N is illustrated by Bailey (1957, Figures 4.2 and 4.3). For large N the same conclusion follows from a comparison of the means of two epidemic curves. In the deterministic case

$$E(T') = \frac{1}{N\beta}\log_e\left(\frac{N + \omega}{\omega}\right). \tag{9.26}$$

Williams (1965) showed that this is always less than the stochastic version given by (9.18). In order for the deterministic solution to be equivalent to the asymptotic stochastic solution it is necessary for ω to be large. It may easily be verified that (9.17) and (9.25) become identical in the limit as $\omega \to \infty$. For fixed ω the difference between the deterministic and stochastic means is of the order of N^{-1}. This may appear to be insignificant until it is recalled that the duration of the whole epidemic is of order $(\log N)/N$. On this time scale a difference of order N^{-1} can be of practical importance for moderate values of N. We must therefore be on the alert for this kind of occurrence when we come to more complex models.

Models with imperfect mixing

The principal assumptions of the birth process model are (a) that all members are equally exposed to the source and (b) that all pairs of members have equal likelihood of communicating. Assumption (a) is not crucial unless α is large compared to β because the main contribution to λ_n, once the process has started, comes from the term $\beta(N - n)n$. The second assumption is certainly invalid in most human populations. It is therefore necessary to investigate the effect of relaxing (b).

We begin by going to an extreme and suppose that there is no communication at all between members of the population. The diffusion is thus entirely attributable to the source. This case is, in fact, covered by our model and is obtained by setting $\beta = 0$. Expressions for the epidemic curve and duration can be obtained from those already given by letting $\omega \to \infty$ with N fixed.† However, they can easily be obtained from first principles along with other results which cannot be found in the general case. When $\beta = 0$ we have what is called the pure death process (see Bailey, 1964, Section 8.5). The exact distribution of $n(T)$ turns out to be binomial with

$$Pr\{n(T) = n\} = \binom{N}{n} e^{-\alpha(N-n)T}(1 - e^{-\alpha T})^n \qquad (n = 0, 1, \ldots, N). \quad (9.27)$$

Hence

$$\bar{n}(T) = N(1 - e^{-\alpha T}) \qquad (9.28)$$

and the epidemic curve is

$$\frac{d\bar{n}(T)}{dT} = N\alpha\, e^{-\alpha T}. \qquad (9.29)$$

There is thus a marked qualitative difference between this case and that illustrated on Figure 9.1. The rate of diffusion declines continuously with time instead of first rising to a maximum. It should therefore be possible, in practice, to form some judgement about the relative importance of inter-personal and source-personal diffusion by an inspection of the empirical epidemic curve. In a study of the diffusion of information about a new drug reported in Coleman (1964b, Figure 17.2) it was found that the growth of knowledge was roughly exponential. This can be interpreted to mean that advertising rather than personal recommendation led to adoption of the new drug.

The time taken for a proportion p of the population to hear the news is also easy to obtain. Since the T_n's are partial sums of the τ's it follows at once that

$$E(T_n) = \frac{1}{\alpha} \sum_{i=1}^{n} \frac{1}{N - i + 1}. \qquad (9.30)$$

If N is large and $p\,(= n/N)$ is not near to one

$$E(T_{Np}) \sim -\frac{1}{\alpha} \log_e (1 - p). \qquad (9.31)$$

This result stands in marked contrast to (9.13) and (9.14). In the present model the time taken to reach a given proportion does not depend on N.

† Note that this is not the same set of conditions which led to the equivalence of the deterministic and stochastic epidemic curves in the last section. In that case N was allowed to tend to infinity.

By combining the results obtained from our two extreme assumptions we may conclude, in general, that the time taken to reach a proportion p cannot be an increasing function of population size. This argument does not cover the limiting case $p = 1$. In this case (9.31) must be replaced by

$$E(T_N) \sim \frac{1}{\alpha} \log_e N. \tag{9.32}$$

Thus the total duration does increase with size but only slowly.

In order to chart the territory between the two extreme degrees of mixing we shall consider the case of a stratified population. One such model was discussed by Haskey (1954). He supposed that the population was composed of two strata with different rates of contact between and within groups. A deterministic model with k strata was analysed by Rushton and Mautner (1955); we return to this later. A similar kind of model for a population with $N/3$ strata was solved semi-deterministically by Coleman (1964b, Chapter 17). We shall first consider a simpler model of the same kind.

Suppose that the population is made up of k strata of equal size. The members of all strata are equally exposed to the source and the rate of contact between members of the *same* stratum is β, as before. However, there is no contact at all between the members of different strata. Under these conditions the diffusion in each stratum develops according to the theory of the preceding sections. The diffusion in the system as a whole is then found by pooling the results; no new theory is required.

Coleman (1964b) developed the theory for the case $N = 2k$. In this case each stratum is of size 2 and there is no difficulty in obtaining the exact distribution of $n(T)$ from (9.6). Let $n_i(T)$ denote the number who have heard at time T in the ith stratum ($i = 1, 2, \ldots, N/2$); then

$$\lambda_0 = 2\alpha, \qquad \lambda_1 = \alpha + \beta, \qquad \lambda_2 = 0.$$

Hence

$$\left.\begin{aligned} Pr\{n_i(T) = 0\} &= e^{-2\alpha T} \\ Pr\{n_i(T) = 1\} &= \frac{2\alpha(e^{-2\alpha T} - e^{-(\alpha+\beta)T})}{\beta - \alpha} \\ Pr\{n_i(T) = 2\} &= 1 - Pr\{n_i(T) = 0 \text{ or } n_i(T) = 1\} \end{aligned}\right\}. \tag{9.33}$$

The limiting forms appropriate when $\alpha = \beta$ are easily deduced. Since

$$n(T) = \sum_{i=1}^{N/2} n_i(T)$$

it follows that

$$\bar{n}(T) = \frac{N}{2}\bar{n}_i(T) \quad \text{and} \quad \operatorname{var} n(T) = \frac{N}{2}\operatorname{var} n_i(T),$$

and that $n(T)$ is approximately normal. The moments of $n_i(T)$ are readily found from (9.33); in particular,

$$\left.\begin{aligned} \bar{n}_i(T) &= 2 - \frac{2(\omega\, \mathrm{e}^{-(\alpha+\beta)T} - \mathrm{e}^{-2\alpha T})}{\omega - 1} & \omega \neq 1 \\ \bar{n}_i(T) &= 2\{1 - (1 + \beta T)\,\mathrm{e}^{-2\beta T}\} & \omega = 1 \quad (\text{i.e. } \alpha = \beta) \end{aligned}\right\}. \tag{9.34}$$

The epidemic curve in the case $\alpha = \beta$ has the form

$$\frac{1}{N}\frac{\mathrm{d}\bar{n}(T)}{\mathrm{d}T} = \beta\{1 + 2\beta T\}\,\mathrm{e}^{-2\beta T}. \tag{9.35}$$

The density on the right-hand side of (9.35) is decreasing with mean $3/4\beta$. This may be compared with the approximate value of $(\log_e N)/N\beta$ for a homogeneously mixing population. The mean for a population with no mixing at all and transmission intensity $\alpha = \beta$ from the source is $1/\beta$. With the limited degree of communication permitted by our stratified model the expected time to hear is reduced but is still independent of the population size. If $\beta = \infty$ the second member of each stratum automatically receives the news at the same time as the first member. Our model is then equivalent to a freely mixing population made up of $N/2$ pairs.

The foregoing analysis expresses in quantitative form the obvious conclusion that incomplete mixing reduces the rate of diffusion—at least in a population of small non-communicating strata. This conclusion can be strengthened by considering the duration of the diffusion for general k. Let us denote by $T_{(i)}$ the duration for the ith stratum. Then

$$T_N = \max_i T_{(i)}.$$

The probability distribution of T_N is then that of the largest member of a sample of size k from the distribution of the duration in a stratum. It is possible to make progress with the general theory of the distribution of T_N but an inequality due to Gumbel (1958) provides sufficient information for our purposes. For any random variable x with finite mean μ and variance σ^2 he states that

$$E(x_{\max}) \leq \mu + \sigma\frac{k - 1}{\sqrt{2k - 1}}$$

where k is the sample size. Applying this result to the case of T_N, μ is the average duration for a stratum and σ^2 is its variance. If N/k is large we may use the asymptotic forms and obtain

$$E(T_N) \leq \frac{2\log_e(N/k)}{\beta(N/k+\omega)} + \frac{\sqrt{\phi^{(1)}(\omega-1)}}{\beta(N/k+\omega)}\frac{(k-1)}{\sqrt{2k-1}}$$

$$= k\frac{2\log_e N}{N\beta} + O(N^{-1}) \tag{9.36}$$

for fixed ω. This result suggests but does not prove, because of the inequality, that division into k strata multiplies the duration by a factor k.

The question of imperfect mixing can be investigated in somewhat greater generality by reverting to the deterministic approximation and following Rushton and Mautner (1955). Suppose there are k strata; let the size of the ith stratum be N_i and denote by $n_i(T)$ the number of hearers at time $T (i = 1, 2, \ldots, k)$. As before, let α be the intensity of transmission of the source; let β_i be the transmission intensity within the ith stratum and let γ_{ij} be the corresponding quantity between members of the ith and jth strata. The set of differential equations corresponding to (9.24) is then

$$\frac{dn_i(T)}{dT} = \{N_i - n_i(T)\}\{\alpha + \beta_i n_i(T)\} + \{N_i - n_i(T)\}\sum_{j\neq i}\gamma_{ij}n_j(T)$$

$$(i = 1, 2, \ldots, k). \tag{9.37}$$

The first term on the right-hand side of (9.37) is the contribution to the rate of increase of $n_i(T)$ from the source and contact within the ith stratum. The second term accounts for the change resulting from between stratum contact. The initial conditions will be $n_i(0) = 0$ for all i. Rushton and Mautner (1955) provided a general method of solving the system (9.37), but considerable insight into the effects of imperfect mixing can be had by considering the following special case. Let $N_i = N/k$, $\beta_i = \beta$ and $\gamma_{ij} = \gamma$ for all i and j; then, because of the deterministic nature of the process, $n_i(T) = n(T)/k$. Under these simplifying assumptions all of the equations in (9.37) have the same form and can therefore be treated as a single equation, since

$$\frac{d}{dT}\left\{\frac{n(T)}{k}\right\} = \left\{\frac{N}{k} - \frac{n(T)}{k}\right\}\left\{\alpha + \frac{\beta}{k}n(T)\right\} + \left\{\frac{N}{k} - \frac{n(T)}{k}\right\}\gamma\left\{\frac{N}{k} - \frac{n(T)}{k}\right\}$$

or

$$\frac{dn(T)}{dt} = \{N - n(T)\}\left\{\alpha + \left(\frac{\beta + (k-1)\gamma}{k}\right)n(T)\right\}. \tag{9.38}$$

In order to see the effect of imperfect mixing we must compare this equation with (9.24). The difference lies in the coefficient $\{\beta + (k - 1)\gamma\}/k$ which replaces the β of the original equation. The epidemic therefore develops at the same rate as in a population with homogeneous mixing and with transmission intensity $\{\beta + (k - 1)\gamma\}/k$. This expression is a weighted average of β and γ and since $\gamma < \beta$ in any meaningful application it will be less than β. In the extreme case when $\gamma = \beta$, (9.38) reduces to (9.24). When $\gamma = 0$, meaning that there is no communication at all between strata, the epidemic spreads as in a homogeneously mixing population with transmission intensity β/k. In effect this amounts to a scaling of time by a factor k so that it takes k times as long to attain a given degree of spread in the stratified population as in a homogeneously mixing one of the same total size. This conclusion provides an interesting confirmation of the stochastic result reached in (9.36).

The foregoing conclusions emphasize that the assumption of homogeneous mixing is crucial and they provide a caution against undue reliance on the model in cases where it is known to be suspect. In spite of this rather severe limitation a careful analysis of the pure birth model yields valuable insight into the mechanism of diffusion and so provides a preparation for the general epidemic models of Chapter 10.

9.3 THE DISTRIBUTION OF HEARERS BY GENERATION

Introduction

As we have already observed, the random variables $\{n_g\}$ have a particular interest in the study of the diffusion of rumours which are subject to distortion as they pass from person to person. In this section we shall investigate the distribution of hearers by generation when diffusion takes place according to the pure birth model. The case $g = 1$ is of particular importance as it relates to those who hear directly from the source and so the distribution of n_1 will be treated in some detail. As before, we shall proceed as far as possible with the stochastic version of the model and then revert to the deterministic approximation.

The joint distribution of $\{n_g\}$

Consider, first, the problem of finding the joint distribution of the variables $\{n_g(T)\}$, giving the generation distribution at time t. Denote the number of ignorants (the term used for those who have not heard) at time T by $m(T)$. Then if the total population size is N,

$$m(T) = N - \sum_{g=1}^{N} n_g(T).$$

The infinitesimal transition probabilities associated with the process are as follows:

$$\begin{aligned} Pr\{m(T + \delta T) &= m - 1, n_g(T + \delta T) = n_g + 1| \\ &\times m(T) = m, n_{g-1}(T) = n_{g-1}, n_g(T) = n_g\} \\ &= \beta m n_{g-1}\, \delta T \quad \text{if } g > 1. \end{aligned} \tag{9.39}$$

These equations express the fact that the size of the gth generation increases by one as a result of contact between a member of the $(g - 1)$th generation and an ignorant. The case $g = 1$ has to be treated separately because the size of the first generation can only increase by contact with the source so that

$$\begin{aligned} Pr\{m(T + \delta T) = m - 1, n_1(T + \delta T) = n_1 + 1|m(T) = m, n_1(T) = n_1\} \\ = \alpha m\, \delta T \end{aligned} \tag{9.40}$$

where α is the intensity of the source. In principle, these probabilities, with the initial conditions, are sufficient to deduce the complete development of the diffusion process. Our main interest is in the limiting behaviour of the generation distribution. This can be found by taking advantage of an embedded Markov chain.

Let $P(n_1, n_2, \ldots, n_N)$ be the probability that the system is in the state $\mathbf{n} = (n_1, n_2, \ldots, n_N)$ at some time during the course of the diffusion (that is that $n_h(T) = n_h$, $h = 1, 2, \ldots, N$ for some T). The method is to find a difference equation for the probabilities using the fact that the process is a random walk on the set of states $\mathbf{n}$. The state vector $\mathbf{n}$ is subject to certain important constraints. The sum of its elements obviously cannot exceed N and it will attain this value at the end of the epidemic. Secondly, the only vectors $\mathbf{n}$ which have a non-zero probability of occurring are those having the form $(X, X, \ldots X, 0, 0, \ldots 0)$ where X represents a non-zero entry. This property follows from the fact that there cannot be any members belonging to the gth generation unless there is first at least one member of the $(g - 1)$th generation. For example, when $N = 3$ the possible $\mathbf{n}$-vectors are

$$(3, 0, 0), (2, 1, 0), (1, 2, 0), (1, 1, 1).$$

Let us call all $\mathbf{n}$-vectors satisfying the foregoing conditions *admissible* vectors; then the states of our Markov chain will be the set of admissible vectors. To simplify the exposition it will be convenient to define $P(\mathbf{n}) = 0$ for all inadmissible states so that ranges of summation can be kept as simple as possible.

Immediately prior to being in the state $(n_1, n_2, \ldots, n_N)$ the system must have been in one of the admissible states among $(n_1, n_2, \ldots n_h - 1, n_{h+1} \ldots n_N)$

$(h = 1, 2, \ldots, N)$. The probability of the transition

$$(n_1, n_2, \ldots n_h - 1, n_{h+1} \ldots n_N) \rightarrow (n_1, n_2, \ldots n_h, n_{h+1}, \ldots n_N)$$

is easily obtained from the infinitesimal transition probabilities as

$$\frac{\beta n_{h-1} m}{\beta m N - \beta m + \alpha m} = \frac{n_{h-1}}{N + \omega - 1} \qquad (h > 1)$$

and

$$\frac{\alpha m}{\beta m N - \beta m + \alpha m} = \frac{\omega}{N + \omega - 1} \qquad (h = 1)$$

where $\omega = \alpha/\beta$. Hence

$$P(n_1, n_2, \ldots, n_N) = \sum_{h=1}^{N} \frac{n_{h-1} P(n_1, n_2, \ldots, n_h - 1, \ldots n_N)}{N - 1 + \omega} \tag{9.41}$$

if we define $n_0 \equiv \omega$. The initial condition is

$$P(1, 0, 0, \ldots, 0) = 1.$$

Equation (9.41) expresses the required probability for a population of size N in terms of a set of probabilities for a population of size $N - 1$. In building up the distribution for a given N we therefore obtain the distribution for all sizes smaller than N. The equation has been programmed for a computer, but as the number of admissible vectors increases as the square of N the demands made by the program on computer storage space limit the size of population which can be considered. We shall give some numerical results later but first we shall pursue the analysis by means of generating functions. Let

$$\Pi_N(s) = E(s_1^{n_1} s_2^{n_2} \ldots s_N^{n_N}) \tag{9.42}$$

where $\sum_{h=1}^{N} n_h = N$. Then, multiplying both sides of (9.41) by $s_1^{n_1} s_2^{n_2} \ldots s_N^{n_N}$ and summing over all admissible **n**-vectors, we obtain

$$\left.\begin{aligned} \Pi_N(\mathbf{s}) &= \frac{1}{N - 1 + \omega}\left[\omega s_1 \Pi_{N-1}(\mathbf{s}) + \sum_{h=1}^{N-1} s_h s_{h+1} \frac{\partial \Pi_{N-1}(\mathbf{s})}{\partial s_h}\right] \quad (N > 1) \\ \Pi_1(s_1) &= s_1 \end{aligned}\right\} \tag{9.43}$$

The dimension of the vector **s** in such formulae is indicated by the subscript of Π. An alternative form of (9.43) which has certain advantages can be derived by an application of Euler's theorem about homogeneous functions. Since $\Pi_N(\mathbf{s})$ has the property that all its terms are of degree N it is a homogeneous function, and the same applies to $\Pi_{N-1}(\mathbf{s})$, which is of degree $N - 1$.

Hence, by Euler's theorem,

$$\sum_{h=1}^{N-1} s_h \frac{\partial \Pi_{N-1}(\mathbf{s})}{\partial s_h} = (N-1)\Pi_{N-1}(\mathbf{s}). \tag{9.44}$$

Adding the right-hand side of (9.44) to that of (9.43) and subtracting the left-hand side yields

$$\Pi_N(\mathbf{s}) = \left(\frac{N-1+s_1\omega}{N-1+\omega}\right)\Pi_{N-1}(\mathbf{s}) + \frac{1}{N-1+\omega}\sum_{h=1}^{N-1} s_h(s_{h+1}-1) \times \frac{\partial \Pi_{N-1}(\mathbf{s})}{\partial s_h}. \tag{9.45}$$

Neither of the expressions for $\Pi_N(\mathbf{s})$ is very helpful for determining the probability distribution unless N is small, and it does not appear easy to deduce a solution for $\Pi_N(\mathbf{s})$ in closed form. The value of the equations lies in their usefulness for the study of special cases and for deriving such things as moments. However, the joint distribution is interesting even for very small N and we list the results for $N \leq 5$ and $\omega = 1$ below. The choice of $\omega = 1$ makes the source equivalent to one spreader introduced into the group.

$$\left.\begin{aligned}
\Pi_1(s_1) &= s_1 \\
\Pi_2(s_1, s_2) &= \tfrac{1}{2}s_1^2 + \tfrac{1}{2}s_1 s_2 \\
\Pi_3(s_1, s_2, s_3) &= \tfrac{1}{6}(s_1^3 + 3s_1^2 s_2 + s_1 s_2^2 + s_1 s_2 s_3) \\
\Pi_4(s_1, s_2, s_3, s_4) &= \tfrac{1}{24}(s_1^4 + 6s_1^3 s_2 + 7s_1^2 s_2^2 + 4s_1^2 s_2 s_3 + s_1 s_2^3 \\
&\quad + 3s_1 s_2^2 s_3 + s_1 s_2 s_3^2 + s_1 s_2 s_3 s_4) \\
\Pi_5(s_1, s_2, s_3, s_4, s_5) &= \tfrac{1}{120}(s_1^5 + 10s_1^4 s_2 + 25s_1^3 s_2^2 + 10s_1^3 s_2 s_3 \\
&\quad + 15s_1^2 s_2^3 + 25s_1^2 s_2^2 s_3 + 5s_1^2 s_2 s_3^2 \\
&\quad + 5s_1^2 s_2 s_3 s_4 + s_1^2 s_2^3 + 6s_1 s_2^3 s_3 \\
&\quad + 7s_1 s_2^2 s_3^2 + 4s_1 s_2^2 s_3 s_4^3 + s_1 s_2 s_3^3 \\
&\quad + 3s_1 s_2 s_3^2 s_4 + s_1 s_2 s_3 s_4^2 + s_1 s_2 s_3 s_4 s_5)
\end{aligned}\right\} \tag{9.46}$$

The marginal distributions for the individual generations are given for $\omega = 1$ and $N = 10$ in Table 9.2. The corresponding distributions for $N \leq 5$ can easily be computed direct from the generating functions given in (9.46). For small N, at least, a large proportion hear at first- or second-hand—4·2 on average out of a population of 5 and 6·4 out of a population of 10.

***Table 9.2** The distribution of n_g for $N = 10$ with $\omega = 1$ for the pure birth model*

	j											
g	0	1	2	3	4	5	6	7	8	9	10	*Mean*
1	0·0000	0·1000	0·2829	0·3232	0·1994	0·0742	0·0174	0·0026	0·0002	0·0000	—	2·9282
2	0·0000	0·0399	0·1635	0·2969	0·2941	0·1583	0·0424	0·0048	0·0002	0·0000	—	3·5151
3	0·0320	0·1979	0·3544	0·2797	0·1110	0·0227	0·0022	0·0001	0·0000	—	—	2·3172
4	0·4000	0·3439	0·1844	0·0594	0·0112	0·0011	0·0000	—	—	—	—	0·9412
5	0·8055	0·1486	0·0387	0·0066	0·0006	0·0000	—	—	—	—	—	0·2482
6	0·9623	0·0325	0·0048	0·0004	0·0000	—	—	—	—	—	—	0·0433
7	0·9954	0·0042	0·0004	0·0000	—	—	—	—	—	—	—	0·0050
8	0·9997	0·0003	0·0000	—	—	—	—	—	—	—	—	0·0003
9	0·9999	0·0001	0·0000	—	—	—	—	—	—	—	—	0·0001
10	1·0000	0·0000	—	—	—	—	—	—	—	—	—	0·0000

For large values of N we shall need the approximate method discussed later.

The number who hear directly from the source

In the case of the first generation the marginal distribution of n_1 can be found explicitly. The probability generating function of n_1 is obtained from (9.45) by putting $s_2 = s_3 = \ldots = s_N = 1$. The second term on the right then vanishes leaving

$$\Pi_N(s_1) = \left(\frac{N - 1 + s_1\omega}{N - 1 + \omega}\right)\Pi_{N-1}(s_1),$$

whence

$$\Pi_N(s_1) = \frac{\omega s_1(1 + \omega s_1)(2 + \omega s_1)\ldots(N - 1 + \omega s_1)}{\omega(1 + \omega)(2 + \omega)\ldots(N - 1 + \omega)}$$

$$= \frac{\Gamma(N + \omega s_1)\Gamma(\omega)}{\Gamma(N + \omega)\Gamma(\omega s_1)} \tag{9.47}$$

where $\Gamma(x)$ denotes the gamma function. Extracting the coefficient of $s_1^{n_1}$ we obtain the probability distribution as

$$P_N(n_1) = \frac{\omega^{n_1}\Gamma(\omega)|S_N^{n_1}|}{\Gamma(N + \omega)} \qquad (n_1 = 1, 2, \ldots, N) \tag{9.48}$$

where $S_N^{n_1}$ is the Stirling number of the first kind, $|S_N^{n_1}|$ being the coefficient of x^{n_1} in $x(x + 1)\ldots(x + N - 1)$. Miles (1959) discussed the distribution in another context when $\omega = 1$ and he tabulated the Stirling numbers for $N = 1(1)12$. A further discussion and tabulation of the distribution is given in Barlow and coworkers (1972, page 142 ff and Table A.5, page 363). In order to give an idea of the shape of the distribution for small N and varying ω it has been tabulated in Table 9.3.

In small groups it is clear that n_1/N is highly variable except for the extreme values of ω. As the last column shows, the mean number informed by the source increases with ω but not as rapidly as might have been expected. The only way to ensure that most people first hear the news from the source is to have a very high value of ω. This can only be achieved by increasing the strength of the source or decreasing the degree of contact between members of the population. It is a characteristic of rumours that the source is weak, consisting, perhaps, of a single person. Under these circumstances it is not surprising that distortion often occurs because almost everyone receives the rumour at second-hand or worse. These conclusions apply also when N is large as we shall now show.

Table 9.3 The distribution $P(n_1|N = 10)$ ***and*** $E(n_1|N = 10)$ ***for various*** ω

	n_1										
ω	1	2	3	4	5	6	7	8	9	10	$E(n_1\|N = 10)$
0	1·0000	—	—	—	—	—	—	—	—	—	1·00
1	0·1000	0·2829	0·3232	0·1994	0·0742	0·0174	0·0026	0·0002	0·0000	0·0000	2·93
2	0·0182	0·1029	0·2350	0·2901	0·2159	0·1014	0·0303	0·0056	0·0006	0·0000	4·04
5	0·0005	0·0071	0·0404	0·1245	0·2317	0·2722	0·2032	0·0936	0·0242	0·0027	5·84
10	0·0000	0·0003	0·0035	0·0216	0·0803	0·1887	0·2819	0·2595	0·1342	0·0298	7·19
∞	—	—	—	—	—	—	—	—	—	1·0000	10·00

The exact mean and variance of n_1 given N can be found directly from the generating function. They are

$$\left.\begin{aligned} E(n_1|N) &= \omega\left\{\frac{1}{\omega} + \frac{1}{\omega+1} + \dots + \frac{1}{N+\omega-1}\right\} \\ &= \omega\{\phi(N+\omega-1) - \phi(\omega-1)\} \\ \operatorname{var}(n_1|N) &= E(n_1|N) - \omega^2 \sum_{i=0}^{N-1} \frac{1}{(i+\omega)^2} \end{aligned}\right\}. \tag{9.49}$$

If ω is fixed and N is large

$$E(n_1|N) \sim \omega \log_e N,$$

and hence the proportion who have heard direct from the source is

$$\frac{E(n_1|N)}{N} \sim \omega \frac{\log_e N}{N}. \tag{9.50}$$

The proportion who hear first-hand from the source under these conditions thus tends to zero as the population size increases. In order for the source to communicate directly with a high proportion of the population it will obviously be necessary to make ω very large. When ω is large

$$\left.\begin{aligned} E(n_1|N) &\sim \omega \log_e\left(\frac{N+\omega-1}{\omega}\right) \\ \text{and} \qquad & \\ \frac{E(n_1|N)}{N} &\sim \frac{\omega}{N}\log_e\left(\frac{1+\omega/N}{\omega/N}\right) \end{aligned}\right\}. \tag{9.51}$$

If $\omega = \frac{1}{2}N$ the expected proportion of hearers is 0·55 and if $\omega = N$ it is 0·69. Thus, for example, if the source consists of N spreaders introduced into a population of size N about 70 per cent would hear the information at first-hand.

The formula for the variance in (9.49) suggests that the limiting distribution of n_1 may be Poisson in form because the mean and variance tend to equality. This is not quite true. The asymptotic form of the probability generating function is

$$\Pi_N(s_1) \sim N^{\omega(s_1-1)}\Gamma(\omega)/\Gamma(\omega s). \tag{9.52}$$

This has the form of the generating function of the convolution of a Poisson variable with mean $\omega \log_e N$ and a random variable with generating function $\Gamma(\omega)/\Gamma(\omega s)$. Since the latter does not depend on N its contribution will be negligible for very large N and the Poisson part should be a good approximation. Barton and Mallows (1961) showed that the limiting form of $P_N(n_1)$ as $N \to \infty$ with n_1 fixed and $\omega = 1$ was Poisson probability with parameter $\log_e N$. This result can be extended to cover the case of a general ω, but as it relates only to the lower tail of the distribution it adds little to the Poisson approximation based on (9.52).

The distribution of n_1 was first considered in the diffusion context by Taga and Isii (1959), who derived the probability distribution from a specialization of the argument leading to (9.41). In fact, if we sum both sides of (9.46) over admissible values of $n_2, n_3, \ldots, n_N$ we obtain the equation

$$P_N(n_1) = \frac{\omega}{N-1+\omega} P_{N-1}(n_1 - 1) + \frac{N-1}{N-1+\omega} P_{N-1}(n_1)$$

$$(1 \leq n_1 \leq N; N \geq 1) \tag{9.53}$$

with initial conditions

$$P(0|N) = 1 \text{ if } N = 0$$
$$= 0 \text{ otherwise,}$$

which is Taga and Isii's result.

The expected number of hearers in the gth generation

The joint generating function of the n_g's provides a convenient means of finding the expectations of the generation sizes. Thus

$$E_N(n_g) = \left.\frac{\partial \Pi_N(\mathbf{s})}{\partial s_g}\right|_{s_1 = s_2 = \ldots = s_N = 1} \quad (g = 1, 2, \ldots, N). \tag{9.54}$$

The case $g = 1$ has to be treated separately from $g > 1$. The result in the former case has already been obtained in the last section and we shall not duplicate the derivation here. Taking $g > 1$ and differentiating both sides of (9.45) with respect to s_g,

$$\frac{\partial \Pi_N(\mathbf{s})}{\partial s_g} = \left(\frac{N-1+s_1\omega}{N-1+\omega}\right)\frac{\partial \Pi_{N-1}(\mathbf{s})}{\partial s_g} + \frac{1}{N-1+\omega}$$

$$\sum_{h=1}^{N-1} s_h(s_{h+1}-1)\frac{\partial^2 \Pi_{N-1}(\mathbf{s})}{\partial s_h^2} + \frac{1}{N-1+\omega}$$

$$\left\{s_{g-1}\frac{\partial \Pi_{N-1}(\mathbf{s})}{\partial s_{g-1}} + (s_{g+1}-1)\frac{\partial \Pi_{N-1}(\mathbf{s})}{\partial s_g}\right\}$$

$$(g = 2, 3, \ldots, N). \tag{9.55}$$

Putting $s_1 = s_2 = \ldots = s_N = 1$ and substituting in (9.54),

$$E_N(n_g) = E_{N-1}(n_g) + E_{N-1}(n_{g-1})/(N - 1 + \omega)$$
$$(g = 2, 3, \ldots, N). \tag{9.56}$$

The advantage of the representation of (9.45) over (9.43) will now be clear from the way terms vanish when we put the s's equal to one. Starting with $E_1(n_1) = 1$ it is easy to build up the complete set of expectations.

Equation (9.56) can also be used to deduce an approximation to the expected generation sizes, valid when N is large. We arrive at this, when ω is fixed, by treating N as a continuous variable and replacing (9.56) by

$$\frac{dE_N(n_g)}{dN} = \frac{E_N(n_{g-1})}{N} \qquad (g = 2, 3, \ldots, N). \tag{9.57}$$

Starting with $E_N(n_1) \sim \omega \log_e \{(N - 1 + \omega)/\omega\}$ we may easily deduce that

$$E_N(n_g) \sim \omega \left\{\log_e \left(\frac{N - 1 + \omega}{\omega}\right)\right\}^g \bigg/ g! \qquad (g = 1, 2, \ldots, N). \tag{9.58}$$

Notice that when these expectations are summed over g we obtain $N - 1$ which is asymptotically the same as in the exact case when the sum is N. The expected sizes of successive generations thus form an exponential series with the first term missing. Typically, this distribution will be unimodal with an upper tail dying away rapidly for large g.

The method of deriving this approximation does not depend on ω being small. The argument holds if ω is of the same order as N. In this case we shall wish to express the expected size of each grade as a proportion of the population size N. As a function of ω/N we then have

$$\frac{E_N(n_g)}{N} \sim \frac{\omega}{N}\left\{\log_e \left(1 + \frac{N}{\omega}\right)\right\}^g \bigg/ g! \qquad (g = 1, 2, \ldots, N). \tag{9.59}$$

Some numerical values for $\omega/N = \frac{1}{2}$ and 1 are given in Table 9.4.

One way in which these results might be used is to see the effect of mixing two populations, one of which possesses information which is passed to

Table 9.4 The expected proportion who hear at gth-hand derived from the asymptotic approximation

g	1	2	3	4	5
$\omega/N = \frac{1}{2}$	0·55	0·30	0·11	0·03	0·01
$\omega/N = 1$	0·69	0·24	0·06	0·01	0·00

members of the other. In this interpretation ω would represent the size of the transmitting population and N the receiving population. The table shows that most people will hear first- or second-hand if the influx is at least half as big as N.

Daley (1967a) derived the approximation of (9.59) as the solution to a deterministic version of the problem. When discussing the generation distribution for a general epidemic model in the next chapter it will not be possible to make any progress with the stochastic model. We shall therefore have to rely on deterministic methods and so to prepare the ground we shall rederive (9.59), starting with the deterministic version of the model.

Let $n_g(T)(g = 1, 2, \ldots, N)$ and $m(T)$ refer to the same quantities as before, but now we treat them as continuous functions of T and interpret them deterministically. The rates of change under the assumptions of the model are then

$$\frac{dn_1(T)}{dT} = \alpha m(T) \tag{9.60a}$$

$$\frac{dn_g(T)}{dT} = \beta n_{g-1}(T)m(T) \qquad (g = 2, 3, \ldots, N) \tag{9.60b}$$

$$\frac{dm(T)}{dT} = -\alpha m(T) - \beta\{N - m(T)\}m(T) \tag{9.60c}$$

with initial conditions

$$m(0) = N, \qquad n_g(0) = 0 \qquad (g = 1, 2, \ldots).$$

The solution of (9.60c) has already been found in (9.25) because $m(T) = N - n(T)$ so

$$m(T) = N\left(1 + \frac{\omega}{N}\right)\bigg/\left(1 + \frac{\omega}{N}e^{\beta(N+\omega)T}\right).$$

Substitution of this in (9.60a) gives $n_1(T)$ by straightforward integration as

$$n_1(T) = \omega \log_e \left\{\left(1 + \frac{\omega}{N}\right)\bigg/\left(\frac{\omega}{N} + e^{-\beta(N+\omega)T}\right)\right\}. \tag{9.61}$$

The limit of $n_1(T)$ as $T \to \infty$ agrees with (9.51) and (9.59). The remaining $n_g(T)$'s are obtained from (9.60b) by putting $g = 2, 3, \ldots$ in turn and finding each $n_g(T)$ by an integration involving its predecessor. It may be verified that this procedure yields

$$n_g(T) = \frac{\omega}{g!}\left\{\log_e \left(1 + \frac{\omega}{N}\right)\bigg/\left(\frac{\omega}{N} + e^{-\beta(N+\omega)T}\right)\right\}^g. \tag{9.62}$$

The deterministic approach gives not only the limiting values of the generation sizes but it also shows how these numbers change throughout the course of the diffusion. The stochastic approach does not allow us to do this without reverting to the much more difficult problem posed by (9.39) and (9.40).

9.4 DIFFUSION IN SPACE

Introduction

When we turn to the study of diffusion in space the theory available is very limited. The emphasis in this section will therefore be on the model-building aspect with relatively little analysis or testing.

Most applications of spatial diffusion theory envisage a geographical context in which hearers are located at points in a two-dimensional plane. However, the theory is often simpler for populations distributed in one dimension on a line. Many of the essential characteristics of spatial diffusion processes are preserved in passing from two dimensions to one, so we shall sometimes take advantage of the simplification which this specialization offers.

The early work on spatial diffusion was carried out principally by Kolmogorov and coworkers (1937) for the spread of an advantageous gene and by Bartlett (1956) and Kendall (1957) in the context of the spread of epidemic diseases. More recently there has been a rapid growth of interest among geographers, much of it stemming from the pioneering work of Hägerstrand (1967). Hägerstrand's work was originally published in Swedish in 1953, but it was not until the publication of the English translation that its main impact was made. His chief interest was in the diffusion of innovations. The study covers agricultural indicators, such as state-subsidized pastures and the control of bovine T.B., together with the spread of things like cars and telephones. Marris (1970) studied the diffusion of consumer demand and Eidem (1968) discussed the diffusion of community-linked innovations such as nursing homes, swimming pools and municipal fluoridation in the urban areas of North Dakota. Morrill (1965) investigated the spread of negro ghettos in Seattle. None of these studies contains a significant amount of theory of the processes they describe and neither do they relate to the earlier work on epidemics. Hägerstrand, however, does formulate several stochastic models of spatial diffusion and it is with this part of his work that we shall be concerned here. A complete study of any of the areas listed above must obviously be concerned with the wider range of geographical, social and economic factors which relate to the situation. Our limited objective is the analysis of the stochastic features which such processes exhibit.

The pattern of diffusion of the simple epidemic in a randomly distributed population

Suppose that members of a population are distributed at random over a plane area. By this we mean that the number of inhabitants per unit area is a Poisson variable and that the numbers in non-overlapping areas are independent. If diffusion takes place according to the assumptions of the birth process then the chance of contact between any pair of individuals will be unrelated to their location. At any stage, therefore, the knowers will form a random subset of the original population and so will themselves be randomly distributed. The pattern revealed if the hearers are plotted on a map will thus be a purely random one. This may seem a somewhat trivial point at which to begin because the spatial aspect only enters through the original scatter of the population. However, even a random distribution of points will show a certain degree of clustering. In order to decide whether clusters arising in practice are due to special local effects or are purely random in origin it is necessary to consider whether they are compatible with the random hypothesis. Statistical methods for this purpose are well known and need not detain us here. The initial assumption of a randomly distributed population is not necessary, of course. If we had any arbitrary distribution we could test the randomness of the spread by comparing the proportion of hearers in non-overlapping regions with the expected value of $n(T)/N$. Hägerstrand (1967) proposed such a model in which the population was assumed to be spread evenly over a plane area with a density of 30 per 25 kilometre square. He used a Monte Carlo technique to study the pattern of diffusion and concluded that such a simple model could not account for the observed patterns of his various indicators. Notice that in this, as in all other spatial spread models, the population members are assumed to have a fixed location. This is reasonable if 'hearing' corresponds to adopting an agricultural innovation which takes place on a particular farm. It may also be reasonable if the 'space' is social rather than geographical, but it will not be appropriate for the study of highly mobile populations.

The pattern of diffusion when the chance of contact depends on distance apart

Stochastic models

When potential hearers are distributed over an area it is natural to expect contact to be more common among people who are close together than those who are far apart. In such circumstances we need to study the number of hearers as a function of both time and location. Several stochastic models have been proposed in which the essential idea is to make the chance of contact a function of distance apart. Little progress has been made on the mathematical analysis of these models but some of their basic qualitative characteristics have been determined from Monte Carlo experiments. We shall first describe two stochastic models and then pursue the analysis by means of deterministic approximations.

Hägerstrand (1967) introduced both one- and two-dimensional versions of a simple discrete model. The one-dimensional model was used simply to aid the exposition, but there is little difficulty in proceeding straight to the more realistic two-dimensional version. The population is again supposed to be evenly distributed over a plane area with a specified density. This area is divided up by a square grid. The diffusion starts with a given distribution of knowers—usually a certain number in a particular central square. Each of the original knowers then communicates with one other person selected in a manner to be described below. Next the original knowers, together with those contacted on the first round, make one further contact. The process continued in this way. At each stage every knower makes one contact; an addition to the total stock of knowers occurs each time a knower makes contact with an ignorant.

The contacts are selected in this model by a two-stage probability process. At the first stage the grid square from which the hearer is to come is chosen and at the second stage one member of that square is chosen. The second stage involves a simple random choice in which each member is equally likely to be chosen. The first stage depends on a probability distribution over the squares in the neighbourhood of the originating knower. In Hägerstrand's investigation he assumed that contacts could be made in any of the 25 squares centred on the originator's square with probabilities illustrated in the diagram below.

0·0096	0·0140	0·0168	0·0140	0·0096
0·0140	0·0301	0·0547	0·0301	0·0140
0·0168	0·0547	0·4431	0·0547	0·0168
0·0140	0·0301	0.0547	0·0301	0·0140
0·0096	0·0140	0·0168	0·0140	0·0096

The probabilities, which had an empirical basis, fall off with increasing distance from the shaded square which marks the origin.

Hägerstrand represented the realizations of this stochastic process on contour diagrams, showing the position at various stages of the development. A typical realization shows an initial cluster of knowers near the original source with several subsidiary clusters derived from a few 'long distance' contacts made in the early stages. As time advances, these clusters tend to thicken and merge to give an irregular concentration of knowers centred near the region where the process started. Hägerstrand used the term 'central stability' to describe this phenomenon. This does not mean that the density of hearers will necessarily be greatest at the origin but that the regions of high concentration will be in its vicinity. The precise way in which the pattern develops will depend, of course, on the form of the contact probability distribution. Hägerstrand's example is symmetrical with probability falling off in all directions. We shall see later that the precise way in which probability of contact falls off with distance is a crucial factor in determining the pattern of spread.

The model described above could be represented as a Markov chain in discrete time. The states of the chain are all possible dispositions of hearers in the area. Each step in the chain involves the making of one contact as a result of which the chain either stays in the same state—if the contact already knows—or moves to a new state. The probabilities of contact are independent and depend only on the present state of the system. In any non-trivial problem the number of states will be enormous and it does not appear possible to use the theory of Markov chains to advantage except, perhaps, to arrive at the otherwise obvious result that the system will ultimately arrive at the only absorbing state—when everyone knows. Other ways in which the Markov chain model can be used in very similar circumstances have been touched on in Chapter 2.

A rather similar model has been investigated by Mollison (1972b) for diffusion in one dimension. Individuals are located on the real line with σ persons at each integer point (positive and negative). Each knower makes contacts at random intervals governed by a Poisson process of rate $\alpha\sigma$. The destination of a given contact is determined by a probability distribution, which for simplicity is assumed to be symmetrical about the source of the contact and is written in the form

$$v(u) = \alpha(|u|)/\alpha \qquad (u = 0, \pm 1, \pm 2, \dots). \tag{9.63}$$

The main differences between this model and the one-dimensional version of Hägerstrand's model is that the former places the 'contact events' in time and so the development of the process in time can be observed. Hägerstrand only ordered his events according to the number of stages which, in practice, would only partly reflect the time scale. Since Hägerstrand was primarily interested in pattern this was adequate for his purposes. Mollison's approach takes us further by enabling us to study the rate of diffusion. The assumption that the rate at which an individual makes contacts is proportional to the number of individuals at his location is not very plausible in the rumour situation, but this only becomes important if we wish to compare processes with different σ's.

Mollison used his model to test the validity of theoretical deductions made from a deterministic model and to suggest lines for further theoretical analysis. We shall treat matters in the reverse order, using the deterministic analysis to illuminate the empirical findings from simulation of the stochastic model.

Mollison (1972b) found two main kinds of behaviour which are related to the form of the contact distribution $v(u)$—in particular to its variance and the nature of its tails. The first kind of behaviour appears to occur when $v(u)$ has exponentially bounded tails (and, hence, finite variance). In such a case contacts are usually made in the neighbourhood of the source and the 'front' of knowers tends to advance outwards from the origin at a steady rate. There is a problem about defining the front but it is clear intuitively what is meant. The profile of knowers might appear at any time in Figure 9.2.

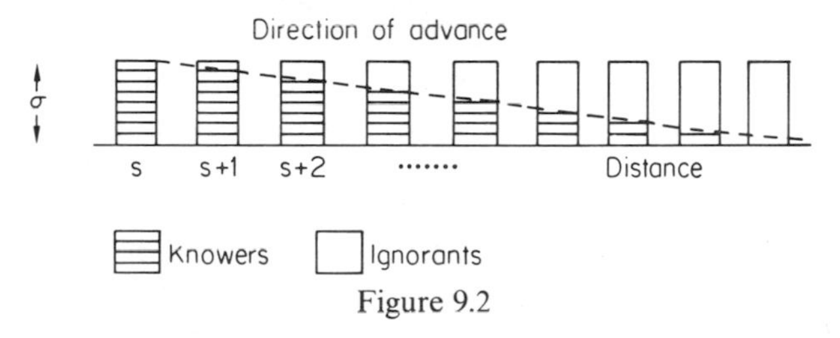

Figure 9.2

The dotted curve suggests a wave travelling outwards as the diffusion develops. At any time there will be an inner region of knowers, an outer region of ignorants and a twilight region in which some know and some are ignorant. In any particular realization there will, of course, be stochastic variation which will blurr the outlines, but the broad outline will be clear. Mollison simulated the case when

$$v(u) = \tfrac{1}{3}(\tfrac{1}{2})^{|u|} \qquad (u = 0, \pm 1, \pm 2, \dots)$$

and found that, on average, the wave advanced at a fairly constant rate. When $v(u)$ has infinite variance, theory suggested and simulation demonstrated a different kind of behaviour. When the position of the wave is plotted as a function of time the steady advance is broken at intervals by what Mollison calls 'great leaps forward'. This describes what happens when a contact is made at a considerable distance ahead of the advancing wave. Such a contact gives rise to a colony of knowers ahead of the wave and so accelerates the overall rate of advance. A similar kind of behaviour can also occur when the variance is finite but when the tails are not exponentially bounded.

We thus have two distinct kinds of behaviour depending on the extent to which chance of contact depends on distance apart. There will either be a steadily advancing wave or a wave preceded by local clusters; as old clusters merge with the main body of knowers, new ones will arise. Mollison suggests that a necessary and sufficient condition for the steady rate of advance may be that $v(u)$ has exponentially bounded tails, but this has not been proved. If $v(u)$ is highly dispersed so as to be almost uniform we are almost back to the case when chance of contact is independent of distance. In this case there will be no discernible wave at all—only a gradually merging set of clusters.

Deterministic models

Some progress can be made with the theory by resorting to a deterministic treatment. In fact, the model of Rushton and Mautner (1955) leading to (9.37) is very similar to Mollison's stochastic model. The k strata can be made to represent the groups of individuals located at points on the line. The between-strata contact rates can be brought into correspondence with the chance of contact for different distances apart. In principle, therefore, the spatial spread can be investigated by solving the set of differential equations (9.37) in the manner described by Rushton and Mautner (1955). In a similar way, spatial diffusion in two dimensions could be handled for populations whose members were located in large groups at discrete points. In either case the problem of computation would be formidable and information about the behaviour of such a process can be obtained with less effort using an approach in which the population is uniformly distributed over the plane.

The following approach was used in an epidemic context by Bartlett (1956) and Kendall (1957, 1965) and extended by Mollison (1972a, 1972b). Let the density of the population be σ per unit area and let $y(s, T)$ be the proportion of knowers located at s at time T. (The argument which follows extends to two dimensions but the essentials can be displayed more simply in one dimension.) Let $v(u)\,\delta u$ be the probability that a given contact is made at a distance in $(u, u + \delta u)$. Then if contacts are made at a rate β the rate of change of knowers at s will be proportional to the product of the number of ignorants there and the number of knowers elsewhere who succeed in making a contact at s. The probability that a contact is made at s from a knower at

$(s - u, s - u + \delta u)$ is proportional to $y(s - u)v(u)\,\delta u$, and the total probability arising from all u may thus be expressed by

$$\bar{y}(s, T) = \int_{-\infty}^{+\infty} y(s - u, T)v(u)\,du. \tag{9.64}$$

This argument thus leads to

$$\frac{\partial y(s, T)}{\partial T} = \beta\sigma\bar{y}(T)\{1 - y(s, T)\}. \tag{9.65}$$

This equation generalizes (9.24), to which it reduces if $v(u)$ is uniform. The boundary conditions will specify the distribution of knowers at $T = 0$.

It does not appear to be possible to find an explicit solution to (9.65), even for densities $v(u)$ of very simple form. Attention has therefore been concentrated on approximate solutions and on the deduction of relevant properties of the solution. A method of approximation used by Kendall (1965) is to replace $\bar{y}(T)$ by the diffusion approximation

$$\bar{y}(T) \doteqdot y(s, T) + K\frac{\partial^2 y(s, T)}{\partial s^2} \qquad (K > 0). \tag{9.66}$$

This can be thought of as resulting from a Taylor expansion of $y(s - u, T)$ as a function of u about $u = 0$ as far as the third term (the second term vanishes because of the assumed symmetry of $v(u)$). The constant K is then half the variance of $v(u)$. For the approximation to be good we shall therefore require $v(u)$ to be concentrated around zero, implying that contacts are usually made in the immediate vicinity of the knower. Substituting (9.66) in (9.65) we have, dropping the arguments of y,

$$\frac{\partial y}{\partial T} = \beta\sigma\left\{y(1 - y) + K(1 - y)\frac{\partial^2 y}{\partial s^2}\right\}. \tag{9.67}$$

Even after this approximation no general solution appears to be available, but progress can be made by looking for a particular sort of solution. We observed with Mollison's stochastic model that the wave of new knowers appeared to advance at a constant rate. It therefore seems reasonable to look for solutions of (9.65) having this property. That is, we search for a solution of the form $y(s, T) = y(s - cT)$. The values of y will be the same at all points (s, T) satisfying $s - cT =$ constant. If, between T_1 and T_2, a fixed point on the wave moves from s_1 to s_2 then

$$s_1 - cT_1 = s_2 - cT_2 \quad \text{or} \quad c = (s_2 - s_1)/(T_2 - T_1),$$

showing that c must be the velocity of the wave. Introducing the new variable

$x = s - cT$ we have

$$\frac{\partial y}{\partial T} = \frac{\partial x}{\partial T}\frac{\mathrm{d}y}{\mathrm{d}x} = -c\frac{\mathrm{d}y}{\mathrm{d}x}$$

and

$$\frac{\partial^2 y}{\partial s^2} = \frac{\partial}{\partial s}\left\{\frac{\partial x}{\partial s}\frac{\mathrm{d}y}{\mathrm{d}x}\right\} = \frac{\partial^2 x}{\partial s^2}\frac{\mathrm{d}y}{\mathrm{d}x} + \frac{\partial x}{\partial s}\frac{\mathrm{d}^2 y}{\mathrm{d}x^2} = \frac{\mathrm{d}^2 y}{\mathrm{d}x^2}.$$

Substituting in (9.67) we obtain

$$\beta\sigma K(1 - y)\frac{\mathrm{d}^2 y}{\mathrm{d}x^2} + c\frac{\mathrm{d}y}{\mathrm{d}x} + \beta\sigma y(1 - y) = 0. \tag{9.68}$$

When x is large we shall be ahead of the wave and so as $x \to \infty$ we require $y \to \infty$. When x is large and negative we shall be behind the wave where $y \to 1$ as $x \to \infty$.

If there is a solution of (9.68) bounded between zero and one, then it will clearly satisfy the original equation with x replaced by $s - cT$ and will establish the possibility of a wave solution. Let us first consider the situation when y is near to zero, in which case (9.68) may be approximated by

$$K\frac{\mathrm{d}^2 y}{\mathrm{d}x^2} + \left(\frac{c}{\beta\sigma}\right)\frac{\mathrm{d}y}{\mathrm{d}x} + y = 0. \tag{9.69}$$

This equation has a general solution of the form

$$y = A\,\mathrm{e}^{r_1 x} + B\,\mathrm{e}^{r_2 x}$$

where r_1 and r_2 are the roots of the quadratic equation

$$Kr^2 + \left(\frac{c}{\beta\sigma}\right)r + 1 = 0.$$

That is,

$$r_1, r_2 = -\frac{1}{2K}\left\{\frac{c}{\beta\sigma} \pm \sqrt{\frac{c^2}{\beta\sigma} - 4K}\right\}.$$

If both roots are real they will be negative, since $K > 0$, and so y will approach zero as $x \to \infty$. If both roots are complex, y will behave in an oscillatory manner and will repeatedly change sign. A solution with negative values of y cannot represent a solution of our problem and therefore we conclude that no waveform solution exists if r_1 and r_2 are complex. The condition for this is

$$c^2 \geq 4K\beta^2\sigma^2,$$
$$c \geq 2\sqrt{K}\,\beta\sigma. \tag{9.70}$$

In other words, there is a minimum velocity, given by (9.70), below which a waveform solution is impossible. At or above that minimum velocity the above analysis leaves the question open, but it may be shown that a waveform solution is possible; it further appears that if the source consists of a concentration of knowers at a point that the wave will move outwards at the minimum velocity.

The foregoing analysis is based on an approximation which assumes that contacts are made by knowers only in their immediate neighbourhood. Mollison (1972a, 1972b) was able to take the analysis further by showing that a similar result held for (9.65). Two kinds of behaviour are possible depending on the degree of dispersion of $v(u)$. If $v(u)$ has two tails which approach zero at least as fast as an exponential then all velocities are possible above a certain minimum. The results derived from the diffusion approximation then give a good indication of the behaviour of the process. On the other hand, if the tails of $v(u)$ are not bounded by an exponential the rate of advance of the epidemic tends to infinity with time.

It is interesting to relate these conclusions to those derived by simulation of the stochastic model. The critical factors governing whether the wave advanced at a steady rate or 'by leaps and bounds' was the behaviour of the tails of $v(u)$. The accelerating rate of advance in the deterministic model corresponds to the phenomenon of the 'great leap forward', and both arise when contacts can be made well beyond a spreader's immediate neighbourhood.

These results, extended into two dimensions, provide some guidance on the interpretation of patterns of diffusion of the kind studied by Hägerstrand (1967). The occurrence of central stability suggests that the chance of a pair making contact depends on their distance apart—and is generally confined to a neighbourhood. An area of stability with occasional clusters at a distance is not necessarily evidence for several sources but may indicate a contact distribution which allows occasional contact at a considerable distance from the main body of knowers.

It is clear that the information we have been able to glean from our analysis of spatial models is fragmentary and further research is needed. However, some broad and useful qualitative characteristics have emerged which will be extended in the following chapter.

Perimeter and hierarchical models

In concluding this chapter we shall briefly mention a group of models in which the contacts made between spreaders and ignorants are constrained by a particular geographical or social structure. The stochastic features of such models have not been explored in any depth in the present context†, and the purpose in introducing the topic is to direct attention to an area in which further research is highly desirable.

Suppose that the members of the population are immobile and are only able to communicate with their immediate neighbours. If information is passed through the population from a single source we may expect the area covered by the knowers to move outwards steadily from the source. New knowers will only be added at the boundary of the region occupied by knowers and for this reason we speak of *perimeter models*. There is an obvious similarity with waveform properties of the models discussed above. The distinction is that in the perimeter model contact is restricted to immediate neighbours; in the models discussed by Kendall and Bartlett contacts can be made with any other member of the population, albeit with small probability in the case of those a long distance apart. The pattern of diffusion in a perimeter model will follow the density of the population—the frontier will advance most rapidly where the population is most sparsely spread. For this reason the analysis cannot be expected to yield much of interest about pattern. We shall, instead, be more concerned with the rate of growth of the total numbers in a situation where contacts are constrained by geography.

A stochastic model of the perimeter variety was proposed by Bailey (1967) and analysed by simulation. According to this model individuals are located at the vertices of a square lattice. To begin with there is a single knower at the centre. This person is able to communicate with any of his eight neighbours, as illustrated in Figure 9.3, but with no-one else. Spreading takes place at discrete points in time and at each such time every knower is able to make contact with his neighbours. The chance that a given ignorant hears the message from a neighbour who knows is p and all contacts are independent. At the first stage the number of new knowers will thus have the binomial distribution

$$\binom{8}{r} p^r(1-p)^{8-r} \qquad (r = 0, 1, 2, \ldots, 8).$$

At the next stage all ignorants who are neighbours of knowers will be exposed to the risk of contact. If a given ignorant has a knowers as neighbours the chance that he will become a knower is $1-(1-p)^a$. Bailey (1967) simulates such a process on grids of size 11×11 and 21×21. To begin

† See, however, Hammersley's (1966) work on 'percolation processes'.

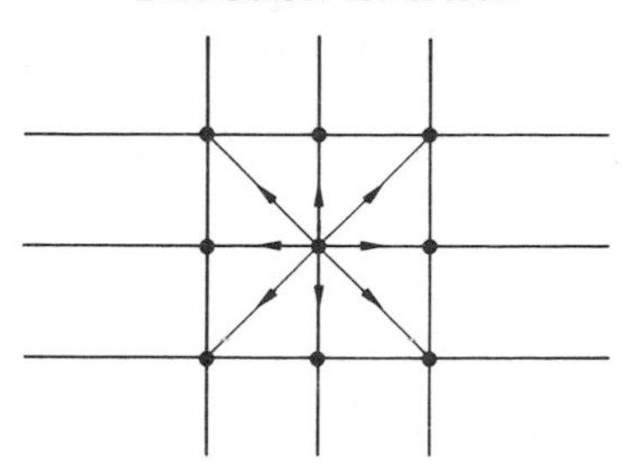

Figure 9.3

with he found that the number of knowers rose as the number of ignorants exposed to risk increased, but as the boundaries were approached an increasing number of knowers had knowers for neighbours so the rate of growth tailed off. The resulting epidemic curve was not unlike that observed in practice with the simple epidemic—rising to a peak and then falling away rather more slowly. The manner of spreading in the perimeter model is, however, quite different from the simple epidemic model. In the one case contact is equally likely between all pairs of individuals. In the other case contacts are severely limited by the structure of the population.

A deterministic investigation of a perimeter model, very similar to Bailey's, has been made by Day (1970). He was interested in the diffusion of agricultural innovations. His model is highly simplified but it provides a pointer for the development of more realistic stochastic models. A large area is divided up into square farms on a checkerboard pattern. Initially, a farm at the centre adopts an innovation. After observing it for one time period all of the four neighbours having a common boundary with the innovator adopt it. These four adopters have eight neighbours who follow suit after one further period has elapsed. At time T there will be $4T - 2$ new adopters and the cumulative number of adopters at that point will be $1 + 2T(T + 1)$. The number of adopters will thus be a quadratic function of time until the boundary eventually stops growth. Such a growth pattern does not correspond very closely to what happens in practice, and Day extended his model by supposing that the innovation starts at farms spaced d units apart in both directions. The initial growth then follows the quadratic pattern, but as soon as $T > \frac{1}{2}(d + 1)$ the clusters begin to overlap and the growth rate slows down. In fact, for $T > \frac{1}{2}(d + 1)$

$$n(T) = \tfrac{1}{2}[2d^2 + 1 - \{1 - 2(d - T)\}^2].$$

The resulting growth curve has the characteristic sigmoid form and so suggests the desirability of constructing stochastic models in which diffusion spreads simultaneously from a number of centres.

A rather similar approach to spatial diffusion involves an extension of the idea on which our original deterministic treatment of the random mixing model was based. Both T and $n(T)$ are treated as continuous variables. Consider first a perimeter model in which news diffuses from a single source through a population of uniform density. Growth will take place on the perimeter of the circle enclosing those who already know. Therefore, the number of knowers who are active at T will be proportional to $\sqrt{n(T)}$ These people will be in contact with an equal number of people on the inner boundary of the ignorants. If any knower on the boundary can communicate with any ignorant on the adjacent boundary we shall have

$$\frac{\mathrm{d}n(T)}{\mathrm{d}T} \propto \sqrt{n(T)} \times \sqrt{n(T)} \tag{9.71}$$

and growth will be exponential. If the knowers on the perimeter can only communicate with their ignorant neighbours then

$$\frac{\mathrm{d}n(T)}{\mathrm{d}T} \propto \sqrt{n(T)} \tag{9.72}$$

and the growth law will be quadratic as in Day's model. We could construct stochastic versions of these models by using a pure birth model with $\lambda_n \propto \sqrt{n}$ or n as the case may be. McNeil (1972) has investigated such a model with $\lambda_n \propto \{n(N - n)\}^{\frac{1}{2}}$, but it is not easy to relate this to a geographical model of diffusion.

Hierarchical models are very similar to perimeter models as far as their mathematical structure is concerned. The 'space' of such models is usually a social or organizational structure through which the information diffuses. Such a system may be viewed as a network in which information is introduced at the highest level and then spreads downwards through the hierarchy defined by the network. We shall not pursue the discussion of these models here. A good introduction to such models may be found in Hudson (1969), who also reviews some of the other models discussed in this chapter, drawing attention to their inadequacies in explaining diffusion.

CHAPTER 10

The General Epidemic Model for the Diffusion of News and Rumours

10.1 INTRODUCTION

The distinguishing feature of the simple epidemic models of the last chapter was that those who had heard the news continued to spread indefinitely. Common experience and experimental work, some of which is described in Section 10.5, suggest that epidemics may die out before everyone has heard. People may cease to be spreaders for a variety of reasons. They may forget, lose interest or gain the impression that 'everyone knows'. In order to achieve the greater degree of realism which these considerations suggest we shall have to consider what we call 'general' epidemic models. This term is used in epidemic theory for Kermack and McKendrick's model, to be discussed in the next section. Since its generality lies in the provision of a mechanism for the cessation of spreading it seems appropriate to extend the usage to cover other models which incorporate the same feature.

The theory of general epidemic models is difficult and it is not possible to make very much progress with the stochastic aspects of the rate of diffusion in time or space. Greater reliance has to be placed on deterministic methods. There is, however, an important new feature of general epidemics which can be treated stochastically and to which the greater part of this chapter will be devoted. This concerns the number who will ultimately hear. Since people eventually stop being active spreaders the epidemic may cease before everyone has heard. Indeed, it is possible that spreading will die out very quickly so that hardly anyone hears. This leads us to the consideration of threshold effects concerning whether or not an epidemic, in the usual sense of the word, is likely to develop.

At time T the state of the system can be described by three random variables as follows:

$m(T)$ persons who have not heard—the ignorants;
$n(T)$ persons who have heard and are actively spreading—the spreaders;
$l(T)$ persons who, having heard the news, have ceased to spread it.

Because the size of the population is assumed to be constant it follows that

$$N = m(T) + n(T) + l(T);$$

any two of the three random variables are sufficient to describe the state of the system. In the simple epidemic $l(T)$ is zero for all T. We shall be particularly interested in the limiting behaviour of the process $T \to \infty$. It is obvious that $n(T)$ must ultimately take the value zero with probability one; hence $m(\infty) = N - l(\infty)$ will be the random variable of particular interest.

10.2 KERMACK AND McKENDRICK'S MODEL

Background

The earliest contribution to mathematical epidemic theory appears to be the celebrated paper by Kermack and McKendrick (1927), in which the authors formulated a deterministic version of what has subsequently become known as the general epidemic model. Most of the basic theory is well known and may be found in Bailey (1957). Our initial formulation is slightly different from Bailey's and we shall give the topic a slant more suited to the diffusion of news.

The model is the same as the pure birth model with the addition that spreaders are only active for a random period of time. More precisely, we assume that the period of spreading for each individual is an exponential random variable with mean μ^{-1}. We further assume that cessations are independent and that once a person has ceased spreading they do not resume their activity. An essential characteristic of the model is that cessation as specified above is independent of the state of knowledge in the population. This means that an individual will spread the information with the same zeal whether many or few of his hearers have heard before. The plausibility of this assumption must be judged by the success with which the model accounts for observed diffusion. It seems most reasonable if the item of news is fairly trivial so that cessation of spreading on the part of an individual is due to forgetfulness. Obviously there will be individual variation in the time taken to forget, but the choice of the exponential distribution to describe that variability is more questionable. We shall see later that the form of this distribution may not be crucial.

The distribution of $m(T)$ and $n(T)$

We shall use $m(T)$ and $n(T)$ to describe the system at time T and shall say that the system is in state (m, n) if $m(T) = m$ and $n(T) = n$. One way of studying the development of the process in time is to consider the joint probability distribution of $m(T)$ and $n(T)$. Let us write this as

$$Pr\{m(T) = m, n(T) = n\} = P_{m,n}(T).$$

Another way would be to study the time taken for the system to reach the state (m, n). However, the system need not reach a given state (m, n) and so the simple inverse relation which existed in the analogous situation in the pure birth model is lost and there are no compensating mathematical advantages.

We shall not be able to obtain explicit formulae for the joint distribution of $m(T)$ and $n(T)$. It is nevertheless a worthwhile exercise to set up equations for them for the light which they throw on the process. From the state (m, n) two transitions are possible. They are set out below with the probabilities that they take place in $(T, T + \delta T)$.

$$\text{(a)} \quad (m, n) \to (m - 1, n + 1) : m(\alpha + \beta n)\, \delta T$$

for $m = 1, 2, \ldots, N, n = 0, 1, \ldots, N - 1$ such that $0 \leq n + m \leq N$.

$$\text{(b)} \quad (m, n) \to (m, n - 1) : n\mu\, \delta T$$

for $n = 1, 2, \ldots, N - 1$.

The transition (a) takes place when a spreader meets an ignorant; transition (b) occurs when a spreader forgets and ceases to spread.

Using these transition probabilities we can relate the joint probability at time $T + \delta T$ to that at T in the usual way and obtain the following bivariate differential-difference equation for $P_{m,n}(T)$:

$$\left.\begin{aligned} P'_{m,n}(T) &= -\{m(\alpha + \beta n) + n\mu\}P_{m,n}(T) + (m + 1) \\ &\qquad (\alpha + \beta(n - 1))P_{m+1,n-1}(T) + (n + 1)\mu P_{m,n+1}(T) \\ P'_{N,0}(T) &= -N\alpha P_{N,0}(T) \\ P_{N,0}(0) &= 1 \\ &\qquad (0 \leq m \leq N, 0 \leq n < N, 0 \leq m + n \leq N) \end{aligned}\right\} \quad (10.1)$$

where it is to be understood that probabilities with subscripts not satisfying $0 \leq m + n \leq N, n, m \geq 0$ are zero. It may be deduced that $P_{m,n}(T)$ can be expressed as a series of descending exponentials but the quadratic coefficients in (10.1) make further progress difficult.

We can draw certain general conclusions about the process by noting that the pure birth model is a special case. If $\mu = 0$ there is no forgetting and it is intuitively obvious that the rate of spread must therefore be greater than when $\mu > 0$. Thus, for example, $\bar{n}(T)$ for the pure birth model provides an upper bound for the same function in our present model. Secondly, if $\beta = 0$ or $\mu = \infty$ forgetting is irrelevant since then no-one is ever actively spreading the news and hence $n(T)$ is always zero. By putting $\mu = \infty$ our model thus becomes identical with the pure birth model with $\beta = 0$. The rate of diffusion will be greater if $\beta > 0$ than if $\beta = 0$ so, this time, we can obtain a lower bound for $\bar{n}(T)$. The two bounds provided by considering the extreme values of μ will usually be rather wide but no further progress has been made with the stochastic model in this form.

The foregoing model does not allow for the possibility, which we envisaged at the beginning of the section, of the epidemic dying out. This is because the source continues to transmit indefinitely and thus, ultimately, all people will be informed. An interesting variant is obtained by supposing that the source transmits for a limited period only. One way in which this could happen is if the source consists of a group of a individuals with the same law of forgetting as the other members of the population. Under these circumstances the infinitesimal transition probabilities become

$$\text{(a)} \quad \beta mn \quad \text{and} \quad \text{(b)} \quad \mu n$$

where n now refers to the total number of spreaders whether they originate from inside or outside the population. The differential-difference equations for the probabilities $P_{m,n}(T)$ are now

$$\left.\begin{aligned} P'_{m,n}(T) &= -(\beta mn + \mu n)P_{m,n}(T) + \beta(m+1)(n-1)P_{m+1,n-1}(T) \\ &\quad + \mu(n+1)P_{m,n+1}(T) \\ P'_{N,a}(T) &= -\{\beta aN + \mu a\}P_{N,a}(T) \\ P_{N,a}(0) &= 1 \\ &\quad (0 \le m \le N, \quad 0 \le n < N + a, \quad 0 \le m + n \le N + a) \end{aligned}\right\} . \qquad (10.2)$$

We again define probabilities to be zero if their subscripts are outside the stated ranges. The equations (10.2) are those that arise in epidemic theory and a considerable body of information has accumulated about their solution. Gani (1965) and Siskind (1965) have obtained methods for finding explicit solutions using a generating function technique. Their methods are extremely unwieldly unless N is very small and so are of little immediate practical value. Two other approaches remain open. One is to concentrate on finding partial solutions, in particular, for the limiting distribution of the number of ignorants. The second is to use a deterministic approximation for the system of equations (10.2). We shall follow both of these courses but, before doing so, we point out a second way in which the present model can arise.

Suppose that the source transmits the information to exactly a people before the diffusion starts and then ceases to operate. From that point onwards the system behaves like one of size $N' = N - a$ into which a

spreaders are introduced. It only requires trivial modifications of notation to make the theory cover a situation of this kind.

The terminal state of the system

Although the equations (10.2) are difficult to solve it is relatively easy to find the limiting values of the probabilities $P_{m,n}(T)$ and $T \to \infty$. After a sufficiently long period the diffusion will cease either because everyone has heard or because the spreaders have ceased to be active. In either event $n(\infty)$, the final number of spreaders, is zero with probability 1. Consequently

$$P_{m,n}(\infty) = 0 \quad \text{if } n > 0. \tag{10.3}$$

When $n = 0$, $P_{m,0}(\infty)$ will be the probability distribution of the terminal number of ignorants. We shall determine this distribution by exploiting the existence of an embedded random walk.

Let $P_{m,n}$ denote the probability that, *at some time* during the diffusion, there are m ignorants and n spreaders. Then clearly

$$P_{m,0} = P_{m,0}(\infty). \tag{10.4}$$

If we consider the process only at those points in time when a change of state takes place we may represent it as following a random walk over the lattice points (m, n). The situation is illustrated in Figure 10.1. We imagine a

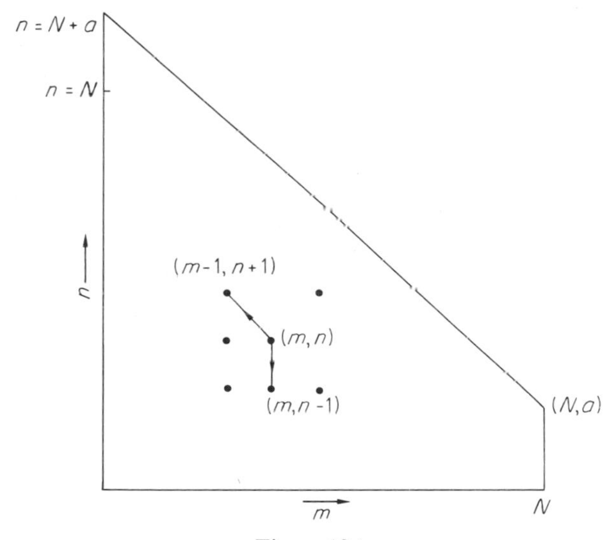

Figure 10.1

particle starting at the point (N, a) and moving, at each step, either diagonally upwards or vertically downwards as shown. When the system is in state (m, n) the two transitions which it can make and their associated probabilities are

$$\text{(a)} \quad (m, n) \to (m - 1, n + 1): \frac{m}{m + \rho}$$

$$\text{(b)} \quad (m, n) \to (m, n - 1): \frac{\rho}{m + \rho}$$

except that every state $(m, 0)$ is an absorbing state. These probabilities are the relative values of the infinitesimal transition probabilities given in (a) and (b) earlier in this section. The random walk is Markovian because the transition probabilities depend only on the present state of the system. Here, $\rho = \mu/\beta$ and is often called the *relative removal rate*. In words, it may be expressed as

$$\rho = \frac{\text{Average time taken for a randomly chosen pair to communicate}}{\text{Average length of time for which a spreader is active}}.$$

A large value of ρ indicates that forgetting takes place relatively rapidly and a small value the converse. The m-axis is an absorbing barrier corresponding to the complete elimination of spreaders from the population. If the particle reaches a point on the n-axis it descends and is absorbed at the origin.

If the particle passes through the point (m, n) it must previously have passed through either $(m + 1, n - 1)$ or $(m, n + 1)$. This enables us to set up a difference equation, with the aid of (a) and (b) above, as follows:

$$\left.\begin{aligned} P_{m,n} &= \left(\frac{m+1}{m+1+\rho}\right)P_{m+1,n-1} + \left(\frac{\rho}{m+\rho}\right)P_{m,n+1} \\ &\qquad (m \geq 0, \quad n > 1, \quad m + n < N + a) \\ P_{m,0} &= P_{m,1}\left(\frac{\rho}{m+\rho}\right) \qquad (0 < m \leq N) \\ P_{m,1} &= P_{m,2}\left(\frac{\rho}{m+\rho}\right) \qquad (0 < m \leq N) \end{aligned}\right\}. \qquad (10.5)$$

The initial condition is $P_{N,a} = 1$. For $m = N$ and $1 \leq n \leq a$ the probabilities are easily seen to be

$$P_{N,a-i} = \left(\frac{\rho}{N+\rho}\right)^i \qquad (i = 0, 1, \ldots, a), \qquad (10.6)$$

while those on the diagonal $m + n = N + a$ are given by the recurrence formula

$$P_{N-i,a+i} = P_{N-i+1,a+i-1}\left(\frac{N-i+1}{N-i+1+\rho}\right) \qquad (i = 1, 2, \ldots, N). \quad (10.7)$$

These results can be used to compute the complete probability distribution from (10.5). Bailey (1957) has given an explicit formula for $P_{m,0}$ (equation 5.53) and Siskind (1965) gave an alternative expression. The probability distribution was computed by Bailey for $a = 1$ and $N = 10, 20$ and 40. Some further calculations have been made for $N = 100, 200$ and 400 and various values of a. These form the basis of Figure 10.2 and Tables 10.1, 10.2 and 10.3. We follow Bailey and express the results in terms of the number $n_H = N - m$.

The distribution of $n_H = N - m$ has a variety of shapes depending on the values of N and ρ. Figure 10.2 illustrates the three principal forms for $a = 1$. If $N \leq \rho$ the distribution is J-shaped, indicating that the information seldom reaches more than a handful of people. If N is a little greater than ρ a mode appears in the upper tail. As N/ρ increases the mode becomes larger and moves to the end of the range until the distribution is U-shaped. Further increase in N/ρ results in a reduction in the probability concentrated near the origin. Finally, in the limit, the process degenerates into a pure birth model with all of the probability at $n_H = N$. The development of the diffusion thus depends critically on the relative sizes of N and ρ. There may be no epidemic at all, there may be an epidemic of uncertain size or there may be an epidemic in which everyone is almost certain to receive the information. These conclusions hold good in broad outline for any fixed a, but the position will be investigated quantitatively in more detail below.

The discussion given above is based on extensive calculations of the exact distribution and on Whittle's (1955) stochastic threshold theorem. Before giving some sample calculations and stating the theorem it is instructive to give an intuitive discussion of the threshold effect. The assumptions of our model imply that any individual communicates, on average, with $N\beta$ others per unit time irrespective of his own state. He himself is actively engaged in spreading for an average length of time μ^{-1}. Hence the expected number of tellings for each person will be $N\beta/\mu = N/\rho = d$, say. We shall see that if $d \leq 1$ the epidemic does not develop and if $d > 1$ it may do so. We might reasonably have expected the diffusion to peter out if the average number of tellings was less than one per head. However, the position can be clarified by comparing our process with a branching process such as that discussed in Chapter 2. Any member of the gth generation of hearers gives

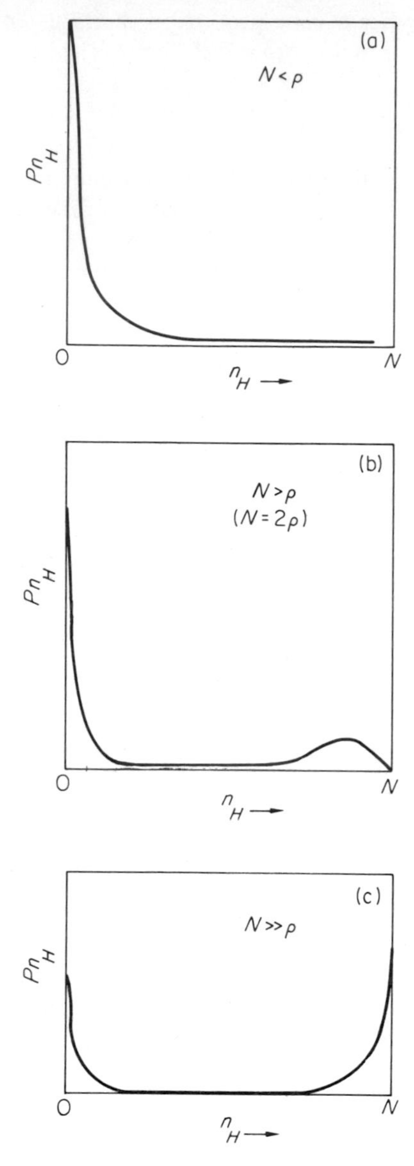

Figure 10.2 Forms of the distribution of n_H, the ultimate number who hear

rise to a random number of *new* hearers in the $(g + 1)$th generation. The expected value of this random number must be less than or equal to 1 if $d \leq 1$ (d is the expected total number of hearers—new and old). Under these circumstances extinction of the branching process is certain. In our application this means that, if N is large and $d \leq 1$, only a small proportion will hear the news. It is very important to notice that this argument does not depend on the distribution of the period of spreading. This is why the kind of threshold effect which we have observed for this model also occurs in a much larger class of models.

The relationship with the branching process can also be used to establish the following theorem.

Theorem *(Stochastic threshold theorem)*

If N is sufficiently large the probability of the epidemic exceeding any arbitrarily chosen size tends to zero for $d \leq 1$. If $d > 1$ the probability of the epidemic exceeding any arbitrarily small proportion of N tends to $1 - d^{-a}$.

This theorem formalizes the conclusions expressed above. The extent to which it holds for moderate N can be judged from the following tables. They give computations for the case $N = 200$. Other calculations for $N = 100$ and 400 have been made and lead to essentially the same conclusions. In particular, the proportions given in Tables (10.1) and (10.2) for $N = 200$ are almost identical with those for $N = 400$.

Table 10.1 ***Probabilities that the ultimate number of hearers will be small or large for various values of d and $N = 200, a = 1$***

	d							
n_H	0	$\frac{1}{2}$	$\frac{2}{3}$	1	2	4	20	∞
0–1	1·000	0·815	0·745	0·626	0·408	0·232	0·050	0·000
0–9	1·000	0·985	0·950	0·830	0·496	0·251	0·050	0·000
0–19	1·000	0·998	0·986	0·889	0·506	0·251	0 050	0·000
181–200	0·000	0·000	0·000	0·000	0·006	0·749	0·950	1·000
191–200	0·000	0·000	0·000	0·000	0·000	0·704	0·950	1·000
199–200	0·000	0·000	0·000	0·000	0·000	0·026	0·950	1·000

The threshold effect at $d = 1$ is very obvious in these tables. By increasing a it is possible to increase the number informed but the overall characteristics of the process remain the same. From Table 10.3 it can be seen that

***Table 10.2** Probabilities that the ultimate number of hearers will be small or large for various values of* d *and* $N = 200, a = 5$

n_H	d 0	$\frac{1}{2}$	$\frac{2}{3}$	1	2	4	20	∞
0–1	1·000	0·279	0·172	0·071	0·009	0·001	0·000	0·000
0–9	1·000	0·860	0·670	0·322	0·028	0·001	0·000	0·000
0–19	1·000	0·981	0·885	0·495	0·034	0·001	0·000	0·000
181–200	0·000	0·000	0·000	0·000	0·017	0·999	1·000	1·000
191–200	0·000	0·000	0·000	0·000	0·000	0·954	1·000	1·000
199–200	0·000	0·000	0·000	0·000	0·000	0·044	1·000	1·000

***Table 10.3** Means and variances of* n_H *for* $N = 200$. *Upper figure is the mean and lower figure the variance*

a	d 0	$\frac{1}{2}$	$\frac{2}{3}$	1	2	4	20	∞
1	0·0	1·0	1·8	7·0	78·2	146·8	190·0	200·0
	0·0	5·3	19·7	265·0	6258·0	7191·5	1900·0	0·0
2	0·0	1·4	3·6	13·3	118·7	183·6	199·5	200·0
	0·0	10·4	37·7	449·0	4849·8	2260·0	99·8	0·0
5	0·0	4·8	8·7	28·8	154·4	196·0	200·0	200·0
	0·0	24·6	83·2	711·2	1011·9	44·2	0·1	0·0
10	0·0	9·3	16·6	47·4	162·9	196·6	200·0	200·0
	0·0	45·1	137·2	747·5	183·6	5·1	0·0	0·0

the expected number of hearers changes smoothly on each side of the threshold value $d = 1$. The variances show very large differences because of the changing shape of the distribution. For $d \leq 1$ it always has a high degree of positive skewness; for $d > 1$ there is a concentration near the origin and a unimodal portion centred in the upper part of the range. As d increases this hump moves to the right until the distribution appears U-shaped. When a is increased the concentration near the origin diminishes rapidly.

Daniels (1966) has shown that if $d > 1$ then $N - n_H$ is asymptotically distributed in the Poisson form. Inspection of Table 10.3 suggests that this will only be a satisfactory approximation in practice for $N = 200$ if a and d are reasonably large. Cane (1966) arrived at a similar result by an approximate method.

The deterministic approximation

The distribution theory for the terminal state of the system tells us nothing about the duration or rate of growth of the epidemic. Except in degenerate cases, the mathematical problems of obtaining this information from the stochastic formulation are formidable. We shall therefore treat the process deterministically. The adequacy of this approach can be judged by comparing its terminal predictions with those already obtained for the stochastic case.

Instead of supposing that the number of transitions from (m, n) to $(m - 1, n + 1)$ in $(T, T + \delta T)$ is a random variable taking the values 0 or 1 with expectation $\beta m(T) n(T)\, \delta T$, we now assume that N is large enough for the expectation to be treated as the actual increase in $n(T)$ during the interval. Similarly, the decrease in $n(T)$ due to cessation of spreading will be assumed to be exactly $\mu n(T)\, \delta T$. Thus $n(T), m(T)$ and $l(T)$ are treated as continuous variables no longer being restricted to integer values. As $T \to 0$ the change in $n(T)$ may then be represented by the differential equation

$$\frac{\mathrm{d}n(T)}{\mathrm{d}T} = n(T)\{\beta m(T) - \mu\}. \tag{10.8}$$

Likewise, the derivatives of $m(T)$ and $l(T) = N + a - m(T) - n(T)$ are given by

$$\frac{\mathrm{d}m(T)}{\mathrm{d}T} = -\beta m(T) n(T) \tag{10.9}$$

and

$$\frac{\mathrm{d}l(T)}{\mathrm{d}T} = \mu n(T). \tag{10.10}$$

The initial conditions are $n(0) = a, m(0) = N, l(0) = 0$ and throughout the diffusion we must have $n(T) + m(T) + l(T) = N + a$.

An important result concerning the behaviour of the system can be found without actually solving the equations. First we note that $\mathrm{d}n(T)/\mathrm{d}T$ is negative or zero if $m(T) \leq \rho$ for any T. Since the number of ignorants cannot increase, this condition will certainly be satisfied if $m(0) = N \leq \rho$. Thus no epidemic occurs when $N \leq \rho$, while if $N > \rho$, $\mathrm{d}n(T)/\mathrm{d}T > 0$ at $T = 0$, the number of spreaders rises initially and an epidemic occurs. This is essentially the same threshold result which we obtained for the stochastic version of the model. The deterministic approximation has therefore been successful in reproducing this important characteristic of the epidemic.

If $N > \rho$, we can find the ultimate size of the epidemic and compare it with the stochastic values given in Table 10.3. To do this we first establish a simple relationship between $m(T)$ and $l(T)$. Dividing each side of (10.9) by the corresponding side of (10.10) we have

$$\frac{\mathrm{d}m(T)}{\mathrm{d}l(T)} = -\frac{\beta}{\mu}m(T) = -\frac{m(T)}{\rho}. \tag{10.11}$$

Integrating this equation and substituting the initial conditions we find that

$$\frac{m(T)}{N} = \exp\left[-l(T)/\rho\right]. \tag{10.12}$$

When the diffusion has ceased $n(T)$ will be zero and hence $m(T) = N + a - l(T)$. We have earlier denoted the ultimate number of hearers by n_H, which is the value taken by $l(T)$ when $n(T) = 0$. Hence from (10.12) we have

$$1 + \frac{a}{N} - p = \mathrm{e}^{-dp}$$

where $p = n_H/N$ and $d = N/\rho$. If a is fixed and N large, p satisfies

$$1 - p = \mathrm{e}^{-dp} \tag{10.13}$$

approximately. This equation can easily be solved by using the tables of Barton and coworkers (1960). A short table is given below (Table 10.4).

Table 10.4 Deterministic approximation to the ultimate number of hearers in the general epidemic model

	d			
	1	2	4	20
p	0	0·797	0·980	1·000
$200p$	0	159·4	196·0	200·0

The last row of the table is included to facilitate the comparison of the deterministic predictions with the values of the stochastic mean given in Table 10.3. In the stochastic case the expected size depends on the value of a.

The agreement between the deterministic and stochastic solutions is reasonably good when $a = 5$ or 10 but is very poor for $a = 1$ or 2. It might thus appear that the deterministic approximation is of little value. However, further investigation shows that the agreement is only poor when the distribution of n_H is bimodal or U-shaped. This happens when there is an appreciable probability that no epidemic will develop. According to the deterministic model an epidemic *always* occurs if $N > \rho$. We are therefore prompted to ask whether the agreement can be improved by omitting those cases where only a few people hear. There is, of course, some degree of arbitrariness about where we draw the line, but in Table 10.5 we have given the mean of n_H *given that* $n_H > 20$.

Table 10.5 ***Values of*** $E(n_H | n_H > 20)$ ***for*** $N = 200$

	d		
a	2	4	20
1	157·3	195·9	200·0
2	157·7	195·9	200·0
5	159·7	196·2	200·0
10	163·1	196·6	200·0

It is clear that the overall agreement between the stochastic and deterministic predictions as given by Tables 10.4 and 10.5 is greatly improved. The minor exception to this rule at $a = 10$, $d = 2$ may be accounted for by the omission of the term a/N from (10.13). It thus appears that the deterministic approach provides a satisfactory terminal description of the process *when an epidemic occurs*. The probability of an occurrence is, of course, given by the stochastic threshold theorem.

The number who have heard at time T is $N - m(T)$; the 'epidemic curve', which is defined as the rate of increase of the number of hearers, is thus obtained by plotting $(-\mathrm{d}m(T)/\mathrm{d}T)$ against T. (In the theory of epidemics the epidemic curve is defined by $\mathrm{d}l(T)/\mathrm{d}T$. This is because only the $l(T)$ members have recognizable symptoms although a further $n(T)$, who are incubating the disease, are actively spreading it.) Combining (10.9) and (10.12) we find

$$\frac{-\mathrm{d}m(T)}{\mathrm{d}T} = \beta m(T)\{N + a - m(T)\} + \mu m(T) \log_e (m(T)/N). \qquad (10.14)$$

We may compare this with the corresponding equation for the pure birth model in Chapter 9 which is obtained by putting $\mu = 0$. Since the last term on the right-hand side of (10.14) is always negative the epidemic curve for $\mu > 0$ lies everywhere below that for $\mu = 0$. Solving the equation with the appropriate initial conditions we have

$$T = \frac{1}{\beta}\int_{m(T)}^{N} \frac{\mathrm{d}x}{x\{N + a - x + \rho \log_e x/N\}} \qquad (m(\infty) \leq m(T) \leq N). \tag{10.15}$$

Using numerical integration this equation may be used to plot $m(T)$, and hence $-\mathrm{d}m(T)/\mathrm{d}T$, against T. Some calculations are given in Kendall (1956). It will be recalled that in the birth model the stochastic and deterministic epidemic curves approached one another as a increased. We may expect that the same will be true in the present case, but there are no results available for the stochastic model to test this conjecture.

The distribution of the number of hearers by generation

We can classify the hearers at time T according to the generation to which they belong. To this end let $n_g(T)$ and $l_g(T)$ denote the numbers of spreaders and inactive knowers, respectively, who first heard the news at gth hand ($g = 1, 2, \ldots, N$). Since the $n_g(T)$'s will be zero in the limit as $T \to \infty$ we shall be primarily interested in the joint distribution of the $l_g(\infty)$'s. Daley (1967b) showed that it was possible to calculate this distribution as a set of absorption probabilities for a certain random walk on a lattice. The argument is of the same kind as that leading up to (9.41), but the greater complexity arising from the two kinds of knower makes the calculations prohibitive in this case.

Instead of pursuing this approach we shall therefore turn to the deterministic treatment—also due to Daley (1967b). For this purpose we introduce $n_0(T)$ to denote the number out of the original a source members who are still active at time T. Then, on the assumptions of the model, the $n_g(T)$'s and the $l_g(T)$'s will change at the following rates:

$$\frac{\mathrm{d}n_0(T)}{\mathrm{d}T} = -\mu n_0(T), \tag{10.16a}$$

$$\frac{\mathrm{d}n_g(T)}{\mathrm{d}T} = \beta m(T) n_{g-1}(T) - \mu n_g(T) \qquad (g = 1, 2, \ldots), \tag{10.16b}$$

$$\frac{\mathrm{d}l_g(T)}{\mathrm{d}T} = \mu n_g(T) \qquad (g = 0, 1, 2, \ldots). \tag{10.16c}$$

If (10.16b) is compared with (9.60b) it will be seen that it differs by the introduction of the term $-\mu n_g(T)$, which reflects the loss to $n_g(T)$ through cessation of spreading. The size of $l_g(T)$ only changes by the addition of gth generation spreaders who cease activity, and this fact is expressed in (10.16c). Since changes in $m(T)$ are determined only by the total number of spreaders we have, as before,

$$\frac{\mathrm{d}m(T)}{\mathrm{d}T} = \beta m(T)n(T). \tag{10.17}$$

To facilitate the solution of this system of equations we introduce

$$w_g(T) = \mathrm{e}^{\mu T} n_g(T), \tag{10.18}$$

leading to

$$\frac{\mathrm{d}w_g(T)}{\mathrm{d}T} = \mu\,\mathrm{e}^{\mu T} n_g(T) + \mathrm{e}^{\mu T}\frac{\mathrm{d}n_g(T)}{\mathrm{d}T}.$$

Making this change of variable in (10.16a) and (10.16b) we have

$$\frac{\mathrm{d}w_0(T)}{\mathrm{d}T} = 0, \tag{10.19a}$$

$$\frac{\mathrm{d}w_g(T)}{\mathrm{d}T} = \beta m(T) w_{g-1}(T) \qquad (g = 1, 2, \ldots). \tag{10.19b}$$

We have already shown how to obtain $m(T)$ from (10.17), its inverse being given in (10.15). Hence, by substituting in (10.19b) the successive $w_g(T)$'s can be obtained by integration. In principle, therefore, the problem is solved. In practice, it helps to express the equations as functions of m instead of T. Using (10.15)

$$\frac{\mathrm{d}T}{\mathrm{d}m} = -\frac{1}{m\beta\{N + a - m + \rho\log_e(m/N)\}} = -\frac{1}{m\beta\psi(m)}, \text{ say,}$$

and, letting $\omega_g(m) = w_g(T(m))$,

$$\begin{aligned}\frac{\mathrm{d}\omega_g}{\mathrm{d}m} &= \beta m\omega_{g-1}(m)\frac{\mathrm{d}T}{\mathrm{d}m}\\ &= -\omega_{g-1}(m)/\psi(m).\end{aligned} \tag{10.20}$$

Integrating both sides and making use of the initial conditions, $m = N$ at $T = 0$ giving $\omega_0(N) = a$, $\omega_g(N) = 0$, we find

$$\omega_g(m) = \int_m^N \frac{\omega_{g-1}(u)\,\mathrm{d}u}{\psi(u)}, \tag{10.21}$$

Repeated application of this relationship yields

$$\omega_g(m) = a \int_R \prod_{i=0}^{g-1} \frac{du_i}{\psi(u_i)} \tag{10.22}$$

where R denotes the region $m \leq u_{g-1} \leq u_{g-2} \leq \ldots \leq u_0 \leq N$. Because of the symmetry in the u_i's of the integrand it is obvious that the integral would have the same value for all $g!$ permutations of the u_i's in the inequalities defining R. Hence, the integral has the same value as that taken over $m \leq u_i \leq N (i = 0, 1, \ldots, g - 1)$ divided by $g!$. Thus

$$\omega_g(m) = \frac{a}{g!}\left\{\int_m^N \frac{du}{\psi(u)}\right\}^g \qquad (m(\infty) < m \leq N) \tag{10.23}$$

and, from this, the $n_g(T)$'s can be determined. Our main interest is in the $l_g(T)$'s because it is their limiting values which will give the generation distribution when the epidemic has ended. These can be obtained from (10.16c) by integration of the $n_g(T)$'s derivable from (10.23), but it will again pay to approach the problem indirectly by changing from T to m.

From (10.16c),

$$\begin{aligned} \frac{dl_g}{dm} &= \frac{dl_g}{dT}\frac{dT}{dm} = \mu n_g(T(m))\{-\beta m\psi(m)\}^{-1} \\ &= -\rho\omega_g(m)\, e^{-\mu T(m)}/m\psi(m). \end{aligned} \tag{10.24}$$

Now

$$\begin{aligned} \mu T(m) &= -\rho \int_m^N \frac{du}{u\psi(u)} = -\int_m^N \frac{(1 + d\psi/du)}{\psi(u)}du \\ &= -\int_m^N \frac{du}{\psi(u)} - \log a + \log \psi(m). \end{aligned}$$

Substituting back in (10.24)

$$\frac{dl_g}{dm} = -\frac{\rho}{g!m}\left\{\int_m^N \frac{du}{\psi(u)}\right\}^g \exp\left[-\int_m^N \frac{du}{\psi(u)}\right]. \tag{10.25}$$

Finally, integrating with respect to m and using the initial condition $l_g = 0$ when $m = N$

$$l_g = \frac{\rho}{g!}\int_m^N \left\{\int_u^N \frac{dv}{\psi(v)}\right\}^g \exp\left[-\int_u^N \frac{dv}{\psi(v)}\frac{du}{u}\right] \qquad (g = 0, 1, 2, \ldots). \tag{10.26}$$

This equation enables us to find l_g as a function of m. It shows, incidentally, that l_g can be represented as a mixture of Poisson distributions but it does not seem possible to turn this observation to good effect. When $T \to \infty$, $m(T) \to m(\infty)$ and so

$$\lim_{T \to \infty} l_g = \lim_{m \to m(\infty)} l_g = \frac{\rho}{g!} \int_{m(\infty)}^{N} \left\{ \int_u^N \frac{dv}{\psi(v)} \right\}^g \exp\left[-\int_u^N \frac{du}{\psi(u)} \right] \frac{dv}{v}$$

$$(g = 0, 1, 2, \ldots,). \qquad (10.27)$$

No simple approximations to this integral have been discovered, but there is no difficulty about evaluating the integrals numerically if required.

The spatial spread of a general epidemic model

We can formulate a spatial version of the general epidemic model discussed above by an extension of the arguments used for the simple epidemic. Most of the theory available is deterministic, but we shall refer to a simulation study by Mollison (1972b) which adds a note of caution to our deterministic analysis. The method followed is to formulate the partial differential equations governing the development of the process in time and then, in a heuristic fashion, to deduce some relevant properties of their solution.

As before, suppose the population to be uniformly spread over a line of unlimited extent with density σ. At time T and location s the members can be divided into three exhaustive and mutually exclusive categories, as follows:

$x(s, T)$ is the proportion of ignorants;

$y(s, T)$ is the proportion of active spreaders;

$z(s, T)$ is the proportion of spreaders who have ceased activity, called passive knowers.

These three proportions add up to one for all s and T. Contacts are made at a rate β and the distance at which a contact is made is governed by the probability density function $v(u)$, $s + u$ being the location of the receiver ($-\infty < u < +\infty$). The new feature is the introduction of the rate μ at which active spreaders become passive. The rates of change of x, y and z now become

$$\frac{\partial x}{\partial T} = -\beta\sigma x\bar{y}, \qquad (10.28a)$$

$$\frac{\partial y}{\partial T} = \beta\sigma x\bar{y} - \mu y, \qquad (10.28b)$$

$$\frac{\partial z}{\partial T} = \mu y. \qquad (10.28c)$$

The arguments of x, y and z have been suppressed for simplicity but we shall reintroduce them whenever they are needed to make the reasoning clear. As before, $\bar{y}$ is a space average of y around the point s.

We begin with the question of threshold effects. This was very easily answered in the non-spatial case by considering the sign of $\mathrm{d}y/\mathrm{d}T$ at $T = 0$. The situation is now more complicated. An epidemic will certainly be impossible if $\partial y/\partial T$ is non-positive at every point s on the line. To take the easiest case first, suppose that the news is injected into the population by informing a small proportion ε uniformly distributed over the line. This means that $y(s, 0) = \bar{y}(s, 0) = \varepsilon$. Initially $x(s, 0) = 1 - \varepsilon$ and so growth will be impossible if

$$\beta\sigma(1 - \varepsilon)\varepsilon \leq \mu\varepsilon. \tag{10.29}$$

Thus we have the result that a small injection of the news will not start an epidemic if $\sigma < \rho$. In other words there must be a sufficient density of ignorants for diffusion to take place.

The symmetry of the problem in this particular case ensures that the proportions x, y and z will be independent of s throughout the development of the process. That is, the epidemic pattern will be the same at all locations and the differential equations become identical in form to those given in (10.8), (10.9) and (10.10). This result should not surprise us because by introducing the information uniformly over the region we have effectively eliminated the spatial element from the problem.

It would be more natural to suppose that the original spreaders are concentrated around a particular location. If they are confined to a small interval then some of their initial contacts will be with ignorants outside the region and so, initially at least, there will be no circumstances under which growth is impossible at all locations. This does not, however, imply that continuing growth throughout the region is certain since growth at some points does not ensure that the total level of knowledge will rise above the initial level. The growth outside the initial area may be more than matched by a decline within it.

The position becomes clearer if we suppose those who are initially informed are distributed over the whole line according to an initial smooth density $y(s, 0)$ chosen to be small for all s. This allows us to have an initial concentration and can be made to approximate very closely to the situation in which the source is confined to a restricted part of the space. Suppose also that $v(u)$ is concentrated in a neighbourhood of $u = 0$ then

$$\bar{y}(s, T) = \int_{-\infty}^{+\infty} y(s - u, T)v(u)\,\mathrm{d}u \doteqdot y(s, T).$$

Having made this approximation we observe that $\partial y/\partial T/T = 0$ is non-positive if

$$\beta\sigma \leq \mu \quad \text{or} \quad \beta \leq \rho. \tag{10.30}$$

Thus if (10.30) is satisfied there will be no growth in the number of spreaders and if it is not satisfied the number of active spreaders will rise above the initial number at all locations. These results indicate a threshold density which must be exceeded if there is to be any diffusion at all. Kendall (1965) investigated the condition under which the epidemic would be propagated as a wave using the diffusion approximation to the space average $\bar{y}$ and arrived at the same threshold condition, which he called the wave threshold theorem.

If diffusion does take place, the next thing is to consider the terminal state and to enquire whether the epidemic will die out before the whole population has been informed. The result obtained earlier for Kermack and McKendrick's model suggests that the proportion who ultimately hear will depend on the amount by which the population density exceeds the threshold. We now show that this expectation is fulfilled. From (10.28c) we have

$$\frac{\partial z(s-u, T)}{\partial T} = \mu y(s-u, T).$$

Multiplying by $v(u)$ and integrating between $-\infty$ and $+\infty$ gives

$$\frac{\partial \bar{z}}{\partial T} = \mu \bar{y} \tag{10.31}$$

where the bar indicates the space average with respect to $v(u)$. Dividing (10.28a) by (10.28c) gives

$$\frac{\mathrm{d}x}{\mathrm{d}\bar{z}} = -x\frac{\sigma}{\rho}, \tag{10.32}$$

which may be compared with (10.11). Solving this equation with the initial condition $\bar{z} = 0$, $x = 1$ we have

$$x = \mathrm{e}^{-\sigma\bar{z}/\rho}. \tag{10.33}$$

This equation expresses the relationship between $x(s, T)$ and $\bar{z}(s, T)$ at each point s on the line. As $T \to \infty$ the epidemic must die out at such a point s either because everyone knows or because the epidemic has 'passed on', leaving no active spreaders in the neighbourhood of s. In either event this implies $y(s, \infty) = 0$ and so, in the limit, we have, from (10.33),

$$1 - z(s, \infty) = \mathrm{e}^{-\sigma\bar{z}(s,\infty)/\rho}. \tag{10.34}$$

If the ultimate number of hearers at each location were to be the same then we should have $\bar{z} = z$ and (10.34) would determine the terminal proportion of hearers. It is not obvious that this condition will hold, at least near to the original source, but it does seem likely that $z(s, \infty) \to z(\infty)$, say, as s becomes increasingly distant from the source. This limiting value will then satisfy

$$1 - z(\infty) = e^{-\sigma z(\infty)/\rho}, \tag{10.35}$$

which is the same as (10.13) with $d = \sigma/\rho$. This equation has a non-zero root in (0, 1) only if $\sigma > \rho$, which is the condition for there to be an epidemic at all. Some values of $z(\infty)$ satisfying (10.35) can be read off from Table 10.4 by entering it with $d = \sigma/\rho$. The 'interesting' values of σ/ρ lie between 1 and 4. Below this interval there will be no epidemic and above it practically everyone will hear. Kendall (1957) uses the term 'pandemic' to describe the situation when $\sigma > \rho$ since then the effects of the epidemic are felt equally at all locations however distant from the source. A fuller discussion of the theory underlying these results may also be found in Kendall's paper.

Kendall (1965) also investigated the existence of waveform solutions to (10.28a, b and c) by making the diffusion approximation used in Chapter 9. After deducing the threshold condition he showed that if there is a waveform solution it cannot have a velocity less than

$$c_{\min} = 2\sqrt{K}(1 - \rho/\sigma)^{\frac{1}{2}}. \tag{10.36}$$

Kendall conjectures that all speed at or above this minimum are possible but that if the epidemic begins with a local concentration of initial spreaders then the wave will move at the minimum velocity. The effect of introducing the cessation of spreading is to slow down the rate of advance by the factor $(1 - \rho/\sigma)^{\frac{1}{2}}$. In a simulation study of a stochastic version of the model Mollison (1972b) found that in an example with $\rho/\sigma = \frac{1}{2}$ the speed was reduced to about one-fifth of $2\sqrt{K}$ instead of the factor 0·7 as Kendall's deterministic theory suggested. It is clear that there is ample scope for further research in this area if the behaviour of spatial spread is to be fully understood.

Having reached this point the reader may feel that the simplifications which have had to be made in order to make the models tractable are so drastic as to make the results almost meaningless. The assumption of a population uniformly distributed over an infinite line, or plane, is far removed from the highly irregular distribution of human populations. Similarly, the assumption of a concentrated symmetrical distribution $v(u)$ supposed to be the same for all people at every location ignores the inevitable individual and topographical variation. Such arguments have considerable force if the object is to predict diffusion in some detail. Detailed prediction is not, however, a reasonable objective in the present state of the theory. The value of our highly simplified models lies in their ability to suggest broad qualitative features like threshold effects. From the manner of their derivation one might reasonably hope such effects would survive a good deal of change in the assumptions in the direction of greater realism.

10.3 RAPOPORT'S MODELS

The basic model

During the period 1948–1954 a number of models for the diffusion of information were proposed by Rapoport and his coworkers (see, for example, Rapoport, 1948, 1951, 1953a, 1953b, 1954; Rapoport and Rebhun, 1952; Solomanoff and Rapoport, 1951. Their theory was originally developed for the study of the random net which arises in neurophysiological problems, but the relevance to diffusion of information and disease was soon apparent. This work pre-dates much of that described by Bailey (1957) in his book. The close relationship between the two bodies of theory seems to have passed largely unnoticed. Rapoport's model is, in fact, a simple variant of Kermack and McKendrick's model and both can be regarded as special cases of a more general model, discussed on page 366.

Rapoport's assumptions were the same as those of the general epidemic model except for the one governing the cessation of spreading. He supposed that each spreader told the news to exactly d other people and then ceased activity. The choice of person to be told was supposed to be random and hence independent of whether or not they had received the information. It is difficult to imagine a real-life situation in which this assumption would be true. The chain-letter in which each recipient is asked to write to d other persons is perhaps the nearest approximation. The value of studying the model is best seen by considering it as a special case of a more general model in which the number told is a random variable. If this random variable is denoted by $\tilde{d}$, Rapoport stated that his results would hold by taking $d = E(\tilde{d})$. Later in this section we shall discuss the conditions under which this statement can be justified.

The present model thus arises as an extreme case when we take the distribution of $\tilde{d}$ as concentrated at the point d. Kermack and McKendrick's model is another special case since then $\tilde{d}$ has a geometric distribution. This may be seen as follows. If a spreader is active for length of time x the number of people he communicates with, excluding the initial spreaders, will have a Poisson distribution with mean $\beta N x$. But x has an exponential distribution with parameter μ. Therefore,

$$
\begin{aligned}
Pr\{\tilde{d} = j\} &= \int_0^\infty e^{-\mu x} \frac{(\beta N x)^j}{j!} e^{-\beta N x} \, dt \\
&= \frac{\mu}{\mu + N\beta}\left(1 - \frac{\mu}{\mu + N\beta}\right)^j \\
&= \frac{1}{1 + d}\left(\frac{d}{1 + d}\right)^j \qquad (j = 1, 2, \ldots,). \qquad (10.37)
\end{aligned}
$$

As another example, take the case where the individual spreads the news for a fixed length of time; $\tilde{d}$ would then have a Poisson distribution. By comparing the results obtained under a variety of fairly extreme assumptions we may hope to establish our results on a broader basis.

Except for the case $d = 1$ we shall not be able to obtain any results about the rate of diffusion. Instead we shall study the random variables $\{n_g\}$—the numbers in the different generations of hearers. Some additional notation is required as follows. Let

$$N_g = \sum_{i=1}^{g} n_i.$$

For some $g \leq N$, N_g will attain its maximum value at which it remains constant. This limiting value is the ultimate size of the epidemic. In the last section we denoted this quantity by n_H but here it is more natural to use N_∞ because it is obtained by allowing g rather than T to tend to infinity. The equivalence between n_H and N_∞ is established by noting that

$$N_\infty = \lim_{g \to \infty} \sum_{i=1}^{g} n_i = \lim_{T \to \infty} \{N - m(T)\} = n_H. \qquad (10.38)$$

Finally, let

$$p_g = E(n_g)/N.$$

This may be interpreted as the probability that a randomly chosen member of the population belongs to the gth generation.

Some exact distribution theory

The exact distribution theory for the random variables $\{n_g\}$ and hence of $\{N_g\}$ for fixed d can be obtained by combinatorial methods. This fact is a consequence of the formal equivalence between our diffusion process and the classical occupancy problem in which balls are distributed randomly into boxes. The precise way in which this correspondence is established depends on what assumptions are made about the operation of the source. Two points which must be considered are the following.

(a) The a persons who introduce the news may or may not be able to communicate with one another. Which of the alternatives is adopted will depend on whether or not the initial spreaders are able to recognize one another.

(b) The members of the population may or may not be able to communicate with the initial spreaders. In practice, communication may be impossible because the initial spreaders withdraw or unnecessary because they can be recognized. The four variants of the basic model to which these categories give rise are set out below.

		Initial spreaders	
		May communicate with one another	May not communicate with one another
Members of population	May communicate with initial spreaders	Ia	IIa
	May not communicate with initial spreaders	Ib	IIb

If N is large and a is small the difference between the four models is negligible. When $a = 1$ the classifications I and II coincide. For convenience, we shall develop the theory for model IIb. Each of the others can be treated by essentially similar arguments which we shall indicate at various points.

Consider first the distribution of n_1 under model IIb. We think of the N members of the population as empty boxes. The $a \times d$ contacts made by the source can then be thought of as balls distributed randomly among the boxes. The distribution of first generation hearers is thus the same as the distribution of the number of non-empty boxes. This is given, for example, by Barton and David (1962, page 242) and is

$$Pr\{n_1 = i\} = \binom{N}{i} \sum_{j=0}^{i-1} (-1)^j \binom{i}{j} \left(\frac{i-j}{N}\right)^{ad} \qquad (i = 1, 2, \ldots, ad). \quad (10.39)$$

The conditional distribution of n_{g+1} given $n_g, n_{g-1}, \ldots, n_1$ may be found by an extension of the same argument. Of the $n_g d$ contacts made by the spreaders of the gth generation suppose that r are with ignorants. Then, from (10.39),

$$Pr\{n_{g+1} = i|n_g, n_{g-1}, \ldots, n_1; r\}$$

$$= \binom{N - N_g}{i} \sum_{j=0}^{i-1} (-1)^j \binom{i}{j} \left(\frac{i-j}{N-N_g}\right)^r \qquad (i = 1, 2, \ldots, r; r > 0)$$

$$= 0 \quad (r = 0, i > 0) \qquad (10.40)$$

$$= 1 \quad (r = 0, i = 0).$$

To find the required distribution we need the distribution of r. This is clearly binomial because each contact is made independently with probability $(N - N_g)/(N - 1)$. The denominator here is $N - 1$ rather than N because a spreader cannot communicate with himself. Combining this result with (10.40), the conditional distribution of n_{g+1} is then

$$Pr\{n_{g+1} = i|n_g, n_{g-1}, \ldots, n_1\}$$

$$= \binom{N - N_g}{i} \sum_{j=0}^{i} (-1)^j \binom{i}{j} \left(\frac{i-j+N_g-1}{N-1}\right)^{n_g d}$$

$$(i = 0, 1, \ldots, \min(n_g d, N - N_g)). \qquad (10.41)$$

In principle our problem is now solved, but the joint distribution of the n_g's obtained from (10.41) is extremely cumbersome. However, the exact result provides a starting point for, and check on, the approximations given below.

Modifying the foregoing theory for the other models is straightforward. In the case of model IIa the only difference is that the binomial probability $(N - N_g)/(N - 1)$ above must be replaced by $(N - N_g)/(N + a - 1)$. Equation (10.41) gives the conditional probability distribution for model Ia for a population of size $(N - a)$ if, instead of (10.39), we have

$$Pr\{n_1 = i\} = 1 \quad \text{if } i = a$$

$$= 0 \quad \text{otherwise.}$$

The treatment of model Ib is slightly more complicated and is left as an exercise for the interested reader.

The distributions of n_g and N_g, and hence of N_∞, can be found using the theory of Markov chains. This fact was noticed and used by Solomanoff (1952). For moderate or large N the arithmetic is very heavy but can easily be programmed for a computer. For any g we can describe the state of the system by a pair of numbers (x, X) where $x = n_g$ and $X = N_g$. The process defined on these states is a Markov chain because the probabilities of transition between states from one generation to the next depend only on x and X. The numbers x and X must obviously satisfy

$$1 \leq X \leq N, \qquad 0 \leq x \leq X$$

so there are $\frac{1}{2}N(N + 3)$ states in all. When $N = 3$ the nine possibilities are as listed in Table 10.6. For the present purpose the process begins when the

***Table 10.6** Calculations required to find the distributions of n_g, N_g and N_∞ for Rapoport's model when $N = 3$, $d = 2$, $a = 1$*

	Transition probabilities									*State probabilities*			
States	01	11	02	12	22	03	13	23	33	$\mathbf{p}'_3$	$\mathbf{p}'_2$	$\mathbf{p}'_1$	$\mathbf{p}'_0$
01	1	0	0	0	0	0	0	0	0	0	0	0	0
11	0	0	0	$\frac{1}{2}$	0	0	0	$\frac{1}{2}$	0	0	0	0	$\frac{1}{3}$
02	0	0	1	0	0	0	0	0	0	$\frac{2}{24}$	$\frac{2}{24}$	$\frac{1}{24}$	0
12	0	0	$\frac{1}{4}$	0	0	0	$\frac{3}{4}$	0	0	0	0	$\frac{1}{6}$	0
22	0	0	$\frac{1}{16}$	0	0	0	$\frac{15}{16}$	0	0	0	0	0	$\frac{2}{3}$
03	0	0	0	0	0	1	0	0	0	$\frac{22}{24}$	$\frac{19}{24}$	0	0
13	0	0	0	0	0	1	0	0	0	0	$\frac{3}{24}$	$\frac{15}{24}$	0
23	0	0	0	0	0	1	0	0	0	0	0	$\frac{1}{6}$	0
33	0	0	0	0	0	1	0	0	0	0	0	0	0

members of the source have communicated with n_1 members of the population. The system is then in the state ($x = i$, $X = i$) and the probability of being in this state is $Pr\{n_1 = i\}$ given by (10.39). The only transitions with non-zero probabilities are those of the form $(x, X) \rightarrow (i, X + i)$ for $i = 0, 1, \ldots, \min(xd, N - x)$. In these cases the transition probability is obtained from (10.41) with $x = n_g$, $X = N_g$. Let us agree to list the states in some particular order and denote the $1 \times \frac{1}{2}N(N + 3)$ vector of initial probabilities by $\mathbf{p}_0$. If $\mathbf{P}$ denotes the matrix of transition probabilities then the state distribution after i generations will be given by

$$\mathbf{p}_i = \mathbf{p}_0\mathbf{P}^i. \tag{10.42}$$

The states of the form $(0, X)$ are absorbing states. Since the number of generations cannot exceed N we shall find that $\mathbf{p}_i$ contains non-zero entries for the absorbing states when i is at most N. These probabilities give the distribution of N_∞, the number who ultimately receive the information. The intermediate vectors $\mathbf{p}_i$ can be used to find the distribution of n_g and N_g for all g. An illustrative example is given in Table 10.6 for $N = 3$, $d = 2$ and $a = 1$.

It follows at once from the column headed $\mathbf{p}'_3$ that

$$Pr\{N_\infty = 2\} = 1/12, \qquad Pr\{N_\infty = 3\} = 11/12$$

and hence that $E(N_\infty) = 35/12$. Solomanoff (1952) made similar calculations for models Ia and IIa for the same values of N, d and a. His version of the model includes the somewhat unrealistic assumption that a person can communicate with himself and thus his calculations are not directly comparable with ours. As a further illustration of the use of Table 10.6 let us find the distribution of n_2. From the $\mathbf{p}'_1$ column

$$Pr\{n_2 = 0\} = Pr\{n_2 = 0, N_2 = 1 \text{ or } 2 \text{ or } 3\} = 1/24,$$

$$Pr\{n_2 = 1\} = Pr\{n_2 = 1, N_2 = 1 \text{ or } 2 \text{ or } 3\} = 1/6 + 15/24 = 19/24,$$

$$Pr\{n_2 = 2\} = Pr\{n_2 = 2, N_2 = 2 \text{ or } 3\} = 1/6.$$

For larger and more realistic values of N the calculations follow the same pattern but the size of the matrix is proportional N^2.

Methods for finding the distribution of N_∞ and its expectation were given by Landau (1952). These methods take advantage of the Markov property to construct difference equations which can then be solved to give the required probabilities or expectations. In particular, Landau showed that, under model Ia when a member can communicate with himself and $a = 1$,

$$E(N_\infty) = \sum_{i=1}^{N-1} b_i \binom{N-1}{i} \left(1 - \frac{i}{N}\right)^{(N-i)d} \tag{10.43}$$

where the b_i satisfy

$$w = \sum_{i=1}^{w} b_i \binom{w}{i} \left(1 - \frac{i}{n}\right)^{(w-i)d} \qquad (w = 1, 2, \ldots, N-1). \tag{10.44}$$

Exact theory when $a = d = 1$

In this case it is possible to derive a simple explicit expression for the distribution of N_∞. This may be obtained as a special case of the general theory above, but we shall proceed from first principles. Since each spreader tells only one individual who is either an ignorant or former spreader, it follows that $n_g = 0$ or 1 for all g. The process will continue until a spreader tells a former spreader. Suppose that this happens at the ith generation. Then

$$\begin{aligned} n_g &= 1, \quad i \leq g \\ &= 0, \quad i > g. \end{aligned}$$

The process can be represented as a random walk on the integers $1, 2, \ldots, N$ as follows:

$$\frac{i-1}{N-1} \longleftarrow \qquad \longrightarrow \frac{N-i}{N-1}$$

$$1 \quad 2 \quad 3 \quad \ldots \quad i-1 \quad i \quad i+1 \quad \ldots \quad N$$

A particle starts at the left-hand end. It moves to the right whenever a spreader tells an ignorant and to the left if a spreader tells a former spreader. The probabilities of each kind of transition are $(N-i)/(N-1)$ and $(i-1)/(N-1)$ respectively and the process terminates as soon as the first move to the left is about to occur. The point of termination is equal to N_∞. A direct argument thus gives

$$Pr\{N_\infty = i\} = \left(\frac{i-1}{N-1}\right) \prod_{j=1}^{i-1} \left(\frac{N-j}{N-1}\right)$$
$$= \frac{(N-2)!(i-1)}{(N-i)!(N-1)^{i-1}} \qquad (i = 2, 3, \ldots, N). \qquad (10.45)$$

For small N the distribution is easily tabulated; the result for $N = 11$ is given in Table 10.7.

Table 10.7 Distribution of N_∞ *for* $N = 11$, $d = 1$, $a = 1$

i	$Pr\{N_\infty = i\}$	i	$Pr\{N_\infty = i\}$
2	0·1000	7	0·0907
3	0·1800	8	0·0423
4	0·2160	9	0·0145
5	0·2016	10	0·0033
6	0·1512	11	0·0004

The distribution is positively skewed with mean value 4·660. If communication with the source is allowed the above table would apply to the case $N = 10$ with i reduced by one.

For large N an approximation to $Pr\{N_\infty = i\}$ can be found as follows. Let K be a constant such that $i = K\sqrt{N}$ and let N be large. Then

$$\prod_{j=1}^{i-1} \left(1 - \frac{j-1}{N-1}\right) = \exp\left[\sum_{j=1}^{i-1} \log_e \left(1 - \frac{j-1}{N-1}\right)\right]$$
$$\sim e^{-K^2/2}.$$

Therefore

$$Pr\{N_\infty = K\sqrt{N}\} \sim \frac{K}{\sqrt{N}} e^{-K^2/2}. \tag{10.46}$$

Omitting the factor $1/\sqrt{N}$, (10.46) gives the probability density function of the continuous approximation to the distribution of N_∞ at $K\sqrt{N}$. As a check, we may note that the density integrates to one. It may be shown that the median of the distribution of N_∞ is at $\sqrt{2 \log_e 2N} = 1{\cdot}177\sqrt{N}$. The asymptotic moments may be deduced by observing that $K = N_\infty/\sqrt{N}$ is distributed like χ^2 with 4 degrees of freedom. Hence

$$\left.\begin{aligned} E(N_\infty) &\sim \sqrt{\frac{\pi N}{2}} = 1{\cdot}253\sqrt{N} \\ \text{and} \qquad \operatorname{var}(N_\infty) &\sim \frac{4-\pi}{2} N = 0{\cdot}4292N \end{aligned}\right\}. \tag{10.47}$$

It thus follows that the expected proportion of the population who eventually hear tends to zero as N increases. When $N = 11$ the asymptotic formula for the mean give $E(N_\infty) = 4{\cdot}1557$ as compared with the exact value of $4{\cdot}6604$.

In the case $a = d = 1$ it is also possible to find the distribution of the duration of the diffusion. On the assumption of random mixing the length of time taken for a spreader to communicate with another person has an exponential distribution with mean value $1/\beta(N-1)$. Given that $N_\infty = i$ the distribution of the duration will then be like that of $\chi^2_{2(i-1)}/2\beta(N-1)$. We have already found the distribution of N_∞ so that an expression for the unconditional distribution of the duration can be written down at once. It is simpler to work with the moments, for which we find

$$\begin{aligned} E(T_{N_\infty}) &= \sum_{i=2}^{N} Pr\{N_\infty = i\}(i-1)/\beta(N-1) \\ &= \frac{E(N_\infty)-1}{\beta(N-1)} \sim \sqrt{\frac{\pi}{\beta^2 N}} = \frac{1{\cdot}253}{\beta\sqrt{N}}. \end{aligned} \tag{10.48}$$

Similarly,

$$\begin{aligned} \operatorname{var}(T_{N_\infty}) &= \sum_{i=2}^{N} Pr\{N_\infty = i\}E(T_i^2|i) - \{E(T_{N_\infty})\}^2 \\ &\sim \left(\frac{2}{\beta^2} - \frac{\pi}{2\beta^2}\right)\Big/ N = \frac{0{\cdot}4292}{\beta^2 N}. \end{aligned} \tag{10.49}$$

The duration of the epidemic thus decreases as the population size increases, as we found with the pure birth model.

An approximation to $E(N_\infty)$

The conditional distribution of n_{g+1} given by (10.41) can be made the basis of an approximation to $E(N_\infty)$. By elementary methods it can be shown that

$$E(n_{g+1}|n_g, N_g) = (N - N_g)\left\{1 - \left(1 - \frac{1}{N-1}\right)^{n_g d}\right\} \qquad (g \geq 1). \tag{10.50}$$

The same result may be derived by the following direct argument. The probability that a randomly chosen individual does not receive *any* of the $n_g d$ transmissions made by members of the gth generation is

$$\left(1 - \frac{1}{N-1}\right)^{n_g d}.$$

The probabilitity that he receives at least one such transmission is thus

$$1 - \left(1 - \frac{1}{N-1}\right)^{n_g d}.$$

A randomly chosen hearer is an ignorant with probability

$$\left(1 - \frac{N_g - 1}{N-1}\right).$$

Therefore the expected numbers of new hearers in the gth generation is

$$E(n_{g+1}|n_g, N_g) = (N-1)\left(1 - \frac{N_g - 1}{N-1}\right)\left\{1 - \left(1 - \frac{1}{N-1}\right)^{n_g d}\right\},$$

which is identical with (10.50). A simple rearrangement gives

$$E\left\{\frac{N - N_{g+1}}{N - N_g}\Big| n_g, N_g\right\} = \left(1 - \frac{1}{N-1}\right)^{n_g d} \qquad (g \geq 1).$$

Suppose that we now attempt to take the expectations of both sides of this equation with respect to all the random variables appearing. There is no simple expression for the result but an approximation to the answer can be obtained by replacing all the random variables by their expectations. For this operation we need

$$E(n_g) = Np_g \quad \text{and} \quad E(N_g) = N \sum_{i=1}^{g} p_i.$$

Substitution now yields

$$\frac{N - N\sum_{i=1}^{g+1} p_i}{N - N\sum_{i=1}^{g} p_i} = \left(1 - \frac{1}{N-1}\right)^{Np_g d}. \tag{10.51}$$

Let g^* be the smallest value of g for which $N_{g+1} = N_g$. We now take the product of both sides of this equation for $g = 1$ to $g = g^*$. On the left-hand side we have

$$N = \sum_{i=1}^{g^*} p_i = N - E(N_\infty)$$

and on the right-hand side

$$\left(1 - \frac{1}{N-1}\right)^{dE(N_\infty)}.$$

Setting $p = E(N_\infty)/N$ the equation thus becomes

$$p = 1 - \left(1 - \frac{1}{N-1}\right)^{Ndp} \sim 1 - \mathrm{e}^{-dp}. \tag{10.52}$$

This result was first obtained by Solomanoff and Rapoport (1951) but it is identical with that found for Kermack and McKendrick's model in (10.13). This should occasion no surprise. The two models differ only in the stochastic aspects of the cessation of spreading and both (10.13) and (10.52) were found by deterministic methods involving only averages. Table 10.4 can thus be used for Rapoport's model also.

In this case it is possible to judge the accuracy of the approximation given by (10.52) since we also have some exact theory. When $d = 2$ Table 10.4 gives $p = 0{\cdot}797$ and thus $E(N_\infty) = 0{\cdot}797N$. For the example discussed in Table 10.6 we found $E(N_\infty) = 3 \times 0{\cdot}972$, which is considerably larger than the approximation, but $N = 3$ is much too small a number for a satisfactory test. A more realistic check can be obtained when $a = d = 1$. In the limit as $N \to \infty$ we have $p = 0$, but we can investigate the situation when N is finite. To do this we rewrite (10.52) in the form

$$p_{g+1} = \left\{1 - \left(1 - \frac{1}{N-1}\right)^{Npd_g}\right\}\left\{1 - \sum_{i=1}^{g} p_i\right\} \qquad (g \geq 1) \tag{10.53}$$

and add

$$p_1 = 1 - \left(1 - \frac{1}{N-1}\right)^{ad}. \tag{10.54}$$

These equations may be solved recursively to give the p_g's and hence p. A comparison of exact and approximate values is given in Table 10.8.

Table 10.8 Comparison of Rapoport's recursive approximation to n_H/N ***with the exact value for*** $a = d = 1$

	N					
	10	20	50	100	200	300
Exact	0·366	0·265	0·171	0·122	0·087	0·071
Approximate	0·292	0·237	0·168	0·125	0·092	0·076

The agreement is reasonably good and suggests that the approximation may be used with confidence when N is moderate or large. Further evidence confirming this conclusion for $a = 1$ and $d = 2$ and 4 has been obtained from the simulation studies summarized in Table 10.9.

Table 10.9 Average values and standard deviations of n_H/N ***in*** 100 ***simulations of Rapoport's model***

		N				Rapoport's asymptotic approximation
		20	100	300	3000	
$d = 2$	Average	0·784	0·799	0·794	0·796	0·797
	S.D.	0·138	0·049	0·031	0·010	—
$d = 4$	Average	0·978	0·981	0·981	0·980	0·980
	S.D.	0·035	0·015	0·010	0·0077	—

By means of (10.53) we can also compute the expected number in each generation. Some calculations for the case $N = 300$, $d = 4$, $a = 1$ are given later in Table 10.11, where a comparison is made with results for a stratified population. Some further results are summarized in Table 10.10, which gives the most common value of g and a deterministic approximation to the duration of the epidemic. The latter is obtained as the largest value of g for which $Np_g \geq 1$. Perhaps the most surprising feature of this table is the small number of generations taken for the news to spread. Nevertheless, even with d as large as 10, most people receive the news at third or fourth hand.

***Table 10.10 The most common value of* g *and estimated duration of the epidemic for* N = 300 *and* 3,000**

		d		
		2	4	10
$N = 300$	Mode	7	4	3
	Duration	13	7	4
$N = 3000$	Mode	11	6	4
	Duration	19	9	5

It is convenient at this point to consider the more realistic general model in which each person tells a random number $\tilde{d}$ of persons, where $E(\tilde{d}) = d$. We have already seen that (10.52) still holds in one such case when $\tilde{d}$ has a geometric distribution (see equation 10.37). That it holds in general may be demonstrated as follows. Let n_{gi} be the number of people informed by the ith member of the gth generation. Assume that the n_{gi}'s are independent. By definition

$$n_{g+1} = \sum_{i=1}^{n_g} n_{gi} \qquad (g = 1, 2, \ldots),$$

$$n_1 = \sum_{i=1}^{a} n_{1i}.$$

As before, we can find the conditional distribution of n_{g+1} given $\{n_{gi}\}$; it has the same form as (10.41). The distribution of n_{g+1} thus depends on the previous n_{gi}'s only through the two sums n_g and N_g. The conditional expectation of n_{g+1} is likewise still given by (10.50), and hence equation (10.51) may be arrived at as before because $E(n_g)$ and $E(N_g)$ are unchanged if we interpret d as an expectation. The general result then follows.

This model, together with the simpler version of Rapoport's model and that of Kermack and McKendrick, can be regarded as a special case of the chain-letter model discussed by Daley (1967b). According to this model spreaders act independently and are active for a random length of time. During their active period (or at the end of it) they pass on the news to a random number of other persons. Those among the hearers who were previously ignorant become spreaders. The same deterministic approximation serves for all the special cases of this model, at least so far as the ultimate state of the population is concerned. In particular, the equations which determine the ultimate number of hearers by their generation will be valid for this wider class of models. The results given in the last part of this section therefore also provide an approximation for the models here. Since Rapoport's models are expressed in terms of generations rather than in terms of time it may not be immediately clear how the connexion can be made. The difficulty is overcome by observing that the terminal state in Rapoport's models would be unaffected by the introduction into the model of a random variable representing the time for which each spreader was active. The essential parameter on which the terminal state depends is the average number of contacts made by an active spreader.

Imperfect mixing

One of the most crucial assumptions of Rapoport's model is that communication between all possible pairs of persons is equally likely. As it seems highly improbable that this will be the case in most social groups Rapoport extended his theory to systems in which there is 'distance bias'. This means that the likelihood of communication between two people will depend on their distance (social, geographical, etc.) apart. In the present section we shall discuss a simple special case of the general model in which the group is stratified. This will serve to indicate the broad qualitative effects of distance bias.

Rapoport supposed that the population was divided into large homogeneous strata with different probabilities of communication between and within strata. We shall take the case where there are only two strata of sizes N' and N'' with $N = N' + N''$. The notation is the same as before except that we add one or two primes according to whether we are referring to the first or second stratum. Thus, for example, $p'_g = E(n'_g)/N'$ is the expected proportion of gth generation hearers in stratum 1.

In the case of a single homogeneous population the probability of contact between a given spreader and any other member of the population is $1/(N - 1)$. For the purposes of obtaining asymptotic results we can replace this by $1/N$. Since we shall only be concerned with asymptotic theory it will be notationally convenient to make this kind of alteration in what follows without further comment. Consider now the case of a system with two strata. If there were no communication barrier the probability that a spreader in stratum 1 makes his next communication with a member of stratum 2 would be N''/N. When there is imperfect mixing this probability will be reduced. Let us therefore suppose that

$$\left.\begin{aligned} &Pr\{\text{next communication of a spreader in stratum 1 is to stratum 2}\}\\ &\qquad = \frac{N''}{(cN' + N'')} \qquad (1 \leq c < \infty).\\ &\text{Similarly, let}\\ &Pr\{\text{next communication of a spreader in stratum 2 is to stratum 1}\}\\ &\qquad = \frac{N'}{cN'' + N'} \qquad (1 \leq c < \infty).\end{aligned}\right\} \quad (10.55)$$

When $c = 1$ we have the case of homogeneous mixing; as $c \to \infty$ the probability of communication between strata approaches zero. We continue to assume that contacts are made at random within any stratum.

We are now in a position to find the conditional expectation of n'_{g+1} by a generalization of the argument which led to (10.50). The probability that a randomly chosen member of stratum 1 receives the communication being made by a given spreader in the same stratum is, by (10.55),

$$\frac{cN'}{cN' + N''} \times \frac{1}{N'} = \frac{c}{cN + N''}.$$

The probability that he does not receive any of the $n'_g d$ transmissions emanating from stratum 1 is thus

$$\left(1 - \frac{c}{cN' + N''}\right)^{n'_g d}.$$

By a similar argument the probability of his not receiving any of the $n''_g d$ transmissions originating from stratum 2 is

$$\left(1 - \frac{c}{cN'' + N'}\right)^{n''_g d}.$$

The probability that he receives at least one communication, regardless of its source, is therefore

$$1 - \left\{\left(1 - \frac{c}{cN' + N''}\right)^{n'_g d}\left(1 - \frac{c}{cN'' + N'}\right)^{n''_g d}\right\}. \qquad (10.56)$$

Finally, since he is an ignorant with probability $(1 - (N'_g/N'))$ we have

$$E(n'_{g+1}|n'_i, n''_i; 1 \leq i \leq g)$$
$$= (N' - N'_g)\left[1 - \left\{\left(1 - \frac{c}{cN' + N''}\right)^{n'_g d}\left(1 - \frac{c}{cN'' + N'}\right)^{n''_g d}\right\}\right]. \qquad (10.57)$$

Dividing both sides of (10.57) by N', replacing the random variables by their expectations and allowing N' to tend to infinity we find

$$p'_{g+1} = \left\{1 - \sum_{i=1}^{g} p'_i\right\}\left\{1 - \exp\left(-\frac{p'_g dcN'}{cN' + N''} - \frac{p''_g dN'}{cN'' + N'}\right)\right\}. \quad (10.58)$$

A similar expression for p''_{g+1} follows by interchanging the single and double primes in (10.58). The initial conditions will depend on the way in which information is first introduced into the system. If it is introduced by a persons into stratum 1 we shall have

$$p'_1 = 1 - \left(1 - \frac{1}{N'}\right)^{ad} \quad \text{and} \quad p''_1 = 0.$$

Using these, or any other appropriate initial conditions, the approximations to the expected proportions can be computed recursively from (10.58) and its counterpart in p''_{g+1}.

The foregoing theory can be used to derive approximations to the expected proportion who eventually receive the news. The method is exactly similar to that leading up to (10.52). In this case it gives the following equations for $p' = N'_\infty/N'$ and $p'' = N''_\infty/N''$:

$$\left.\begin{aligned} 1 - p' &= \exp\left[-p'd\left(\frac{cN'}{cN' + N''}\right) - p''d\left(\frac{N'}{cN'' + N'}\right)\right] \\ 1 - p'' &= \exp\left[-p''d\left(\frac{cN''}{cN'' + N'}\right) - p'd\left(\frac{N''}{cN' + N''}\right)\right] \end{aligned}\right\}. \quad (10.59)$$

The expected proportion who receive the information in the whole population is then

$$p = \frac{p'N' + p''N''}{N}. \quad (10.60)$$

In the case $N' = N''$ the solution of (10.59) is easily obtained. The equations are simultaneously satisfied by $p' = p'' = p_0$, where p_0 is the non-zero root of the equation

$$1 - p = e^{-dp},$$

which is identical with (10.52). This means that the proportion who ultimately hear the information is unaffected by the stratification. The same result holds if $N'/N'' \to 0$ or ∞ with N' and N'' both large. If would be surprising if p depended strongly upon N'/N''. Calculations which we have made show that p is almost independent of c which only affects it in the third decimal. We can easily extend the argument used for two strata to the case of many strata and the conclusions reached are essentially the same. It seems likely that the rate of diffusion will be affected by stratification. Some light is thrown on this question by the results in Table 10.11.

***Table 10.11** Expected proportion of hearers, p_g, for* $N' = 200$, $N'' = 100$, $d = 4$

	g						
c	1	2	3	4	5	6	7
1	0·051	0·172	0·380	0·300	0·058	0·005	0·000
20	0·050	0·160	0·310	0·268	0·142	0·033	0·004

We have computed the expected proportions of hearers in each generation for a stratified population with $N' = 200$, $N'' = 100$ and $d = 4$. The case $c = 1$ corresponds to homogeneous mixing and shows that most people hear at second, third or fourth hand. When $c = 20$ there is very little communication between strata. The effect on the values of p_g is not great but the tendency is for people to receive the news at a greater remove from the source.

The main pre-supposition of our analysis in this and the previous chapter is that the information is spread by means of chance contacts between individuals. A quite different situation arises if there are recognized channels of communication. In most human organizations such channels exist in order to ensure that information reaches every member of the system as quickly as possible. Perhaps the most important general conclusion we have reached from this study is that, even when such channels do not exist, *casual contact is often sufficient to ensure that most people are informed*. The most striking practical illustration of this fact is the way in which rumours spread in human socieities. The models which we shall describe in the next section were specifically constructed for studying the diffusion of rumours.

10.4 DALEY AND KENDALL'S MODEL

The basic model and some exact theory

One important feature of the diffusion of information in human populations has been disregarded when constructing the models given earlier in the chapter. On telling the news the spreader is likely to discover whether his hearer has already heard it. This knowledge may very well influence his enthusiasm for continuing to spread the news. In other words, the cessation of spreading may depend on the state of knowledge in the population. This interaction between spreader and hearer is peculiar to this application and has no counterpart in the theory of epidemics. It is thus desirable to develop new models to assess the importance of this characteristic. A class of models having the property that cessation of spreading depends on the state of knowledge in the population was proposed by Daley and Kendall (1965). Their work forms the basis of the present section.

We retain all the assumptions of Kermack and McKendrick's model except the one which governs the cessation of spreading. In the simplest version of the model the diffusion proceeds as follows. If a spreader meets an ignorant the news is transmitted and the ignorant becomes a spreader. If a spreader meets someone who has previously been informed he ceases to spread the news. Those who have heard but are no longer spreading are called 'stiflers' because, on contact with a spreader, they cause him to cease spreading. As before, $m(T)$ and $n(T)$ denote the number of ignorants and spreaders, respectively, at time T. The theory for this model has only been worked out for the case when the a persons who initiate the diffusion process are indistinguishable from the other members of the population. Thus they may communicate with one another and they remain present as stiflers throughout the process. The number of stiflers at time T, denoted by $l(T)$, is thus $N + a - m(T) - n(T)$.

When the system is in state (m, n) three transitions are possible as follows:

(a) $(m, n) \to (m - 1, n + 1)$. This transition occurs whenever a spreader meets an ignorant. The probability that such a meeting occurs in $(T, T + \delta T)$ is $\beta mn\delta T$.

(b) $(m, n) \to (m, n - 1)$. The number of spreaders is reduced by one whenever a spreader meets a stifler. This happens with probability $\beta nl\delta T$.

(c) $(m, n) \to (m, n - 2)$. This transition results from contact between two spreaders when both become stiflers. There are $\frac{1}{2}n(n - 1)$ pairs of spreaders so the total probability of contact is $\beta\frac{1}{2}n(n - 1)\delta T$.

It is possible to set up differential-difference equations for the probabilities $P_{m,n}(T)$ in exactly the same way as for Kermack and McKendrick's model. As in that case they are difficult to solve, except in degenerate cases. Some information can be obtained by comparison with the pure birth process model because the expected number of hearers will always be greater in the absence of stifling. Thus the curve for $N - \overline{m}(T)$ for the birth process will provide an upper bound for the corresponding function of the stifling model. However, we shall see later that the deterministic approximation provides sufficient information.

The distribution of the ultimate number of ignorants, or knowers, can be obtained by the method of the embedded random walk. The situation closely parallels the development in Section 10.2 so we shall give a more condensed treatment here. We construct difference equations for the probabilities $\{P_{m,n}\}$, using the transition probabilities which are as follows:

$$(m, n) \to (m-1, n+1): \frac{m}{N+a-\frac{1}{2}(n+1)}$$

$$(m, n) \to (m, n-1): \frac{N+a-n-m}{N+a-\frac{1}{2}(n+1)}$$

$$(m, n) \to (m, n-2): \frac{\frac{1}{2}(n-1)}{N+a-\frac{1}{2}(n+1)}.$$

The equations are:

$$\left.\begin{aligned} P_{m,n} &= \frac{m+1}{N+a-\frac{1}{2}n} P_{m+1,n-1} + \frac{N+a-n-m-1}{N+a-\frac{1}{2}(n+2)} P_{m,n+1} \\ &\quad + \frac{\frac{1}{2}(n+1)}{N+a-\frac{1}{2}(n+3)} P_{m,n+2} \\ &\quad (m \geq 0, n > 1) \\ P_{m,1} &= \frac{N+a-m-2}{N+a-3/2} P_{m,2} + \frac{1}{N+a-2} P_{m,3} \qquad (m \geq 0) \\ P_{m,0} &= \frac{N+a-m-1}{N+a-1} P_{m,1} + \frac{\frac{1}{2}}{N+a-3/2} P_{m,2} \qquad (m \geq 0) \end{aligned}\right\}. \tag{10.61}$$

If we define $P_{m,n} = 0$ whenever $m + n > N + a$ these equations hold for all m and n satisfying $0 \leq m \leq N$ and $0 \leq m + n \leq N + a$. The initial condition is $P_{N,a} = 1$. The m-axis is an absorbing barrier so $\{P_{m,0}\}$ is the probability distribution of ignorants remaining when diffusion ceases. It has not proved possible to obtain an explicit solution of (10.61) for this distribution but numerical values can easily be obtained by using a computer.

The shape of the distribution of n_H is illustrated in Figure 10.3 for $N = 50$, $a = 1$. This shows that the number reached will usually be in the neighbourhood of 40 but on a few occasions the diffusion will fail to develop. If N is increased the hump near the origin diminishes in size while the major hump becomes more nearly symmetrical with decreasing variance. Provided that N is large the value of a has hardly any effect on the main part of the distribution of n_H, for reasons which we shall explain below. Some values of the means and variances of n_H/N are given in Table 10.12.

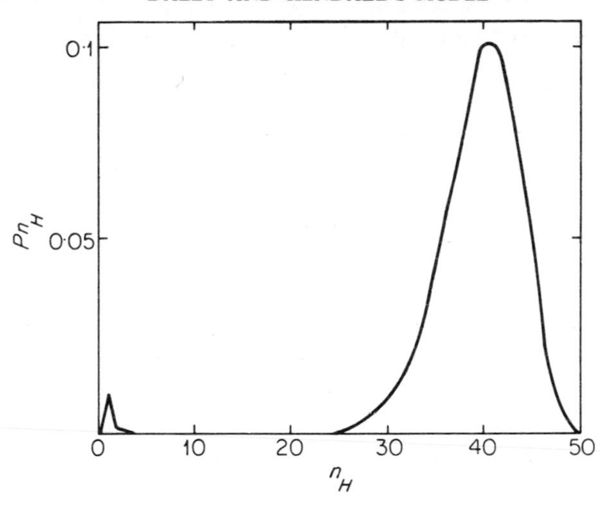

Figure 10.3

***Table 10.12** Means and variances of n_H/N for Daley and Kendall's model*

a		N: 50	100	200	400
1	Mean	0·782	0·790	0·794	0·795
	Variance	0·0139	0·00653	0·00320	0·00158
5	Mean	0·774	0·786	0·792	0·794
	Variance	0·00752	0·00338	0·00165	0·00080

The figures in the table suggest that the mean proportion who hear is approaching an asymptotic value in the neighbourhood of 0·80 independently of a. The variances are nearly proportional to $1/N$ but those for $a = 1$ are almost twice those when $a = 5$. This difference arises because, as Daley and Kendall showed, the hump near the origin accounts for about half of the variance. When $a = 5$ the minor hump makes a negligible contribution to the variance. A more meaningful result can be obtained by omitting the minor hump from the calculations. Daley and Kendall computed the mean and variance of the major part of the distribution only. They investigated the behaviour of these moments for large N by fitting quadratics in N^{-1} to their values when $N = 191, 383$ and 767. The results are expressed in the following formulae:

$$\left.\begin{aligned} E\left(\frac{n_H}{N}\middle| n_H \geq 20\right) &= 0{\cdot}7968 - \frac{0{\cdot}2738}{N} - \frac{1{\cdot}7268}{N^2} \\ \operatorname{var}\left(\frac{n_H}{N}\middle| n_H \geq 20\right) &= \frac{0{\cdot}3107}{N} + \frac{1{\cdot}2327}{N^2} + \frac{19{\cdot}5734}{N^3} \end{aligned}\right\}. \qquad (10.62)$$

At first sight the negligible effect of a on the expectation seems paradoxical. We might expect to achieve a greater coverage of the population by introducing more initial spreaders whereas, in fact, a slightly smaller proportion hear when $a = 5$ than when $a = 1$. The explanation of this phenomenon is quite simple. A system with a initial spreaders is in state (N, a) at time zero. Its future development is in no way different from a process which began in state $(N + a - 1, 1)$ and is now in state (N, a). Since all processes starting in $(N + a - 1, 1)$ must pass through (N, a), unless the unlikely event of premature extinction takes place, we would expect the two processes to terminate with the same number of ignorants. If a is fixed and N is large the eventual expected proportion of people informed will thus be virtually independent of a. The same argument holds for any other characteristic of the distribution of n_H provided that we ignore the minor hump.

The results obtained for the distribution of the ultimate size of the epidemic are in marked contrast to those for Kermack and McKendrick's model. There is no threshold effect, by which we mean that the development of the process does not depend critically on the population size. Daley and Kendall's model is like Rapoport's in that a fixed proportion, which is strictly less than one, ultimately receive the information.

A more general version of the model

In the simple model described above the spreaders are very easily rebuffed. After meeting only one person who has already heard the news they assume that 'everyone knows' and cease to be spreaders. Various generalizations of this model are possible. We shall describe one of the two proposed by Daley and Kendall. Suppose that when a spreader communicates with another spreader or a stifler he becomes a stifler himself with probability ζ. As a further generalization let η be the probability that communication of the news actually takes place when a spreader meets another person. In our previous model both ζ and η were unity. (We could have introduced the second generalization in any of the earlier models but it would have been superfluous. Its only effect would have been to reduce the effective rate of contact to $\eta\beta$. In the stifling model things are a little more complicated.)

The infinitesimal transition probabilities are as follows:

(a) $(m, n) \rightarrow (m - 1, n + 1): \eta\beta mn\, \delta T$

(b) $(m, n) \rightarrow (m, n - 1): \beta[\eta\zeta nl + \zeta(1 - \zeta)\{1 - (1 - \eta)^2\}n(n - 1)]\, \delta T$

(c) $(m, n) \rightarrow (m, n - 2): \beta\{1 - (1 - \eta)^2\}\zeta^2 \tfrac{1}{2}n(n - 1)\, \delta T.$

The transition probabilities for the associated random walk are obtained by expressing (a), (b) and (c) above as proportions of their total. These may then be used to construct difference equations for the probabilities $\{P_{m,n}\}$ after the manner of (10.61). Extensive calculations have been made of the terminal probabilities $\{P_{m,0}\}$ in order to assess the effect of these generalizations on the ultimate state of the system. As we should expect, the value of a has almost no effect on the expected proportion who finally hear the news. The influence of ζ and η can thus be adequately demonstrated by considering the case $a = 1$. Table 10.13 gives the expected proportion of hearers for values of ζ and η intermediate between 0 and 1.

***Table 10.13** The expected proportion of hearers for the general model of Daley and Kendall with $a = 1$*

	$N = 50$			$N = 200$		
	η					
ζ	0·25	0·50	0·75	0·25	0·50	0·75
0·25	0·979	0·983	0·987	0·981	0·985	0·989
0·50	0·892	0·905	0·920	0·899	0·911	0·925
0·75	0·793	0·813	0·834	0·807	0·825	0·844

The general effect of relaxing the assumption that $\zeta = \eta = 1$ is to increase the proportion who ultimately hear. It is clear from the table that η plays a relatively minor role compared with ζ. Inspection of the actual distribution of n_H/N shows that the basic form is preserved under this generalization. That is, it remains true that the diffusion reaches a fixed proportion of the population on average but the actual values of that proportion will depend on the extent to which stifling occurs.

An alternative way of generalizing the simple model is to suppose that a spreader becomes a stifler after k unsuccessful attempts at telling the news.

The quantity k could be fixed or it could be a random variable. Using deterministic methods Daley and Kendall obtained the following asymptotic values for the $E(n_H/N)$ when k is fixed.

k	1	2	3
$E(n_H/N)$	0·797	0·940	0·980

It is thus true to say that 'most people' will hear the news in a homogeneously mixing population unless there is a high degree of stifling. By supplementing our analysis by deterministic methods we shall be able to show that $E(n_H/N)$ cannot be less than 0·63 under any of the models we have considered in this section. The deterministic approach also enables us to obtain information about the rate of diffusion.

The deterministic approximation

If we treat m and n as continuous variables with rates of change determined by the transition probabilities given above, we obtain

$$\left.\begin{aligned}\frac{\mathrm{d}m(T)}{\mathrm{d}T} &= -\eta\beta m(T)n(T)\\ \frac{\mathrm{d}n(T)}{\mathrm{d}T} &= \eta\beta n(T)\{(1+\zeta)m(T) - \zeta(1-\eta)n(T) + \zeta(2-\eta-N-a)\}\end{aligned}\right\}. \tag{10.63}$$

The initial conditions are $m = N$, $n = a$. A direct consequence of (10.63) is that

$$\frac{\mathrm{d}n}{\mathrm{d}m} = -(1+\zeta) + \zeta(1-\zeta)\frac{n}{m} - \frac{\zeta(2-\eta-N-a)}{m} \tag{10.64}$$

where we are now treating n as a function of m. This differential equation may be solved by standard methods to give

$$n = -\frac{(N+a+\eta-2)}{(1-\eta)} - \frac{(1+\zeta)n}{1-\zeta(1-\eta)} + \left(\frac{m}{n}\right)^{\zeta(1-\eta)}\left(\frac{2-\eta}{1-\eta}\right)$$
$$\times\left[\frac{N}{\{1-\zeta(1-\eta)\}} + (a-1)\right] \qquad (\eta \neq 1). \tag{10.65}$$

When $\eta = 1$ the appropriate relationship between $m(T)$ and $n(T)$ can be obtained either directly from (10.64) or by a limiting operation on (10.65). Equation (10.65) enables us to find $n(T)$ in terms of $m(T)$ at any stage of the process. Hence we may eliminate $n(T)$ from (10.63) to obtain a differential equation for $m(T)$. Before doing this we shall check to see whether the deterministic and stochastic methods agree in their predictions of the final number of hearers.

The ultimate number of ignorants is found from (10.65) by setting $n = 0$ and solving for m. To effect the comparison with the stochastic solution let us first consider the case of fixed a and large N. Writing $p = n_H/N = (N - m)/N$ and allowing N to tend to infinity the equation for p is

$$(2 - \eta)(1 - p)^{\zeta(1-\eta)} = \{1 - \zeta(1 - \eta) + (1 + \zeta)(1 - \eta)(1 - p)\} \quad (\eta \neq 1) \tag{10.66}$$

and

$$\zeta \log_e (1 - p) + (1 + \zeta)p = 0 \quad (\eta = 1). \tag{10.67}$$

Equation (10.67) is essentially the same as (10.13), whose solution is given in Barton and coworkers (1960) and in David and coworkers (1966). When $\zeta = 1$, (10.67) is satisfied by $p = 0{\cdot}7968$ which is exactly equal to the asymptotic mean in the stochastic case as estimated from (10.62). Further calculations for $\zeta \neq 1$ and $\eta \neq 1$ confirm the agreement between the two approaches. We may therefore proceed with confidence to use deterministic methods to extend the analysis of the diffusion process.

We have already noted and explained the fact that the terminal behaviour does not depend on a provided that a is fixed and N is large. This conclusion no longer holds if a is of the same order as N. Suppose, for example, that $a = CN$ $(C > 0)$ and, for simplicity, let $\zeta = \eta = 1$. Equation (10.65), with $n = 0$, is then asymptotically equivalent to

$$2p + (1 + C)\log_e (1 - p) + C = 0. \tag{10.68}$$

When $C = 0$ we have shown above that $p = 0{\cdot}7968$. When $C = 1, p = 0{\cdot}6983$ and as $C \to \infty, p \to 1 - e^{-1} = 0{\cdot}6321$. Here we have a clearer demonstration of a conclusion revealed in Table 10.12, namely, that the proportion who hear is a decreasing function of the number of initial spreaders. Since decreasing either ζ or η increases p, the value $p = 0{\cdot}6321$ represents the minimum expected proportion of hearers under the general model of Daley and Kendall.

The rate at which the news spreads can be investigated by solving the first member of (10.63) for $m(T)$. A solution can be obtained numerically after using (10.65) to express $-\eta\beta m(T)n(T)$ as a function of $m(T)$. The method is exactly the same as the one used for the model discussed in Section 10.2. We shall merely point out an interesting connexion between the two models in this respect. Let $a = 1, \zeta = \eta = 1$; then $m(T)$ satisfies

$$\frac{\mathrm{d}m(T)}{\mathrm{d}T} = -\beta m(T)\left\{2N + 1 - 2m(T) + N \log_e \frac{m(T)}{N}\right\}, \tag{10.69}$$

with initial condition $m(0) = N$. If we refer to (10.14), the corresponding equation for Kermack and McKendrick's model, we see that the two equations are very similar. In fact, the rate of spread, $dm(T)/dT$, in the present model is the same as it is in Kermack and McKendrick's model with a population size of $2N$ and $d = 2$. This fact was also noted by Cane (1966).

10.5 EMPIRICAL DATA ON THE DIFFUSION OF INFORMATION

At the beginning of Chapter 9 we referred to empirical evidence on the growth of the number of hearers and gave some references. Further references will be found in Chapter 17 of Coleman (1964b). The data given in these publications, mainly relating to the adoption of innovations, were obtained by observing diffusion processes arising naturally. On the whole, such data are not sufficiently refined to enable us to examine the models discussed here in detail or to discriminate effectively between them. For such purposes we either need more detailed observation of a naturally occurring diffusion process or controlled experimentation. A well-documented study was carried out by Coleman and coworkers (1957) on the diffusion of knowledge about a new drug among physicians. This work permitted a comparison between doctors who shared an office and those who did not, but it was not really possible to meaningfully estimate the parameters of the simple diffusion model which the authors proposed.

The most extensive collections of experimental data on the diffusion of news have been made by the Washington Public Opinion Laboratory, Seattle by Dodd and his coworkers. Much of the material was obtained as part of 'Project Revere', which was carried out in the early part of the 1950's. Some of the experiments were made using small populations of school children or college students; others used the populations of small towns. There was a particular interest in the diffusion of information contained in leaflets dropped from the air. Dodd devoted a good deal of effort to the fitting of logistic curves to the growth of knowledge in a population. He found that the fit could often be improved by allowing β, the contact rate, to be a function of T. This assumption implies either that meetings become more or less frequent as time passes or that the frequency of telling changes with the state of knowledge in the population. Although generalizations of this kind provide closer agreement between theory and observation they have little explanatory value. The fit of almost any model can be improved by the addition of new parameters and it requires more than improved agreement to justify their arbitrary introduction.

Rapoport had also noted that the fits of his models could be improved by allowing d, the number of people each spreader tells, to depend on g, but the same objections apply. In the diffusion application it seems pointless to adopt this kind of refinement when the fault is much more likely to lie with the assumption of homogeneous mixing. In fact, Dodd showed that in a small community of 184 seventh-grade pupils at a school in Washington state, pairing off in the diffusion process was far from random. Members of the same sex, the same group of friends or the same living group were much more likely to communicate among themselves than with members outside such groupings. If the mixing assumption is invalid in such apparently homogeneous groups it is most unlikely that it holds in larger populations. It is for this reason that we have devoted considerable attention to the situation in stratified populations.

We conclude this chapter by quoting some data from the 'coffee slogan' experiment which formed part of Project Revere. This relates to the distribution of hearers by generation and illustrates the kind of data needed if the models described in this chapter and the previous are to be adequately tested and developed.

In a United States village of 210 housewives, 42 were told a slogan about a particular brand of coffee. Each housewife thus informed was asked to pass on the information to other housewives. As an incentive it was explained that there would be a reward—a free pound of coffee—for all people who knew the slogan when an interviewer called later. Data were obtained forty-eight hours afterwards by interviewers calling at each household. It was possible to trace the route by which the slogan had been passed to each hearer so that they could be classified by generations. The results are given in Table 10.14.

Table 10.14 **Numbers of hearers n_g in successive generations g of the 'coffee slogan' experiment, with the expected values $E(n_g)$ on Rapoport's model with $d = 2$, $a = 42$, $N = 168$ (*from* equations 10.53 and 10.54)**

	Generation g					
	1	2	3	4	5	*Total*
n_g	69	53	14	2	4	142
$E(n_g)$	66·6	55·8	22·3	5·5	1·1	151·3

When the experiment ended, 142 out of the original 168 housewives (84·5 per cent) had been informed. It seems unlikely that very much, if any, diffusion would have taken place after forty-eight hours. It thus appears that diffusion ceased before all persons had been informed. This at once rules out models such as the pure birth process. In other experiments the terminal percentage of hearers varied between 60 per cent and 100 per cent. Some of these figures must be accepted with caution because evidence of hearing was sometimes provided by handing in a card containing a completed message. In spite of incentives offered to participants there is no guarantee that all members of the population did take part. Nevertheless, it seems reasonable to infer that some kind of forgetting or stifling process did occur. The figure of 84·5 per cent in the 'coffee slogan' experiment is very close to the figure predicted by Daley and Kendall's simple model. Alternatively, Rapoport's or Kermack and McKendrick's model would lead to this result with d in the neighbourhood of 2.

Another convenient way of comparing the models is by means of the number who heard directly from the 42 source members. On Rapoport's model with $d = 2$ the expected number of first generation hearers is

$$E(n_1) = 168\left\{1 - \left(1 - \frac{1}{168}\right)^{84}\right\} = 66{\cdot}6$$

as given in the table. The agreement with observation is remarkably good. However, equally good agreement can be obtained using the pure birth model. We have already ruled this out as a complete explanation because it predicts 100 per cent coverage in the long run. However, both Kermack and McKendrick's and Daley and Kendall's models will be very similar to the birth model near to the beginning of the process. If we use the theory of Section 9.3 and, in particular (9.51), we have $\omega = 42$, $N = 168$ and

$$E(n_1) \sim \omega \log_e \frac{N + \omega - 1}{\omega} = 42 \log_e \frac{209}{42} = 67{\cdot}4.$$

There is thus no empirical basis for choice between the various models in this comparison. Both methods give excellent predictions of what actually happened.

A more detailed comparison is less favourable to our models. This is evident from Table 10.14, where the observed and expected n_g's are less close for $g > 1$. Similar results were obtained with other data. It is encouraging to find some agreement between our predictions and the course of actual epidemics, but it is also clear that further refinement is necessary.

Bibliography and Author Index

The bibliography contains all the books, papers and reports referred to in the text; the pages on which they are cited are given in brackets. It also contains references to other work relevant to the theme of the book. Some of this is to general discussions on the use of mathematics and statistics in the social sciences, some relates directly to the topics treated in the book and the remainder has to do with stochastic modelling in other fields of application. No claim is made to completeness, especially in respect of the last group, but I think the list includes most published work on stochastic models for social processes.

Aaker, D. A. (1971a). 'The new-trier stochastic model of brand choice'. *Man. Sci.*, Application Series **17,** B435–B450.

Aaker, D. A. (1971b). 'A new method for evaluating stochastic models of brand choice'. *J. Marketing Research*, **7,** 300–306.

Adelman, I. G. (1958). 'A stochastic analysis of the size distribution of firms'. *J. Amer. Statist. Ass.*, **53,** 893–904 [15, 16].

Agnew, R. A. (1971). 'Counter examples to an assertion concerning the normal distribution and a new stochastic price fluctuation model'. *Rev. Econ. Studies*, **38,** 381–383.

Aitchison, J. (1955). Contributions to the discussion on Lane and Andrew (1955) [189].

Alling, D. W. (1958). 'The after history of pulmonary tuberculosis: a stochastic model'. *Biometrics*, **14,** 527–547.

Almond, G. *See* Young and Almond (1961).

Alper, P. *See* Armitage, Smith and Alper (1969).

Anderson, T. W. (1954). 'Probability models for analyzing time changes in attitudes'. In P. F. Lazarsfeld (1954), 17–66 [15].

Andrew, J. E. *See* Lane and Andrew (1955).

Armacost, R. L. *See* Oliver, Hopkins and Armacost (1972).

Armitage, P. (1959). 'The comparison of survival curves'. *J. R. Statist. Soc.*, **A122,** 279–300.

Armitage, P. H., C. M. Phillips, and J. Davies (1970). 'Towards a model of the upper secondary school system (with discussion)'. *J. R. Statist. Soc.*, **A133,** 166–205 [56].

Armitage, P. H., and C. S. Smith (1972). 'Controllability: an example'. *Higher Education Review*, **5,** 55–66.

Armitage, P. H., C. S. Smith, and P. Alper (1969). *Decision Models for Educational Planning*, Allen Lane, The Penguin Press, London [56, 93].

Arrow, K. J., S. Karlin, and P. Suppes (Eds.), (1961). *Mathematical Methods in the Social Sciences*, Stanford University Press.

Bailey, N. T. J. (1957). *The Mathematical Theory of Epidemics*, Griffin, London [297, 299–301, 307, 336, 341, 355].

Bailey, N. T. J. (1963). 'The simple stochastic epidemic: a complete solution in terms of known functions'. *Biometrika*, **50,** 235–240 [301].

Bailey, N. T. J. (1964). *The Elements of Stochastic Processes with Applications to the Natural Sciences*, John Wiley, New York [13, 17, 60, 81, 308].

Bailey, N. T. J. (1967). 'The simulation of stochastic epidemics in two dimensions'. *Proc. Fifth Berkeley Symp. Math. Statist. Prob.*, **4,** 237–257 [332].

Bailey, N. T. J. (1968a). 'Stochastic birth, death and migration processes for spatially distributed populations'. *Biometrika*, **55,** 189–198.

Bailey, N. T. J. (1968b). 'A perturbation approximation to the simple stochastic epidemic in a large population'. *Biometrika*, **55,** 199–209 [301].

Balinsky, W., and A. Reisman (1972). 'Some manpower planning models based on levels of educational attainment'. *Man. Sci.*, **18,** B691–B705.

Balinsky, W., and A. Reisman (1973). 'A taxonomy of manpower-educational planning models'. *Socio-Economic Planning Sciences*, **7,** 13–18.

Balleer, M. (1968). 'The exclusion of sets of persons in sickness insurance considered as a Markov process'. *Bla. Dtsch. Ges. Versichmath*, **8,** 611–632.

Barankin, E. W. (1969). 'Towards the mathematics of a general theory of behaviour I'. *Ann. Inst. Statist. Math.*, **21,** 421–456.

Barlow, R. E., D. J. Bartholomew, J. M. Bremner, and H. D. Brunk (1972). *Statistical Inference Under Order Restrictions*, John Wiley, Chichester [184, 215, 317].

Barron, F. H. *See* Mackenzie and Barron (1970).

Bartholomew, D. J. (1959). 'Note on the measurement and prediction of labour turnover'. *J. R. Statist. Soc.*, **A122,** 232–239 [185, 187, 214, 215].

Bartholomew, D. J. (1963a). 'A multistage renewal process'. *J. R. Statist. Soc.*, **B25,** 150–168 [249].

Bartholomew, D. J. (1963b). 'An approximate solution of the integral equation of renewal theory'. *J. R. Statist. Soc.*, **B25,** 432–441 [214, 217].

Bartholomew, D. J. (1969a). 'Renewal theory models for manpower systems'. In N. A. B. Wilson (1969), 120–128.

Bartholomew, D. J. (1969b). *A Mathematical Analysis of Structural Control in a Graded Manpower System*, Paper P.3. Research Program in University Administration, University of California [98, 110].

Bartholomew, D. J. (1971). 'The statistical approach to manpower planning'. *Statistician*, **20,** 3–26 [41, 209, 241].

Bartholomew, D. J. (1972). 'The effect of changes in quits and hires on the length of service composition of employed workers: a comment on Stoikov's paper'. *Brit. J. Indust. Rel.*, **10,** 130–133.

Bartholomew, D. J., and E. E. Bassett (1971). *Let's Look at the Figures: The Quantitative Approach to Human Affairs*, Penguin Books, London.

Bartholomew, D. J., and A. D. Butler (1970). 'The distribution of the number of leavers for an organization of fixed size'. In A. R. Smith (1970), 417–426 [222, 223].

Bartholomew, D. J., and B. R. Morris (Eds.) (1971). *Aspects of Manpower Planning*, English Universities Press, London.

Bartholomew, D. J., and A. R. Smith (Eds.) (1971). *Manpower and Management Science*, English Universities Press, London, and D. C. Heath and Co., Lexington, Mass. [2].

Bartholomew, D. J. *See* also Barlow, Bartholomew, Bremner and Brunk (1972).

Bartlett, M. S. (1955). *An Introduction to Stochastic Processes*, Cambridge University Press [13, 289, 300].

Bartlett, M. S. (1956). 'Deterministic and stochastic models for recurrent epidemics'. *Proc. Third Berkeley Symp. Math. Statist. Prob.*, **4,** 81–109 [323, 328].

Barton, D. E., and F. N. David (1962). *Combinatorial Chance*, Griffin, London [357].

Barton, D. E., F. N. David and M. Merrington (1960). 'Tables for the solution of the exponential equation, $\exp(-a) + ka = 1$'. *Biometrika*, **47,** 439–445 [271, 346, 377].

Barton, D. E., and C. L. Mallows (1961). 'The randomization bases of the problem of the amalgamation of weighted means'. *J. R. Statist. Soc.*, **B23,** 423–433 [320].

Barton, D. E., and C. L. Mallows (1965). 'Some aspects of the random sequence'. *Ann. Math. Statist.*, **36,** 236–260.

Barton, D. E. *See also* David, Kendall and Barton (1966).

Bartos, O. J. (1967). *Simple Models of Group Behaviour*, Columbia University Press, New York and London.

Bassett, E. E. *See* Bartholomew and Bassett (1971).

Becker, N. G. (1968). 'The spread of an epidemic to fixed groups within the population'. *Biometrics*, **24,** 1007–1014.

Bell, E. J. *See* Preston and Bell (1961).

Benjamin, B. (1972). 'Stochastic processes as applied to life tables'. *Bull. Inst. Math. Appl.*, **8,** 12–16.

Benjamin, B., and J. Maitland (1958). 'Operational research and advertising: some experiments on the use of analogies'. *Operat. Res. Quart.*, **6,** 207–217.

Bernhardt, I. (1970). 'Diffusion of catalytic techniques through a population of medium-size petroleum refining firms'. *J. Indust. Economics*, **19,** 50–64.

Bernhardt, I., and K. D. Mackenzie (1970). ' Some problems in using diffusion models for new products'. In G. Fisk (Ed.), *Essays in Marketing Theory*, Allyn and Bacon, Boston, Mass.

Bernhardt, I., and K. D. Mackenzie (1972). 'Some problems in using diffusion models for new products'. *Man. Sci.*, **19,** 187–200.

Berry, B. J. L. (1971). 'Monitoring trends, forecasting change and evaluating goal achievement in the urban environment'. In M. Chisholm, A. E. Frey and P. Haggett (1971), 93–117 [15].

Bhargava, T. N., and L. Katz (1964). 'A stochastic model for a binary dyadic relation with applications to social and biological science'. *Bull. Int. Statist. Inst.*, **40,** II, 1055–1057.

Bharucha-Reid, A. T. (1960). *Elements of the Theory of Markov Processes and their Applications*, McGraw-Hill, New York [139].

Bibby, J. (1970). 'A model to control for the biasing effects of differential wastage'. *Brit. J. Indust. Rel.*, **8,** 418–420 [241].

Birnbaum, Z. W., and S. C. Saunders (1969). 'A new family of life distributions'. *J. Appl. Prob.*, **6,** 319–327.

Blumen, I., M. Kogan, and P. J. McCarthy (1955). *The Industrial Mobility of Labour as a Probability Process*, Cornell University Press, Ithaca, New York [16, 38, 43, 47–49, 51, 53, 54].

Boag, J. (1949). 'Maximum likelihood estimates of the proportion of patients cured by cancer therapy'. *J. R. Statist. Soc.*, **B11,** 15–53 [147].

Boudon, R. (1973). *Mathematical Structures of Social Mobility*, American Elsevier Co., New York [23].

Bowers, R. V. (1937). 'The direction of the intra-societal diffusion'. *Amer. Soc. Review*, **2,** 826–836.

Bowey, A. M. (1969). 'Labour stability curves and labour stability'. *Brit. J. Indust. Rel.*, **7,** 69–84 [241].

Box, G. E. P., and G. M. Jenkins (1971). *Times Series Analysis Forecasting and Control*, Holden-Day, San Francisco, Cambridge, London and Amsterdam [4].

Boyce, A. J. *See* Hiorns, Harrison, Küchemann and Boyce (1969).

Bremner, J. M. *See* Barlow, Bartholomew, Bremner and Brunk (1972).

Brissenden, P. F., and E. Frankel (1922). *Labour Turnover in Industry: A Statistical Analysis*, Macmillan, New York.

Bristow, M. R. *See* Fisher, Henderson and Bristow (1964).

British Association Mathematical Tables: Vol. I, Circular and Hyperbolic Functions (1951). Cambridge University Press [303].

Brunk, H. D. *See* Barlow, Bartholomew, Bremner and Brunk (1972).

Bryant, D. T. (1965). 'A survey of the development of manpower planning policies'. *Brit. J. Indust. Rel.*, **3,** 279–290.

Bryant, D. T. (1972). 'Recent developments in manpower research'. *Personnel Review*, **1,** 14–31.

Burkart, A. J. *See* Lee and Burkart (1960).

Bush, R. R., and C. F. Mosteller (1955). *Stochastic Models for Learning*, John Wiley, New York [1].

Butler, A. D. (1970). *Renewal Theory Applied to Manpower Systems*, Ph.D. thesis, University of Kent [217, 221, 226, 246].

Butler, A. D. (1971). 'The distribution of numbers promoted and leaving in a graded organization'. *Statistician*, **20,** 69–84 [222, 249].

Butler, A. D. *See also* Bartholomew and Butler (1970).

Cane, V. R. (1966). 'A note on the size of epidemics and the number of people hearing a rumour'. *J. R. Statist. Soc.*, **B28,** 487–490 [345, 378].

Canon, M. D., C. D. Cullum, and E. Polack (1970). *Theory of Optimal Control and Mathematical Programming*, McGraw-Hill, New York [99].

Carlsson, G. (1958). *Social Mobility and Class Structure*, CWK Gleerup, Lund, Sweden.

Carman, J. M. (1966). 'Brand switching and linear learning models'. *J. Advertising Res.*, **6,** 23–31.

Casstevens, T. W., and W. Morris (1972). 'The cube law and the decomposed system'. *Canadian J. Pol. Sci.*, **5,** 521–532.

Champernowne, D. G. (1953). 'A model of income distribution'. *Economic J.*, **63,** 318–351.

Charnes, A., W. W. Cooper, and R. J. Niehaus (1971). 'A generalised network model for training and recruiting decisions in manpower planning'. In D. J. Bartholomew and A. R. Smith (1971), 115–130.

Charnes, A., W. W. Cooper, R. J. Niehaus, and D. Sholtz (1970). 'A model for civilian manpower management and planning in the U.S. Navy'. In A. R. Smith (1971), 247–263 [98].

Chatfield, C., and G. J. Goodhardt (1970). 'The beta-binomial model for consumer purchasing behaviour'. *App. Stats.*, **19,** 240–250.

Chiang, C. L. (1968). *Introduction to Stochastic Processes in Biostatistics*, John Wiley, New York [139, 184].

Chisholm, M., A. E. Frey, and P. Haggett (Eds.) (1971). *Regional Forecasting*, Butterworths, London.

Chorley, R. J., and P. Haggett (1967). *Models in Geography*, Methuen, London.

Chow, L. P. *See* Liu and Chow (1971).

Christopher, S. C. *See* Dodd and Christopher (1968).

Clough, D. J., and W. P. McReynolds (1966). 'State transition model of an educational system incorporating a constraint theory of supply and demand'. *Ontario J. Educational Research*, **9,** 1–18 [56].

Clowes, G. A. (1972). 'A dynamic model for the analysis of labour turnover'. *J. R. Statist. Soc.*, **A135**, 242–256 [241].

Cohen, B. (1963). *Conflict and Conformity: A Probability Model and Its Application*, The M.I.T. Press, Cambridge, Mass.

Coleman, J. S. (1964a). *Models of Change and Response Uncertainty*, Prentice-Hall, Englewood Cliffs.

Coleman, J. S. (1964b). *Introduction to Mathematical Sociology*, The Free Press of Glencoe and Collier-Macmillan, London [1, 8, 295, 308, 309, 378].

Coleman, J. S., E. Katz, and H. Menzel (1957). 'The diffusion of an innovation among physicians'. *Sociometry*, **20**, 253–270 [378].

Coleman, J. S., E. Katz, and H. Menzel (1966). *Medical Innovation: A Diffusion Study*, Bobbs-Merrill Co., New York.

Collar, A. R. *See* Frazer, Duncan and Collar (1946).

Collison, P. (1962). 'Career contingencies of English university teachers'. *Brit. J. Sociology*, **13**, 286–293.

Conrath, D. W., and W. F. Hamilton (1971). 'The economics of manpower pooling'. *Man. Sci.*, **18**, B19–B29.

Cooper, W. W. *See* Charnes, Niehaus, Sholtz and Cooper (1970); *and* Charnes, Niehaus and Cooper (1971).

Cox, D. R. (1962). *Renewal Theory*, Methuen, London [210, 219, 226, 232, 265].

Cox, D. R., and H. D. Miller (1965). *The Theory of Stochastic Processes*. Methuen, London [13, 17, 70, 139, 199].

Crain, R. L. (1966). 'Fluoridation: the diffusion of an innovation among cities'. *Social Forces*, **44**, 467–476.

Criswell, J. H., H. Soloman, and P. Suppes (Eds.) (1972). *Mathematical Methods in Small Group Processes*, Stanford University Press.

Cullum, C. D. *See* Canon, Cullum and Polack (1970).

Cutolo, I. (1963). 'A stochastic analysis of the patrimonial structure of a firm'. *Riv. Pol. Econ.*, **8** and **9**, 1197–1218.

Dacey, M. F. (1971). 'Regularity in spatial distributions: a stochastic model of the imperfect central place plane'. *Statistical Ecology*, **1**, 287–309.

Daley, D. J. (1967a) 'Concerning the spread of news in a population of individuals who never forget'. *Bull. Math. Biophysics*, **29**, 373–376 [322].

Daley, D. J. (1967b). *Some Aspects of Markov Chains in Queueing Theory*, Ph. D. thesis, University of Cambridge [348, 366].

Daley, D. J., and D. G. Kendall (1965). 'Stochastic rumours'. *J. Inst. Math. Applns.*, **1**, 42–55 [371].

Daniels, H. E. (1966). 'The distribution of the total size of an epidemic'. *Proc. Fifth Berkeley Symp. Math. Statist. Prob.*, **4**, 281–293 [344].

David, F. N., M. G. Kendall, and D. E. Barton (1966). *Symmetric Functions and Allied Tables*, Cambridge University Press [377].

David, F. N. *See also* Barton and David (1962) *and* Barton, David and Merrington (1960).

David, H. A. (1970). 'On Chiang's proportionality assumption in the theory of competing risks'. *Biometrics*, **26**, 336–339.

Davies, G. S. (1973). 'Structural control in a graded manpower system'. *Man. Sci.*, **20** (to appear) [98, 104, 110, 120, 131].

Davies, J. *See* Armitage, Phillips and Davies (1970).

Day, R. H. (1970). 'A theoretical note on the spatial diffusion of something new'. *Geographical Analysis*, **2**, 68–76 [333].

De Cani, J. S. (1961). 'On the construction of stochastic models of population growth and migration'. *J. Regional Science*, **3,** 1–13.

Dietz, K. (1967). 'Epidemics and rumours: a survey'. *J. R. Statist. Soc.*, **A130,** 505–528.

Dill, W. R., D. P. Gaver, and W. L. Weber (1966). 'Models and modelling for manpower planning'. *Man. Sci.*, **13,** B142–B166.

Dodd, S. C. (1955). 'Diffusion is predictable: testing probability models for laws of interaction'. *Amer. Sociol. Rev.*, **20,** 392–401 [296].

Dodd, S. C., and S. C. Christopher (1968). 'The reactant model'. In *Essays in Honour of George A. Lunenberg*, Behavioural Research Council, Great Barrington, Mass.

Dodd, S. C., and M. McCurtain (1965). 'The logistic diffusion of information through randomly overlapped cliques'. *Operat. Res. Quart.*, **16,** 51–63.

Douglas, J. B. (1971). 'Stirling numbers in discrete distributions'. *Statistical Ecology*, **1,** 69–98.

Downton, F. (1967a). 'A note on the ultimate size of the general stochastic epidemic'. *Biometrika*, **54,** 314–316.

Downton, F. (1967b). 'Epidemics with carriers: a note on a paper of Dietz'. *J. Appl. Prob.*, **4,** 264–270.

Drinkwater, R. W., and O. P. Kane (1971). ' "Rolling up" a number of Civil Service classes'. In D. J. Bartholomew and A. R. Smith (1971), 293–302.

Dryden, M. (1968). 'Short-term forecasting of share prices: an information theory approach'. *Scottish J. Political Economy*, **15,** 227–249.

Duncan, G. T., and L. G. Lin (1972). 'Inference for Markov chains having stochastic entry and exit'. *J. Amer. Statist. Ass.*, **67,** 761–767.

Duncan, O. D. (1966). 'Methodological issues in the analysis of social mobility'. In N. J. Smelsner and S. M. Lipset (Eds.), *Social Structure and Mobility in Economic Development*, Aldine, Chicago, 51–97.

Duncan, W. J. *See* Frazer, Duncan and Collar (1946).

Ehrenberg, A. S. C. (1965). 'An appraisal of Markov brand-switching models'. *J. Marketing Res.*, **2,** 247–363.

Ehrenberg, A. S. C. (1970). 'Models of fact: examples from marketing'. *Man. Sci.*, **16,** 435–445.

Eidem, R. J. (1968). *Innovation Diffusion Through the Urban Structures of North Dakota*, Master's thesis, University of North Dakota [323].

El Agizy, M. (1971). 'A stochastic programming model for manpower planning'. In D. J. Bartholomew and A. R. Smith (1971), 131–146.

Elsterman, G. *See* Menges and Elsterman (1971).

Fararo, T. J., and M. H. Sunshine (1964). *A Study of a Biased Friendship Net*, Syracuse N.Y. Youth Development Centre.

Farquharson, R. (1969). *Theory of Voting*, Oxford University Press.

Farrow, S. C., D. J. H. Fisher, and D. B. Johnson (1971). 'Statistical approach to planning an integrated haemodialysis/transplantation programme'. *British Medical Journal*, **2,** 671–676.

Fase, M. M. G. (1970). *An Econometric Study of Age Income Profiles*. Rotterdam University Press.

Feichtinger, G. (1971). *Stochastiche Modelle Demographischer Prozesse*, Springer-Verlag, Berlin [191, 205].

Feichtinger, G. (1972). 'Stochastic decrement models of demography'. *Biom. Zeit.*, **14,** 106–125.

Fein, E. (1970). 'Demography and thermodynamics'. *Amer. J. Phys.*, **38,** 1373–1379.

Feldman, F. *See* Revelle, Feldman and Lynne (1969).

Feller, W. (1966). *An Introduction to Probability Theory and Its Applications*, Vol. II. John Wiley, New York [302].

Feller, W. (1968). *An Introduction to Probability Theory and Its Applications*, Vol. I (3rd ed.), John Wiley, New York [17, 162, 222].

Fienberg, S. E. (1971). 'Randomization and social affairs: the 1970 draft lottery'. *Science*, **171,** 255–261.

Firey, W. (1950). 'Mathematics and social theory'. *Social Forces*, **29,** 20–25.

Fisher, D. J. H. *See* Farrow, Fisher and Johnson (1971).

Fisher, J. W., M. R. Bristow, and L. Henderson (1964). 'An actuarial procedure for assessing the experience of mental hospital patients'. *Bull. Int. Statist. Inst.*, **40,** II, 1102–1120.

Fisz, M. (1963). *Probability Theory and Mathematical Statistics* (3rd. ed.), John Wiley, New York.

Fix, E., and J. Neyman (1951). 'A simple stochastic model of recovery, relapse, death and loss of patients'. *Human Biology*, **23,** 205–241 [6, 139, 145].

Fleur, Melvin L. de (1956). 'A mass communication model of stimulus response relationships: an experiment in leaflet message diffusion'. *Sociometry*, **19,** 12–36.

Foley, J. D. (1967). 'A Markovian model of the university of Michigan executive system'. *Communications of the A.C.M.*, **10,** 584–588 [16].

Forbes, A. F. (1970). 'Promotion and recruitment policies for the control of quasi-stationary hierarchical systems'. In A. R. Smith (1970), 401–414 [98, 104].

Forbes, A. F. (1971a). 'Markov chain models for manpower systems'. In D. J. Bartholomew and A. R. Smith (1971), 93–113 [56, 57].

Forbes, A. F. (1971b). 'Non-parametric methods of estimating the survivor function'. *Statistician*, **20,** 27–52 [184, 241].

Frank, R. E. (1962). 'Brand choice as a probability process'. *J. Business*, **35,** 43–56.

Frankel, E. *See* Brissenden and Frankel (1922).

Frazer, R. A., W. J. Duncan, and A. R. Collar (1946). *Elementary Matrices*, Cambridge University Press [60, 141].

Frey, A. E. *See* Chisholm, Frey and Haggett (1971).

Freytag, H. L., and C. C. Von Weizsäcker (1968). *Schulwahl und Schulsystem in Baden Württemberg*, Heidelberg.

Freytag, H. L., and C. C. Von Weizsäcker (1969). *Schulwahl and Schulsystem: Modell theoretische Entwürfe-Verlaufsstatistische Analysen*, Verlag, Julius Beltz, Weinheim.

Fuguitt, G. V. (1965). 'The growth and decline of small towns as a probability process'. *Amer. Social Rev.*, **30,** 403–411 [15].

Fuguitt, G. V. *See* Lieberson and Fuguitt (1967).

Funkhauser, G. R., and M. E. McCoombs (1972). 'Predicting the diffusion of information to mass audiences'. *J. Math. Sociology*, **2,** 121–130.

Gani, J. (1963). 'Formulae for projecting enrolments and degrees awarded in universities'. *J. R. Statist. Soc.*, **A126,** 400–409 [55].

Gani, J. (1965). 'On a partial differential equation of epidemic theory, I'. *Biometrika*, **52,** 617–622 [338].

Gart, J. J. (1968). 'The mathematical analysis of an epidemic with two kinds of susceptibles'. *Biometrics*, **24,** 557–566.

Gaver, D. P. *See* Dill, Gaver and Weber (1966).

Gear, A. E., J. S. Gillespie, A. G. Lockett, and A. W. Pearson (1971). 'Manpower modelling: a study in research and development'. In D. J. Bartholomew and A. R. Smith (1971), 147–162.

Gillespie, J. S. *See* Gear, Gillespie, Lockett and Pearson (1971).
Ginsberg, R. B. (1971). 'Semi-Markov processes and mobility'. *J. Math. Sociology*, **1,** 233–262 [54, 172].
Ginsberg, R. B. (1972a). 'Critique of probabilistic models: application of the semi-Markov model to migration'. *J. Math. Sociology*, **2,** 63–82 [54].
Ginsberg, R. B. (1972b). 'Incorporating causal structure and exogenous information with probabilistic models: with special reference to gravity, migration and Markov chains'. *J. Math. Sociology*, **2,** 83–103.
Glass, D. V. (Ed.) (1954). *Social Mobility in Britain*, Routledge and Kegan Paul, London [20].
Goffman, W. (1965). 'An epidemic process in an open population'. *Nature*, **205,** 831–832.
Goffman, W., and V. A. Newill (1964). 'Generalization of epidemic theory—an application to the transmission of ideas'. *Nature*, **204,** 225–228.
Goffman, W., and V. A. Newill (1967). 'Communication and epidemic processes'. *Proc. Roy. Soc. Series A*, **298,** 316–334.
Goodhardt, G. J. *See* Chatfield and Goodhardt (1970).
Goodman, L. A. (1961). 'Statistical methods for the "mover–stayer" model'. *J. Amer. Statist. Ass.*, **56,** 841–868 [54].
Goodman, L. A. (1968). 'Stochastic models for the population growth of the sexes'. *Biometrika*, **55,** 469–487.
Greville, T. N. E. (Ed.) (1973). *Population Dynamics*, Academic Press, New York and London.
Gribbens, W. D., S. Halperin and P. R. Loynes (1966). 'Applications of stochastic models in research and career development'. *J. Counseling Psychology*, **13,** 403–408.
Grigelionis, B. I. (1964). 'Limit theorems for series of renewal processes'. *Cybernetics in the Service of Communism*, **2,** 246–266 [221].
Griliches, Z. (1957). 'Hybrid corn: an exploration in the economics of technical change'. *Econometrics*, **25,** 501–522 [296].
Griliches, Z. (1960). 'Hybrid corn and the economics of innovation'. *Science*, **132,** 275–280 [296].
Gross, N. C. *See* Ryan and Gross (1943).
Gumbel, E. J. (1958). *The Statistics of Extremes*, Columbia University Press, New York [310].
Gupta, S. K., and K. S. Krishman (1967). 'Mathematical models in marketing'. *Operat. Res.*, **15,** 1040–1050.
Gurevitch, M., and Loevy Zipora (1972). 'The diffusion of television as an innovation. The case of the Kibbutz'. *Human Relations*, **25,** 181–197.
Gurley, W. R. *See* Tarver and Gurley (1965).
Hägerstrand, T. (1952). *The Propagation of Innovation Waves*, Lund Studies in Geography, Ser. B, No. 4.
Hägerstrand, T. (1967). *Innovation Diffusion as a Spatial Process*, The University of Chicago Press, Chicago and London [2, 296, 297, 323–325, 331].
Haggett, P. *See* Chisholm, Frey and Haggett (1971); *and* Chorley and Haggett (1967).
Haines, G. H. (1964). 'A theory of market behaviour after innovation'. *Man. Sci.*, **10,** 634–666.
Hajnal, J. (1956). 'The ergodic properties of non-homogeneous finite Markov chains'. *Proc. Camb. Phil. Soc.*, **52,** 67–77.
Hamilton, W. F. *See* Conrath and Hamilton (1971).

Hammersley, J. M. (1966). 'First-passage percolation'. *J. Roy. Statist. Soc.*, **B28,** 491–496 [332].

Harary, F., and B. Lipstein (1962). 'The dynamics of brand loyalty: a Markovian approach'. *Operat. Res.*, **10,** 19–40.

Harary, F., and I. C. Ross (1954). 'The number of complete cycles in a communication network'. *Soc. Psych.*, **40,** 329–333.

Harrison, G. A. J. *See* Hiorns, Küchemann and Harrison (1970); *and* Hiorns, Harrison, Boyce and Küchemann (1969).

Hart, P. E., and S. J. Prais (1956). 'The analysis of business concentration'. *J. R. Statist. Soc.*, **A119,** 150–191 [15].

Haskey, H. W. (1954). 'Stochastic cross-infection between two otherwise isolated groups'. *Biometrika*, **44**, 193–204 [301, 309].

Hawkes, A. G. (1969). 'An approach to the analysis of electroal swing'. *J. R. Statist. Soc.*, **A132,** 68–79 [15].

Hedberg, M. (1961). 'The turnover of labour in industry, an actuarial study'. *Acta Sociologica*, **5,** 129–143 [194].

Henderson, L. F. (1971). 'The statistics of crowd fluids'. *Nature*, **229,** 381–383.

Henderson, L. *See* Fisher, Bristow and Henderson (1964).

Henry, N. W. (1971). 'The retention model: a Markov chain with variable transition probabilities'. *J. Amer. Statist. Ass.*, **66,** 264–267 [16].

Henry, N. W., R. McGinnis, and H. W. Tegtmeyer (1971). 'A finite model of mobility'. *J. Math. Sociology*, **1,** 107–116 [16, 40].

Hensley, C. *See* Mansfield and Hensley (1961).

Herbst, P. G. (1954). 'Analysis of social flow systems'. *Human Relations*, **7,** 327–336.

Herbst, P. G. (1963). 'Organizational commitment: a decision model'. *Acta Sociologica*, **7,** 34–45 [139, 191].

Herniter, J. (1971). 'A probabilistic market model of purchase timing and brand selection'. *Man. Sci.*, **18,** B102–B113.

Hill, B. M. (1970). 'Zipf's law and prior distributions for the composition of a population'. *J. Amer. Statist. Ass.*, **65,** 1220–1232.

Hill, J. M. M. (1951). 'A consideration of labour turnover as the resultant of a quasi-stationary process'. *Human Relations*, **4,** 255–264.

Hill, J. M. M. *See* Rice, Hill and Trist (1950).

Hill, R. J. (1957). 'An experimental investigation of the logistic model of message diffusion'. *Social Forces*, **36,** 21–26.

Hill, R. T. and N. C. Severo (1969). 'The simple epidemic for small populations with one or more initial infectives'. *Biometrika*, **56,** 183–196.

Hiorns, R. W., G. A. J. Harrison, A. J. Boyce, and C. F. Küchemann (1969). 'Mathematical analysis of the effects of movement on the relatedness between populations'. *Ann. Hum. Genet.*, **32,** 237–250.

Hiorns, R. W., G. A. J. Harrison, and C. F. Küchemann (1970). 'Social class relatedness in some Oxfordshire parishes'. *J. Bio. Soc. Sci.*, **2**, 71–80.

Hodge, R. W. (1966). 'Occupational mobility as a probability process'. *Demography*, **3,** 19–34 [16, 32].

Hoem, J. M. (1969). 'Purged and partial Markov chains'. *Skand. Aktuarietidskrift*, **52,** 147–155.

Hoem, J. M. (1971). 'Point estimation of forces of transition in demographic models'. *J. R. Statist. Soc.*, **33,** 275–289 [139, 191, 195].

Hopkins, D. S. P. *See* Oliver, Hopkins and Armacost (1972).

Horden, W. R., and M. T. Tcheng (1971). 'Projection of enrolment distributions with enrolment ceilings by Markov processes'. *Socio-Economic Planning Sciences*, **5**, 467–473.

Horowitz, A. and I. (1968). 'Entropy, Markov processes and competition in the brewing industry'. *J. Indust. Econ.*, **16**, 196–211.

Horvath, W. J. (1966). 'Stochastic models of behaviour'. *Man. Sci.*, **12**, B513–B518.

Horvath, W. J. (1968). 'A statistical model for the duration of wars and strikes'. *Behavioural Science*, **13**, 18–28 [207].

Hudson, J. C. (1969). 'Diffusion in a central place system'. *Geographical Analysis*, **1**, 45–58 [334].

Hyman, R. (1970). 'Economic motivation and labour stability'. *Brit. J. Indust. Rel.*, **8**, 159–178 [241].

Hyrenius, A. H. (1954). *Utredning rörande pråstkärens rekytering*. (Statistical studies in the structure and recruitment of the clergy of the Church of Sweden). *Publications of the Statistical Institute*, University of Gothenburg, Stockholm.

Isii, K. *See* Taga and Isii (1959).

Jenkins, G. M. *See* Box and Jenkins (1971).

Johnson, D. B. *See* Farrow, Fisher and Johnson (1971).

Jones, E. (1946). 'An actuarial problem concerning the Royal Marines'. *J. Inst. Actu. Students Soc.*, **6**, 38–42.

Jones, E. (1948). 'An application of the service-table technique to staffing problems'. *J. Inst. Actu. Students Soc.*, **8**, 49–55.

Jones, J. M. (1971). 'A stochastic model for adaptive behaviour in a dynamic situation'. *Man. Sci.*, **17**, 484–497.

Jones, J. M. (1973). 'A composite heterogeneous model for brand choice behaviour'. *Man. Sci.*, **19**, 499–509.

Judge, G. C. *See* Lee, Judge and Zellner (1968); *and* Lee, Judge and Zellner (1970).

Kadane, J. B., and G. Lewis (1969). 'The distribution of participation in group discussion: an empirical and theoretical appraisal'. *Amer. Soc. Rev.*, **34**, 710–723.

Kamat, A. R. (1968a). 'Estimating wastage in a course of education'. *Sankhya*, **B30**, 5–12.

Kamat, A. R. (1968b). 'Mathematical schemes for describing progress in a course of education'. *Sankhya*, **B30**, 13–24 [56].

Kamat, A. R. (1968c). 'A stochastic model for progress in a course of education'. *Sankhya*, **B30**, 25–32 [56].

Kane, O. P. *See* Drinkwater and Kane (1971).

Kaplan, R. S. (1972). 'Stochastic growth models'. *Man. Sci.*, **18**, 249–264.

Karlin, S. (1966). *A First Course in Stochastic Processes*, Academic Press, New York [13, 17].

Karlin, S. *See* Arrow, Karlin and Suppes (1960).

Katz, E. *See* Coleman, Katz and Menzel (1957); Bhargava and Katz (1964); *and* Coleman, Katz and Menzel (1966).

Keenay, A. G. *See* Morgan, Keenay and Ray (1973).

Kelley, A. C., and L. W. Weiss (1969). 'Markov processes and economic analysis: the case of migration'. *Econometrica*, **37**, 280–297.

Kemeny, J. G., and L. Snell (1960). *Finite Markov Chains*, Van Nostrand, New York [17, 22, 39, 66, 205].

Kemeny, J. G., and L. Snell (1962). *Mathematical Models in the Social Sciences*, Ginn and Co., Boston [1].

Kemp, A. G., and G. C. Reid (1971). 'The random walk hypothesis and the recent behaviour of equity prices'. *Economica*, **38,** 28–51.

Kendall, D. G. (1956). 'Deterministic and stochastic epidemics in closed populations'. *Proc. Third Berkeley Symp. Math. Statist. Prob.*, **4,** 149–165 [348].

Kendall, D. G. (1957). 'La propagation d'une épidémie au d'un bruit une population limitée'. *Publ. de L'Inst. de Statistique de l'Université de Paris*, **6,** 307–311 [302, 323, 328, 354].

Kendall, D. G. (1965). 'Mathematical models in the spread of infection'. In *Mathematics and Computer Science in Biology and Medicine*, H.M.S.O., London, 213–325 [328, 329, 353, 354].

Kendall, D. G. *See* Daley and Kendall (1965).

Kendall, M. G. (1961). 'Natural law in the social sciences'. *J. R. Statist. Soc.*, **A124,** 1–19.

Kendall, M. G. *See* David, Kendall and Barton (1966).

Kermack, W. O., and A. G. McKendrick (1927). 'Contributions to the mathematical theory of epidemics'. *Proc. Roy. Soc.*, **A115,** 700–721 [336].

Keyfitz, N. (1968). *Introduction to the Mathematics of Population*, Addison-Wesley, Reading, Mass. [2].

Khintchine, A. J. (1960). *Mathematical Methods in the Theory of Queueing*, Griffin, London [221, 279].

Kodama, F. (1970). 'An approach to the analysis of vocational education and training requirements'. *Man. Sci.*, **17,** B178–B191.

Kogan, M. *See* Blumen, Kogan and McCarthy (1955).

Kolmogoroff, A. N., I. Petrovsky, and N. Piscounoff (1937). 'Étude de l'équation de la diffusion avec croissance de la quantité de matière et son application à une problème biologique'. *Bull. de l'Univ. d'Etat à Moscou*, **A1,** fasc 6, 1–25 [323].

Kotler, P. (1968). 'Mathematical models of individual buyer behavior'. *Behavioral Science*, **13,** 274–287.

Krenz, R. D. (1964). 'Projection of farm numbers for North Dakota with Markov chains'. *Agricultural Economics Research*, **16,** 77–83.

Krishman, K. S. *See* Gupta and Krishman (1967).

Kruskal, W. (Ed.) (1970). *Mathematical Sciences and Social Sciences*, Prentice-Hall, Englewood Cliffs, N.J. [1].

Kryscio, R. J. (1972). 'The transition probabilities of the extended simple stochastic epidemic model and the Haskey model'. *J. App. Prob.*, **9,** 471–485.

Küchemann, C. F. *See* Hiorns, Harrison, Boyce and Küchemann (1969); *and* Hiorns, Harrison and Küchemann (1970).

Kuehn, A. A. (1961). 'A model of budget advertising'. In Bass and coworkers (Eds.), *Mathematical Models and Methods in Marketing*, R. D. Irwin, Inc., Homewood, Illinois, 302–356.

Kuehn, A. A. (1962). 'Consumer brand choice as a learning process'. *J. Advertising Research*, **2,** 10–17.

Kuhn, A., A. Poole, P. Sales, and H. P. Wynn (1973). 'An analysis of graduate job mobility'. *Brit. J. Indust. Rel.*, **11,** 124–142 [139].

Kukuk, C. R. *See* Wegman and Kukuk (1971).

Kushner, H. J., and A. J. Kleinman (1971). 'Mathematical programming and the control of Markov chains'. *Int. J. Control*, **13,** 801–820.

Kuzma, J. W. (1967). 'A comparison of two life table methods'. *Biometrics*, **23,** 51–64.

Lancaster, A. (1972). 'A stochastic model for the duration of a strike'. *J. R. Statist. Soc.*, **A135,** 257–271 [198, 200].

Land, K. (1969). 'Duration of residence and prospective migration'. *Demography*, **6,** 133–140 [15, 39].

Landau, H. G. (1952). 'On some problems of random nets'. *Bull. Math. Biophysics.*, **14,** 203–212 [360].

Landau, H. G., and A. Rapoport (1953). 'Contribution to the mathematical theory of contagion and spread of information I: through a thoroughly mixed population'. *Bull. Math. Biophysics*, **15,** 173–183.

Lane, K. F., and J. E. Andrew (1955). 'A method of labour turnover analysis'. *J. R. Statist. Soc.*, **A118**, 296–323 [184, 188, 190, 213, 242, 243].

Lawrence, J. R. (Ed.) (1966). *Operational Research and the Social Sciences*, Tavistock Publications, London.

Lazarsfeld, P. F. (Ed.) (1954). *Mathematical Thinking in the Social Sciences*, The Free Press of Glencoe, New York.

Lee, A. M. (1962). 'Decision rules for media scheduling: static campaigns'. *Operat. Res. Quart.*, **13,** 229–242.

Lee, A. M. (1963). 'Decision rules for media scheduling: dynamic campaigns'. *Operat. Res. Quart.*, **14,** 365–372.

Lee, A. M., and A. J. Burkart (1960). 'Some optimization problems in advertising media planning'. *Operat. Res. Quart.*, **11,** 113–122.

Lee, T. C., G. C. Judge, and A. Zellner (1968). 'Maximum likelihood and Bayesian estimation of transition probabilities'. *J. Amer. Statist. Ass.*, **63,** 1162–1179.

Lee, T. C., G. C. Judge, and A. Zellner (1970). *Estimating the Parameters of the Markov Probability Model from Aggregate Time Data*, North Holland Publishing Co.

Lewis, G. *See* Kadane and Lewis (1969).

Lieberson, S., and G. V. Fuguitt (1967). 'Negro–white occupational differences in the absence of discrimination'. *Amer. J. Sociology*, **73,** 188–200 [16].

Lipstein, B. (1965). 'A mathematical model of consumer behaviour'. *J. Marketing Res.*, **2,** 259–265.

Lipstein, B. *See also* Harary and Lipstein (1962).

Lissitz, R. W. (1972). 'Comparison of the small sample power of the chi-square and likelihood ratio tests of the assumptions for stochastic models'. *J. Amer. Statist. Ass.*, **67,** 574–577.

Littell, A. S. (1952). 'Estimation of the *T*-year survival rate from follow-up studies over a limited period of time'. *Human Biology*, **24,** 87–116.

Lockett, A. G. *See* Gear, Gillespie, Lockett and Pearson (1971).

Loevy, Z. *See* Gurevitch and Loevy (1972).

Long, L. H. (1970). 'On measuring geographic mobility'. *J. Amer. Statist. Ass.*, **65,** 1195–1203 [15].

Lui, P. T., and L. P. Chow (1971). 'A stochastic approach to the prevalence of IUD: example of Taiwan, Republic of China'. *Demography*, **8,** 341–353 [15].

Lynne, W. *See* Revelle, Feldman and Lynne (1969).

MacCrimmon, K. R. *See* Vroom and MacCrimmon (1968).

Mackenzie, K. D., and F. H. Barron (1970). 'Analysis of a decision-making investigation'. *Man. Sci.*, **17,** B226–B241.

Mackenzie, K. D. *See also* Bernhardt and Mackenzie (1970); *and* Bernhardt and Mackenzie (1972).

McCall, J. J. (1971). 'A Markovian model of income dynamics'. *J. Amer. Statist. Assoc.*, **66,** 439–447 [38].

McCarthy, P. J. *See* Blumen, Kogan and McCarthy (1955).

McCurtain, M. *See* Dodd and McCurtain (1965).

McFarland, D. (1970). 'Inter-generational social mobility as a Markov process: including a time-stationary Markovian model that explains observed declines in mobility rates over time'. *Amer. Sociological Review*, **35,** 463–476 [16, 34].

McGinnis, R. (1964). *Mathematical Foundations for Social Analysis*, Bobbs-Merrill, Chicago.

McGinnis, R. (1968). 'A stochastic model of social mobility'. *Amer. Soc. Rev.*, **33,** 712–721 [16, 39].

McGinnis, R. *See also* Myers, McGinnis and Masnick (1967); *and* Henry, McGinnis and Tegtmeyer (1971).

McGuire, T. W. (1969). 'More on least squares estimation of the transition matrix in a stationary first-order Markov process from sample population data'. *Psychometrika*, **34,** 335–345.

McKendrick, A. G. *See* Kermack and McKendrick (1927).

McNamara, J. F. (1973). 'Mathematical programming applications in educational planning'. *Socio-Economic Planning Sciences*, **7,** 19–35.

McNeil, D. R. (1972). 'On the simple stochastic epidemic'. *Biometrika*, **59,** 494–497 [302, 334].

McPhee, W. N. (1963). *Formal Theories of Mass Behaviour*, The Free Press, New York.

McReynolds, W. P. *See* Clough and McReynolds (1966).

Maffei, R. B. (1960). 'Brand preferences and simple Markov processes'. *Operat. Res.*, **8,** 210–218.

Maffei, R. B. (1961). 'Brand preferences and simple Markov processes'. In Bass and coworkers (Eds.), R. D. Irwin, Inc., Homewood, Illinois, 103–120.

Mahoney, T. A., and G. T. Milkovich (1971). 'The internal labor market as a stochastic process'. In D. J. Bartholomew and A. R. Smith (1971), 75–91 [57].

Maitland, J. *See* Benjamin and Maitland (1958).

Mallows, C. L. *See* Barton and Mallows (1965).

Manpower Planning. Proceedings of NATO conference, Brussels, 1965. English Universities Press, London.

Mansfield, E. (1961). 'Technical change and the rate of imitation'. *Econometrica*, **29,** 741–766.

Mansfield, E., and C. Hensley (1960). 'The logistic process: tables of the stochastic epidemic curve and applications'. *J. R. Statist. Soc.*, **B22,** 332–337 [301].

Marris, R. (1970). *The Economic Theory of 'Managerial' Capitalism*, Macmillan, New York and London [323].

Marshall, K. T. (1973). 'A comparison of two personnel prediction models'. *Operat. Res.*, **21,** 810–822 [56, 92].

Marshall, K. T., and R. M. Oliver (1970). 'A constant-work model for student attendance and enrollment'. *Operat. Res.*, **18,** 193–206.

Marshall, M. L. (1971a). 'Some statistical methods for forecasting wastage'. *Statistician*, **20,** 53–68 [190].

Marshall, M. L. (1971b). 'The use of probability distributions for comparing the turnover of families in a residential area'. *London Papers in Regional Science*, Vol. 2. (Urban Regional Planning) A. G. Wilson (Ed.), Pion, London, 171–193.

Maslov, P. P. (1962). 'Model building in sociological research'. *Soviet Sociology*, **1,** 11–23.

Masnick, G. *See* Myers, McGinnis and Masnick (1967).

Massy, W. F. (1968). 'Stochastic models for monitoring new product introductions'. In F. M. Bass, C. W. King and E. A. Pessemier (Eds.), *Applications of the Sciences in Marketing Management*, Wiley, New York.

Massy, W. F., D. B. Montgomery, and D. G. Morrison (1970). *Stochastic Models of Buying Behaviour*, The M.I.T. Press, Cambridge, Mass. [2].

Massy, W. F. *See also* Morrison, Massy and Silverman (1971).

Mathematical Model Building in Economics and Industry (1968). Griffin, London.

Mathematics and Social Sciences I (1965). Proceedings of seminars at Menthon-St-Bernard and Gösing, Mouton and Co., Paris and the Hague.

Matras, J. (1960a). 'Comparison of intergenerational occupational mobility patterns: an application of the formal theory of social mobility'. *Population Studies*, **14,** 163–169 [16].

Matras, J. (1960b). 'Differential fertility, intergenerational occupational mobility and change in the occupational distribution: some elementary interrelationships'. *Population Studies*, **15,** 187–197 [16, 23, 33].

Matras, J. (1967). 'Social mobility and social structure: some insights from the linear model'. *Amer. Soc. Rev.*, **32,** 608–614 [16, 19, 31, 34].

Mautner, A. J. *See* Rushton and Mautner (1955).

Meier, P. (1955). 'Note on estimation in a Markov process with constant transition rates'. *Human Biology*, **27,** 121–124.

Menges, G., and G. Elstermann (1971). 'Capacity models in university management'. In D. J. Bartholomew and A. R. Smith (1971), 207–221 [56].

Menzel, H. *See* Coleman, Katz and Menzel (1957); *and* Coleman, Katz and Menzel (1966).

Merck, J. W. (1961). *A Mathematical Model of Personnel Structure of Large Scale Organizations Based on Markov Chains*, Ph.D. Thesis, Duke University, North Carolina.

Miles, R. E. (1959). 'The complete amalgamation into blocks, by weighted means, of a finite set of real numbers'. *Biometrika*, **46,** 317–327 [317].

Milkovich, G. T. *See* Mahoney and Milkovich (1971).

Miller, H. D. *See* Cox and Miller (1965).

Miller, S. *See* Van der Merwe and Miller (1971).

Miller, W. L. (1972). 'Measures of electoral change using aggregate data'. *J. R. Statist. Soc.*, **A135,** 122–142 [15].

Mollison, D. (1972a). 'Possible velocities for a simple epidemic'. *Adv. Appl. Prob.*, **4,** 233–257 [328, 331].

Mollison, D. (1972b). 'The rate of spatial propagation of simple epidemics'. *Proc. Sixth Berkeley Symp. Math. Statist. Prob.*, **3,** 579–614 [327, 328, 331, 351, 354].

Montgomery, D. B. (1967). 'Stochastic modelling of the consumer'. *Industrial Management Review*, Spring 1967, 31–42.

Montgomery, D. B. (1969). 'A stochastic response model with application to brand choice'. *Man. Sci.*, **15,** 323–337.

Montgomery, D. B. (1972). 'Note on a limit distribution arising in certain stochastic response models'. *J. R. Statist. Soc.*, **C21,** 204–207.

Montgomery, D. B. *See* Massy, Montgomery and Morrison (1970).

Moran, P. A. P. (1968). *An Introduction to Probability Theory*, Clarendon Press, Oxford [13, 17, 139].

Morgan, R. W. (1970). 'Manpower planning in the Royal Air Force: an exercise in linear programming'. In A. R. Smith (1970), 317–325 [99].

Morgan, R. W. (1971). 'The use of a steady state model to obtain the recruitment, retirement and promotion policies of an expanding organization'. In D. J. Bartholomew and A. R. Smith (1971), 283–291 [99].

Morgan, R. W., G. A. Keenay, and K. Ray (1973). 'A steady state model for career planning'. To appear in the proceedings of the Cambridge NATO conference on Manpower Planning Models, D. J. Clough and coworkers (Ed.) [263].

Morrill, R. L. (1965). 'The negro ghetto: problems and alternatives'. *The Geographical Review*, **55,** 349–361 [297, 323].

Morrill, R. L. (1968). 'Waves of spatial diffusion'. *J. Regional Science*, **8,** 1–18.

Morrill, R. L. (1970). 'The shape of diffusion in space and time'. *Economic Geography*, **46** (1970 supplement), 259–268.

Morris, B. R. *See* Bartholomew and Morris (1971).

Morrison, D. G. (1966). 'Testing brand-switching models'. *J. Marketing Research*, **3,** 401–409.

Morrison, D. G., W. F. Massy and F. N. Silverman (1971). 'The effect of non-homogeneous populations on Markov steady state probabilities'. *J. Amer. Statist. Ass.*, **66,** 268–274 [35].

Morrison, D. G. *See* Massy, Montgomery and Morrison (1970).

Morrison, P. (1967). 'Duration of residence and prospective migration: the evaluation of a stochastic model'. *Demography*, **4,** 533–561 [15, 39].

Mosteller, C. F. *See* Bush and Mosteller (1955).

Myers, G. C., R. McGinnis and G. Masnick (1967). 'The duration of residence approach to a dynamic stochastic model of internal migration: a test of the axiom of cumulative inertia'. *Eugenics Quarterly*, **14,** 121–126 [39].

Nagaev, A. V., and A. N. Startsev (1970). 'The asymptotic analysis of a stochastic model of an epidemic'. *Theor. Prob. Appl.*, **15,** 98–107.

Näslund, B. (1970). 'Size distribution and the optimal size of firms'. *Zeit. Nat. Okon.*, **30,** 271–282.

Newill, V. A. *See* Goffman and Newill (1964); *and* Goffman and Newill (1967).

Neyman, J., and E. L. Scott (1964). 'A stochastic model of epidemics'. In J. Gurland (Ed.), *Stochastic Models in Medicine and Biology*, University of Wisconsin Press, Madison, 45–83.

Neyman, J. *See* Fix and Neyman (1951).

Nicholson, M. B. (1968). 'Mathematical models in the study of international relations'. *The Year Book of World Affairs*, **22,** 47–63.

Nicosia, F. M. (1967). 'New developments in advertising research: stochastic models'. *Proc. Amer. Ass. Public Opinion Research.*

Niehaus, R. J. *See* Charnes, Cooper, Niehaus and Sholtz (1970); *and* Charnes, Cooper and Niehaus (1971).

OECD (1967). *Mathematical Models in Educational Planning*, OECD, Paris.

Oliver, R. M. (1972). 'Operations research in university planning'. In A. Drake, R. Keeney and P. M. Morse (Eds.), *Analysis of Public Systems*, The M.I.T. Press, Cambridge, Mass.

Oliver, R. M., D. S. P. Hopkins and R. L. Armacost (1972). 'An equilibrium flow model of a university campus'. *Operat. Res.*, **20,** 249–264.

Oliver, R. M. *See* Marshall and Oliver (1970).

Ostry, S., and A. Sunter (1970). 'Canadian job vacancy survey: a measure of labour demand'. *J. Amer. Statist. Ass.*, **65,** 1059–1070.

Padberg, D. I. (1962). 'The use of Markov processes in measuring changes in market structure'. *J. Farm. Economics*, **44,** 189–199.

Parzen, E. (1962). *Stochastic Processes*. Holden-Day, San Francisco [13, 220].

Pearson, A. W. *See* Gear, Gillespie, Lockett and Pearson (1971).

Peltier, R. (Ed.) (1966). *Model Building in the Social Sciences*, Union Européane d'Editions, Monaco.
Pemberton, H. E. (1936). 'The curve of culture diffusion rate'. *Amer. Soc. Review*, **1,** 547–556 [296].
Pemberton, H. E. (1938). 'The spatial order of culture diffusion'. *Sociology and Social Research*, **22,** 246–251.
Perrin, E. B., and M. C. Sheps (1964). 'Human reproduction: a stochastic process'. *Biometrics*, **20,** 28–45.
Pike, M. C. (1970). 'A note on Kimball's paper "Models for the estimation of competing risks from grouped data" '. *Biometrics*, **26,** 579–581.
Pitts, F. R. (1963). 'Problems in computer simulation of diffusion'. *Papers and Proc. Regional Science Assn.*, **11,** 111–119.
Phillips, C. M. *See* Armitage, Phillips and Davies (1970).
Polack, E. *See* Canon, Cullum and Polack (1970).
Pollard, A. H. (1970a). 'Minimum rate of progress formulae at universities'. *Aust. J. Statist.*, **12,** 92–103.
Pollard, A. H. (1970b). 'Some hypothetical models in systems of tertiary education'. *Aust. J. Statist.*, **12,** 78–91.
Pollard, J. H. (1966). 'On the use of the direct matrix product in analysing certain stochastic population models'. *Biometrika*, **53,** 397–415 [27, 29].
Pollard, J. H. (1967). 'A note on certain discrete time stochastic population models with Poisson immigration'. *J. App. Prob.*, **4,** 209–213 [75].
Pollard, J. H. (1969). 'Continuous-time and discrete-time models of population growth'. *J. R. Statist. Soc.*, **A132,** 80–88.
Pollard, J. H. (1973). *Mathematical Models for the Growth of Human Populations*, Cambridge University Press.
Poole, A. *See* Kuhn, Poole, Sales and Wynn (1973).
Praetz, P. D. (1969). 'Australian share prices and the random walk hypothesis'. *Aust. J. Statist.*, **11,** 123–139.
Prais, S. J. (1955a). 'Measuring social mobility'. *J. R. Statist. Soc.*, **A118,** 56–66 [16, 20, 23, 24, 31, 32].
Prais, S. J. (1955b). 'The formal theory of social mobility'. *Population Studies*, **9,** 72–81.
Prais, S. J. *See* Hart and Prais (1956).
Preston, L. E., and E. J. Bell (1961). 'The statistical analysis of industry structure: an application to food industries'. *J. Amer. Statist. Ass.*, **56,** 925–932 [15].
Pyke, R. (1961a). 'Markov renewal processes: definitions and preliminary properties'. *Ann. Math. Statist.*, **32,** 1231–1242 [44].
Pyke, R. (1961b). 'Markov renewal processes with finitely many states'. *Ann. Math. Statist.*, **32,** 1243–1259 [44].
Rainio, Kulleroo (1961). 'Stochastic processes of social contacts'. *Scandinavian J. Psychology*, **2,** 113–128.
Rapoport, A. (1948). 'Cycle distributions in random nets'. *Bull. Math. Biophysics*, **10,** 145–157 [355].
Rapoport, A. (1949a). 'Outline of a probabilistic approach to animal sociology, I'. *Bull. Math. Biophysics*, **11,** 183–196.
Rapoport, A. (1949b). 'Outline of a probabilistic approach to animal sociology, II'. *Bull. Math. Biophysics*, **11,** 273–281.
Rapoport, A. (1951). 'Nets with distance bias'. *Bull. Math. Biophysics*, **13,** 85–91 [355].
Rapoport, A. (1953a). 'Spread of information through a population with socio-

structural bias: I. Assumption of transitivity'. *Bull. Math. Biophysics*, **15,** 523–533 [355].
Rapoport, A. (1953b). 'Spread of information through a population with socio-structural bias: II. Various models with partial transitivity'. *Bull. Math. Biophysics*, **15,** 535–546 [355].
Rapoport, A. (1954). 'Spread of information through a population with socio-structural bias: III. Suggested experimental procedures'. *Bull. Math. Biophysics*, **16,** 75–81 [355].
Rapoport, A., and L. I. Rebhun (1952). 'On the mathematical theory of rumour spread'. *Bull. Math. Biophysics*, **14,** 375–383 [355].
Rapoport, A. *See* Solomanoff and Rapoport (1951); *and* Landau and Rapoport (1953).
Rashevsky, N. (1947). *Mathematical Theory of Human Relations*, The Principia Press Inc., Bloomington, Indiana.
Rashevsky, N. (1951). *Mathematical Biology of Social Behaviour*, University of Chicago Prcss, Chicago.
Rashevsky, N. (1959). *Mathematical Biology of Social Behaviour* (Revised ed.), University of Chicago Press, Chicago.
Ray, K. *See* Morgan, Keenay and Ray (1973).
Rebhun, L. I. *See* Rapoport and Rebhun (1952).
Rees, P. H. (1973). 'A revised notation for spatial demographic accounts and models'. *Environment and Planning*, **5,** 147–155.
Rees, P. H., and A. G. Wilson (1973). 'Accounts and models for spatial demographic analysis I: aggregate population'. *Environment and Planning*, **5,** 61–90.
Reid, G. C. *See* Kemp and Reid (1971).
Reiger, M. H. (1968). 'A two-state Markov model for behavioural chance'. *J. Amer. Statist. Ass.*, **63,** 993–999.
Reisman, A. (1966). 'A population flow feedback model'. *Science*, **153,** 89–91.
Reisman, A. *See also* Balinsky and Reisman (1972).
Revelle, C., F. Feldmann, and W. Lynne (1969). 'An optimization model of tuberculosis epidemiology gives a Markov process model for tuberculosis'. *Man. Sci.*, **16,** B190–B211.
Rice, A. K., J. M. M. Hill, and E. L. Trist (1950). 'The representation of labour turnover as a social proccss'. *Human Relations*, **3,** 349–381 [184].
Rice, A. K., and E. L. Trist (1952). 'Institutional and sub-institutional determinants of social change in labour turnover'. *Human Relations*, **5,** 347–371.
Ridler-Rowe, C. J. (1967). 'On a stochastic model of an epidemic'. *J. Appl. Prob.*, **4,** 19–33.
Roberts, H. V. (1959). 'Stock market "patterns" and financial analysis: methodological suggestions'. *J. Finance*, **14,** 1–10.
Robertson, T. S. (1967). 'The process of innovation and the diffusion of innovation'. *J. Marketing*, **31,** 14–19.
Rogers, A. (1966). 'A Markovian policy model of inter-regional migration'. *Papers of the Regional Science Association*, **17,** 205–224.
Rogers, E. (1962). *Diffusion of Innovations*, Free Press of Glencoe, New York.
Rogoff, N. (1953). *Recent Trends in Occupational Mobility*, The Free Press, Glencoe, Illinois [20].
Rose, H. M. (1970). 'The development of an urban sub-system: the case of the negro ghetto'. *Ann. Ass. of Amer. Geographers*, **60,** 1–17.
Rosenbaum, S. (1971). 'A report on the use of statistics in Social Science research' (with discussion). *J. R. Statist. Soc.*, **A134,** 534–610 [1].

Ross, I. C. *See* Harary and Ross (1954).

Rowe, S. M., W. G. Wagner, and G. B. Weathersby (1970). *A Control Theory Solution to Optimal Faculty Staffing*, Paper P-11, Research Program in University Administration, University of California [98].

Rushton, S., and A. J. Mautner (1955). 'The deterministic model of a simple epidemic for more than one community'. *Biometrika*, **42**, 126–132 [309, 311, 328].

Rutherford, R. S. G. (1955). 'Income distributions: a new model'. *Econometrica*, **23,** 277–294.

Ryan, B., and N. C. Gross (1943). 'The diffusion of hybrid seed corn in two Iowa communities'. *Rural Sociology*, **8,** 15–24 [296].

Sabolo, Y. (1971). 'A structural approach to the projection of occupational categories and its application to South Korea and Taiwan'. *Internat. Labour Review*, **103,** 131–155 [16].

Sales, P. (1971). 'The validity of the Markov chain model for a branch of the Civil Service'. *Statistician*, **20,** 85–110 [56].

Sales, P. *See also* Kuhn, Poole, Wynn and Sales (1973); Wynn and Sales (1973a); *and* Wynn and Sales (1973b).

Saunders, S. C. *See* Birnbaum and Saunders (1969).

Schach, E., and S. Schach (1972). 'A continuous time stochastic model for the utilization of health services'. *Socio-Econ. Plan. Sci.*, **6,** 263–272 [139].

Scott, E. L. *See* Neyman and Scott (1964).

Seal, H. L. (1945). 'The mathematics of a population composed of *k* stationary strata each recruited from the stratum below and supported at the lowest level by a uniform annual number of entrants'. *Biometrika*, **33,** 226–230 [139, 153].

Seal, H. L. (1970). 'Probability distributions of aggregate sickness durations'. *Skand. Aktuarietidskrift*, Parts 3 and 4, 193–204.

Severo, N. C. (1967). 'Generalizations of stochastic epidemic models'. *Bull. Int. Statist. Inst.*, **42,** Book 2, 1064–1066.

Severo, N. C. (1969a). 'Solving non-linear problems in the theory of epidemics'. *Bull. Int. Statist. Inst.*, **43,** Book 2, 226–228.

Severo, N. C. (1969b). 'The probabilities of some epidemic models'. *Biometrika*, **56,** 197–201.

Severo, N. C. *See also* Hill and Severo (1969).

Sheps, M. C. (1971). 'A review of models for population change'. *Rev. Int. Statist. Inst.*, **39,** 185–196.

Sheps, M. C. *See also* Perrin and Sheps (1964).

Sholtz, D. *See* Charnes, Cooper, Niehaus and Sholtz (1970).

Silcock, H. (1954). 'The phenomenon of labour turnover'. *J. R. Statist. Soc.*, **A117,** 429–440 [184, 187, 215, 221, 262].

Silverman, F. N. *See* Morrison, Massy and Silverman (1971).

Simon, H. A. (1957a). 'The compensation of executives'. *Sociometry*, **20,** 32–35 [283].

Simon, H. A. (1957b). *Models of Man, Social and Rational*; *Mathematical Essays on Rational Human Behaviour in a Social Setting*, John Wiley, New York.

Simon, H. A. (1969). *The Sciences of the Artificial*, M.I.T. Press, Cambridge, Mass.

Siskind, V. (1965). 'A solution of the general stochastic epidemic'. *Biometrika*, **52,** 613–616 [338, 341].

Smith, A. R. (1967). 'Manpower planning in management of the Royal Navy'. *J. Management Studies*, **4,** 127–139.

Smith, A. R. (Ed.) (1971). *Models for Manpower Systems*, English Universities Press, London [2].

Smith, A. R. *See also* Bartholomew and Smith (1971).
Smith, C. S. *See* Armitage, Smith and Alper (1969).
Smith, H. *See* Styan and Smith (1964).
Smith, P. E. (1961). 'Markov chains, exchange matrices and regional development'. *J. Regional Science*, **3,** 27–36.
Snell, L. *See* Kemeny and Snell (1960); *and* Kemeny and Snell (1962).
Soloman, H. *See* Criswell, Soloman and Suppes (1962).
Solomanoff, R. (1952). 'An exact method for the computation of the connectivity of random nets'. *Bull. Math. Biophysics*, **14,** 153–157 [358, 360].
Solomanoff, R., and A. Rapoport (1951). 'Connectivity of random nets'. *Bull. Math. Biophysics*, **13,** 107–117 [355, 364].
Spilerman, S. (1970). 'The causes of racial disturbances: a comparison of alternative explanations'. *Amer. Sociol. Rev.*, **35,** 627–649.
Spilerman, S. (1972a). 'The analysis of mobility processes by the introduction of independent variables into a Markov chain'. *Amer. Sociol. Rev.*, **37,** 277–294 [16].
Spilerman, S. (1972b). 'Extensions of the mover–stayer model'. *Amer. J. Sociol.*, **78,** 599–626 [16, 49, 50].
Staff, P. J., and M. K. Vagholkar (1971). 'Stationary distributions of open Markov processes in discrete time with application to hospital planning'. *J. Appl. Prob.*, **8,** 668–680 [75].
Startsev, A. N. *See* Nagaev and Startsev (1970).
Steindhl, J. (1965). *Random Processes and the Growth of Firms*, Griffin, London [1].
Stoikov, V. (1971). 'The effect of changes in quits and hires on the length-of-service composition of employed workers'. *Brit. J. Indust. Rel.*, **9,** 225–233 [241].
Stone, R. (1965). 'A model of the educational system'. *Minerva*, **3,** 172–186.
Stone, R. (1966). *Mathematics in the Social Sciences and Other Essays*, Chapman and Hall, London.
Stone, R. (1972). 'A Markovian educational model and other examples linking social behaviour to the economy' (with discussion). *J. R. Statist. Soc.*, **A135,** 511–543 [67].
Styan, G. P. H., and H. Smith (1964). 'Markov chains applied to marketing'. *J. Marketing Research*, **1,** 50–55 [16].
Sunshine, M. H. *See* Fararo and Sunshine (1964).
Sunter, A. *See* Ostry and Sunter (1970).
Suppes, P. *See* Arrow, Karlin and Suppes (1960); *and* Criswell, Soloman and Suppes (1962).
Suzuki, T. *See* Taga and Suzuki (1957).
Svalagosta, K. (1959). *Prestige, Class and Mobility*, Heinemann, London [20, 139].
Sverdrup, E. (1965). 'Estimates and test procedures in connexion with stochastic models of deaths, recoveries and transfers between different states of health'. *Skand. Aktuar.*, **46,** 184–211 [139, 195].
Sykes, Z. M. (1969). 'Some stochastic versions of the matrix model for population dynamics', *J. Amer. Statist. Ass.*, **64,** 111–130 [16].
Taga, Y. (1963). 'On the limiting distributions in Markov renewal processes with finitely many states'. *Ann. Inst. Statist. Math.*, **15,** 1–10 [44].
Taga, Y., and K. Isii (1959). 'On a stochastic model concerning the pattern of communication–diffusion of news in a social group'. *Ann. Inst. Statist. Math.*, **11,** 25–43 [299, 320].
Taga, Y., and T. Suzuki (1957). 'On the estimation of average length of chains in the communication pattern'. *Ann. Inst. Statist. Math.*, **9,** 149–156.
Takács, L. (1970). 'On the fluctuations of election returns'. *J. Appl. Prob.*, **7,** 114–123.

Tarver, J. D., and W. R. Gurley (1965). 'A stochastic analysis of geographic mobility and population projections of the census divisions in the United States'. *Demography*, **2,** 134–139 [15].

Taylor, G. C. (1971). 'Moments in Markovian systems with lumped states'. *J. Appl. Prob.*, **8,** 599–605.

Tegtmeyer, H. W. *See* Henry, McGinnis and Tegtmeyer (1971).

Theil, H. (1970). 'The cube law revisited'. *J. Amer. Statist. Ass.*, **65,** 1213–1219.

Theodorescu, R., and I. Vaduva (1967). 'A mathematical demographic model concerning the dynamics of specialists'. *Studii Cereetari Mat.*, **19,** 329–337.

Thonstad, T. (1969). *Education and Manpower: Theoretical Models and Empirical Applications*, Oliver and Boyd, Edinburgh and London [1, 56].

Tinbergen, J. (1956). 'On the theory of income distribution'. *Weltwirtschaftliches Archiv. Bd.*, **77,** 10–31.

Todard, M. P. (1971). 'Income expectations, rural-urban migration and employment in Africa'. *International Labour Review*, **104,** 387–413.

Trist, E. L. *See* Rice, Hill and Trist (1950); *and* Rice and Trist (1952).

Tuck, R. H. (1954). *An Essay on the Economic Theory of Rank*, Blackwell, Oxford.

Tweedie, M. C. K. (1957). 'Statistical properties of inverse Gaussian distributions'. *Ann. Math. Statist.*, **28,** 362–377 and 696–705 [200].

Usher, M. B., and M. H. Williamson (1970). 'A deterministic matrix model for handling the birth, death and migration process of spatially distributed populations'. *Biometrics*, **26,** 1–12.

Vaduva, I. *See* Theodorescu and Vaduva (1967).

Vagholkar, M. K. *See* Staff and Vagholkar (1971).

Vajda, S. (1947). 'The stratified semi-stationary population'. *Biometrika*, **34,** 243–254 [139, 173].

Vajda, S. (1948). 'Introduction to a mathematical theory of a graded stationary population'. *Bull. de l'Ass. Actuair, Suisses*, **48,** 251–273 [153, 155].

Van der Merwe, R., and S. Miller (1971). 'The measurement of labour turnover'. *Human Relations*, **24,** 233–253 [241].

Von Weizsäcker, C. C. *See* Freytag and Von Weizsäcker (1968); *and* Freytag and Von Weizsäcker (1969).

Vroom, V. H., and K. R. MacCrimmon (1968). 'Towards a stochastic model of managerial careers'. *Admin. Sci. Quart.*, **13,** 26–46.

Wagner, W. G. *See* Rowe, Wagner and Weathersby (1970).

Walmsley, C. W. (1971). 'A simulation model for manpower planning'. In D. J. Bartholomew and A. R. Smith (1971), 163–170.

Watson, R. K. (1972), 'On an epidemic in a stratified population'. *J. App. Prob.*, **9,** 659–666.

Waugh, W. A. O'N. (1971). 'Career prospects in stochastic social models with time-varying rates'. *Fourth Conference on the Mathematics of Population*, East-West Population Institute, Honolulu.

Weathersby, G. B. *See* Rowe, Wagner and Weathersby (1970).

Weber, W. L. (1971). 'Manpower planning in hierarchical organizations: a computer simulation approach'. *Man. Sci.*, **18,** 119–144.

Weber, W. L. *See also* Dill, Gaver and Weber (1966).

Wegman, E. J., and C. R. Kukuk (1971). 'A time series approach to the life table'. *Ann. Math. Statist.*, **42,** 1491.

Weinberg, C. B. (1971). 'Response curves for a leaflet distribution—further analysis of the De Fleur data'. *Operat. Res. Quart.*, **22,** 177–179.

Weiss, H. K. (1963). 'Stochastic models for the duration and magnitude of a "deadly quarrel" '. *Operat. Res.*, **11,** 101–121 [207].

Weiss, L. W. *See* Kelley and Weiss (1969).

Wellman, G. (1972). 'Practical obstacles to effective manpower planning'. *Personnel Review*, **1,** 32–47.

White, H. C. (1962). 'Chance models of systems of casual groups'. *Sociometry*, **25,** 153–172.

White, H. C. (1963). 'Cause and effect in social mobility tables'. *Behavioral Science*, **8,** 14–27 [19].

White, H. C. (1969). 'Control and evolution of aggregate personnel: flows of men and jobs'. *Admin. Sci. Quart.*, **14,** 4–11.

White, H. C. (1970a). *Chains of Opportunity*, Harvard University Press, Cambridge, Mass. [2, 93, 246].

White, H. C. (1970b). 'Matching, vacancies and mobility'. *J. Pol. Econ.*, **78,** 97–105.

Whittle, P. (1955). 'The outcome of a stochastic epidemic—a note on Bailey's paper'. *Biometrika*, **42,** 116–122. [341].

Whittle, P. *See also* Wold and Whittle (1957).

Widder, D. V. (1946). *The Laplace Transform*, Princeton University Press [165].

Williams, E. J. (1967). 'The development of biomathematical models'. *Bull. Int. Statist. Inst.*, **42,** Book 1, 131–143.

Williams, T. (1965). 'The simple stochastic epidemic curve for large populations of susceptibles'. *Biometrika*, **52,** 571–579 [302, 304, 305, 307].

Williamson, M. H. *See* Usher and Williamson (1970).

Wilson, N. A. B. (Ed.) (1969). *Manpower Research*. English Universities Press, London [2].

Winick, C. (1961). 'The diffusion of an innovation among physicians in a large city'. *Sociometry*, **24,** 384–396.

Winthrop, H. (1958). 'Experimental results in relation to a mathematical theory of behavioural diffusion'. *J. Soc. Psych.*, **47,** 85–100.

Wold, H. (1967). 'Non-experimental statistical analysis from the general point of view of scientific method'. *Bull. Int. Statist. Inst.*, **42,** Book 1, 391–427.

Wold, H. O. A., and P. Whittle (1957). 'A model explaining the Pareto distribution of wealth'. *Econometrica*, **25,** 591–595.

Wood, F. (1969). *An Investigation into the Introduction of Progressive Patient Care*, M.Sc. Dissertation, University of Essex.

Wood, S. (1965). 'A simple arithmetic approach to career planning and recruitment'. *Brit. J. Indust. Rel.*, **3,** 291–300.

Wynn, H. P. (1973). 'Limiting second moments for transient states of Markov chains'. *J. Appl. Prob.*, **10** (to appear).

Wynn, H. P., and P. Sales (1973a). 'A simple model for projecting means and variances of population grade sizes'. In *Stochastic Analysis of National Manpower Problems*, Research Report, University of Kent [30, 31, 92].

Wynn, H. P., and P. Sales (1973b). 'The mover–stayer model and the 1963 labour mobility survey'. In *Stochastic Analysis of National Manpower Problems*, Research Report, University of Kent [16, 38, 54].

Wynn, H. P. *See also* Kuhn, Poole, Sales and Wynn (1973).

Yang, G. Lo (1968). 'Contagion in stochastic models for epidemics'. *Ann. Math. Statist.*, **39,** 1863–1889.

Young, A. (1965). 'Models for planning recruitment and promotion of staff'. *Brit. J. Indust. Rel.* **3,** 301–310.

Young, A. (1971). 'Demographic and ecological models for manpower planning'. In D. J. Bartholomew and B. R. Morris (1971), 75–97 [57, 60, 66, 94, 95, 190, 191].

Young, A., and G. Almond (1961). 'Predicting distributions of staff'. *Comp. J.*, **3**, 246–250 [57, 79].

Zahl, S. (1955). 'A Markov process model for follow-up studies'. *Human Biology*, **27**, 90–120 [139, 145, 151].

Zellner, A. *See* Lee, Judge and Zellner (1968); *and* Lee, Judge and Zellner (1970).

Zemach, R. (1968). 'A state-space model for resource allocations in higher educations'. *IEEE Trans. of Systems Science and Cybernetics*, Vol. SSc-4, No. 2.

Subject Index